Effects of Non-thermal Plasma Treatment on Plant Physiological and Biochemical Processes

Effects of Non-thermal Plasma Treatment on Plant Physiological and Biochemical Processes

Editors

Vida Mildažienė
Božena Šerá

MDPI • Basel • Beijing • Wuhan • Barcelona • Belgrade • Manchester • Tokyo • Cluj • Tianjin

Editors

Vida Mildažienė
Department of Biochemistry,
Faculty of Natural Sciences,
Vytautas Magnus University,
Vileikos str. 8,
44404 Kaunas, Lithuania

Božena Šerá
Department of Environmental Ecology
and Landscape Management,
Faculty of Natural Sciences,
Comenius University in
Bratislava, Ilkovičova 6,
842 15 Bratislava, Slovakia

Editorial Office
MDPI
St. Alban-Anlage 66
4052 Basel, Switzerland

This is a reprint of articles from the Special Issue published online in the open access journal *Plants* (ISSN 2223-7747) (available at: https://www.mdpi.com/journal/plants/special_issues/plasma_plants).

For citation purposes, cite each article independently as indicated on the article page online and as indicated below:

LastName, A.A.; LastName, B.B.; LastName, C.C. Article Title. *Journal Name* **Year**, *Volume Number*, Page Range.

ISBN 978-3-0365-4205-8 (Hbk)
ISBN 978-3-0365-4206-5 (PDF)

Contents

About the Editors

Vida Mildažienė is a Senior Researcher and Professor at the Department of Biochemistry, Faculty of Natural Sciences, Vytautas Magnus University in Kaunas, Lithuania. She graduated in Biochemistry from Vilnius University and completed her PhD in Biochemistry at the University Health Sciences in Kaunas. She has been a visiting scientist at the University of Cambridge (UK), the Independent University of Amsterdam (Netherlands), Thomas Jefferson University (Philadelphia, USA), Bordeaux II University (France), and Charite University in Berlin (Germany). Her research interests included bioenergetics (mitochondrial oxidative phosphorylation), cell biology (stem and cancer cell research), and systems biology. Currently, her main interests are focused on plant biochemistry, seed physiology, molecular mechanisms of plant response to seed stress, and the development of innovative techniques to improve the agronomic properties of plants by using plasma technologies for seed treatment.

Božena Šerá works at the Department of Environmental Ecology and Landscape Management, Comenius University, in Bratislava. She is a researcher with long-term experience in the academic-pedagogical system. She is a botanist with a deep interest in plant ecology and with rich experience in seed science. She also addresses issues of greenery in urban areas, road ecology, the ecology of trees and plant autecology.

Preface to "Effects of Non-thermal Plasma Treatment on Plant Physiological and Biochemical Processes"

The potential of cold plasma-based applications in sustainable agriculture is supported by numerous studies which have gathered experimental evidence that the plasma treatment of seeds, water or plants can be used to improve yields, increase the size and the robustness of plants and to reduce the need of antifungal agents, as well as other chemicals. However, the development of reliable and manageable agro-biotechnologies is ultimately based on the understanding of the molecular mechanisms underlying such effects. Despite considerable efforts, such knowledge still remains elusive. Recent breakthroughs in this area are strongly linked to recent discoveries in plant physiology and biochemistry related to topics of plant plasticity, adaptability, stress response and communication. Short plasma treatments of plant materials can induce various changes in plant development and metabolism that persist for a long time. We are only just beginning to understand how to use very complex molecular mechanisms for the mobilisation of plant resources and for the improvement in agricultural plant performance. It is likely that investigations of plasma-induced changes in plant physiological and biochemical processes may reveal new facts of both fundamental and applied importance. This Special Issue of *Plants* aims to present the most recent findings on changes in plant signal transduction, metabolism, development and physiological processes induced by the exposure of seeds or plants to cold plasma or plasma-activated water, leading to increased plant productivity.

Vida Mildažienė and Božena Šerá

Editors

Editorial

Effects of Non-Thermal Plasma Treatment on Plant Physiological and Biochemical Processes

Vida Mildaziene [1,*] and Bozena Sera [2]

[1] Faculty of Natural Sciences, Vytautas Magnus University, LT-44404 Kaunas, Lithuania
[2] Department of Environmental Ecology and Landscape Management, Faculty of Natural Sciences, Comenius University in Bratislava, 84215 Bratislava, Slovakia; bozena.sera@uniba.sk
* Correspondence: vida.mildaziene@vdu.lt

Citation: Mildaziene, V.; Sera, B. Effects of Non-Thermal Plasma Treatment on Plant Physiological and Biochemical Processes. *Plants* **2022**, *11*, 1018. https://doi.org/10.3390/plants11081018

Received: 30 March 2022
Accepted: 6 April 2022
Published: 8 April 2022

Publisher's Note: MDPI stays neutral with regard to jurisdictional claims in published maps and institutional affiliations.

Plasma, also called the fourth state of matter, is partially or fully ionized gas. A distinction is made between thermal (high temperature, equilibrium) and non-thermal (cold, low temperature, non-equilibrium) plasma. Thermal plasma reaches temperatures of thousands of kelvin and occurs in the Sun, lighting, electric sparks, tokamaks, etc. Non-thermal plasma (NTP) occurs at near-ambient temperatures, and its high kinetic energy is only stored in electrons [1,2].

NTP can be implemented at low (using expensive vacuum equipment) or atmospheric pressure (a simpler and cheaper option). NTP at atmospheric pressure is of considerable practical interest because it allows one to avoid the use of expensive vacuum equipment and can be used for the treatment of exhaust gases and polluted liquids. The latter cannot be realized at low pressure. Another advantage of atmospheric-pressure NTP, over low-pressure NTP, is its ability to intensify plasma treatment and induce rapid changes in treated materials [3,4].

NTP at atmospheric pressure has wide applications in biology, medicine, dermatology, dentistry, agriculture, forestry, and the food industry [5–11]. The possible applications include disinfection processes, the acceleration of blood coagulation, the improvement of wound healing and infection clearance, dental applications, and cancer therapy [12]. Applications of both low-pressure and atmospheric-pressure NTP based on various effects observed in plants have also been widely reported, and the number of publications has been increasing in recent years. The topics include changing seed-surface properties, microbial seed decontamination, the degradation of mycotoxins, the stimulation of seed germination and seedling growth, inducing changes in metabolic plant pathways, modulating the enzymatic activities of enzymes, modulating the contents of phytohormones, inducing stress resistance, and influencing productivity [13–16].

The application of NTP in agriculture includes both direct and indirect means of NTP treatment. A quantity of published data demonstrate mechanisms of direct interactions between plasma and plant material [17]. Indirect NTP treatments are used less often, it is usually the application of plasma-activated water (PAW) to the plant [18,19].

The potential of NTP applications in sustainable agriculture is supported by numerous studies. The experimental evidence gathered to date suggests that the plasma treatment of seeds, water or plants can be used to improve yields, increase the size and robustness of plants, and reduce the need for antifungal agents, as well as other chemicals. However, the development of reliable and manageable agro-biotechnologies is ultimately based on the understanding of the molecular mechanisms underlying such effects. Despite considerable efforts, such knowledge still remains elusive.

This Special Issue of the *Plants* journal aims to present the most recent findings on changes in plant signal transduction, metabolism, development, and physiological processes induced by the exposure of seeds or plants to cold plasma or PAW and leading to increased plant productivity. This Special Issue includes a number of original studies on the effects of NTP on plants that can be divided into three groups: (1) articles presenting experimental studies on the direct action of NTP on seeds or plants; (2) results obtained using PAW; (3) literature reviews.

Direct action of NTP (1). The potential of pre-sowing seed treatment with low-pressure capacitively coupled NTP and radiofrequency electromagnetic fields (EMFs) for the stimulation of the biosynthesis of steviol glycosides, stevioside (Stev) and rebaudioside A (RebA) in stevia plants was studied by Judickaite et al. [20]. In this study, the impact of treatments on seed germination and the content of other secondary metabolites—phenolic compounds and flavonoids—as well as the subsequent effects on antioxidant activity were also investigated. It was demonstrated that short pre-sowing treatments of stevia seeds with low-pressure NTP and EMFs could be a powerful tool for the enhancement of the biosynthesis/accumulation of RebA and Stev. However, the applied treatments decreased the RebA/Stev ratio, phenolic and flavonoid contents, and antioxidant activity.

Starič et al. [21] studied the impact of direct (glow) and indirect (afterglow) radiofrequency (RF) oxygen plasma treatments (a type of NTP) on the germination and early growth of two winter wheat varieties (*Triticum aestivum*, var. Apache and var. Bezostaya 1). The germination rate, number of roots, length of the root system, and fresh weight of the seedlings were measured. Both positive and negative differences between the two wheat varieties were observed. Overall, it can be concluded that, besides the NTP treatment parameters, the plant variety and/or seed characteristics may play a crucial role in optimizing NTP for seed treatment, as has previously been demonstrated for other plant species (e.g., [22–24]).

The authors of [25] studied the effect of NTP on the fresh and dry biomass production of an ornamental gerbera plant (*Gerbera jamesonii* 'Babylon F1') grown in peat or green compost using a standard or a low-fertilization regime. The different types of treatments and their combinations were evaluated to assess possible improvements in plant nutrition, yield, and ornamental traits, as well as variations in the presence of microorganisms in the rhizosphere, when using NTP to treat the nutrient solution. The NTP treatment promoted fresh leaf and flower biomass production in plants grown in peat as well as nutrient adsorption in plants grown in both substrates, except for Fe, while decreasing the dry plant matter. The results revealed that peat, along with an NTP-treated solution containing a high concentration of nutrients, is the most efficient combination of treatments. On the other hand, according to Cannazzaro et al. [25], the combination of NTP and compost may not be suitable for this type of ornamental species.

Using of PAW pre-treatment (2). In an experiment by Cortese et al. [26], the signaling mechanisms behind the effects induced by PAW in *Arabidopsis thaliana* seedlings were investigated. The potential involvement of calcium in the plant's response and signal transduction induced by the molecules contained in PAW was evaluated. By using an *Arabidopsis* line expressing the bioluminescent Ca^{2+} reporter aequorin in the cytosol, the authors demonstrated that PAW evoked rapid and sustained cytosolic Ca^{2+} elevations, characterized by specific signatures. These signatures depend on several parameters: (a) the operational conditions of the torches used to generate the PAW; (b) the duration of H_2O exposure to plasma; (c) the dose of PAW administered to the plants; and (d) the temperature and duration for the PAW storage. The authors concluded that the magnitude of the recorded Ca^{2+} signals was dependent on the torch power or the energy transferred to the water during PAW generation. The obtained data suggest that the PAW-induced Ca^{2+} signature may be attributable to a complex "cocktail" of different reactive chemical species contained in PAW, rather than to a single component. This work provides evidence that Ca^{2+} acts as an intracellular messenger in the signaling pathway triggered by PAW in *Arabidopsis*. Understanding the signaling mechanisms behind the plant's response to PAW is important for its applications in agriculture, with the potential for enabling more sustainable agriculture.

For the first time, Danilejko et al. [27] introduce a manufactured plasma-chemical reactor for the rapid activation of large volumes of liquids (PAW). In their study, the antifungal activity of PAW was investigated. The authors demonstrated that PAW significantly reduced the amount of phytopathogens in sorghum of the Estonskoe variety (*Sorghum bicolor*), wheat (*Triticum aestivum*) seeds, and strawberry (*Fragaria* sp.) fruits. They confirmed a positive effect of PAW on sorghum seed germination and grain biomass production. In addition, a positive effect of PAW on sorghum drought tolerance in a saline semi-desert region was discovered under field conditions.

In the study presented by Lukacova et al. [28], PAW generated by a transient spark discharge operating in ambient air was used on maize (*Zea mays*, hybrid Bielik) corns and seedlings (cultivated for 3 days in paper rolls and 10 days via hydroponics). The roots and shoots were analyzed for guaiacol peroxidase (POX) activity; the root tissues were analyzed for their lignification, and the root cell walls, for in situ POX activity. The activity of POX and catalase after the arsenic-stress treatment of the seedlings was examined, along with the concentration of the photosynthetic pigments in the leaves, and the concentrations in the leaves and roots. The results of this relatively complex experiment suggest that PAW has a positive effect on the physiological responses in maize corns and young plants under arsenic stress.

The work of Kostolani et al. [17] aimed to investigate the effect of PAW on three-day-old pea (*Pisum sativum* cv. Eso) and barley (*Hordeum vulgare* cv. Kangoo) seedlings by estimating the PAW-induced differences in growth and physiological parameters: the germination dynamics; growth parameters; the total soluble protein concentration; the activity of hydrolytic enzymes, antioxidant enzymes, and dehydrogenases; and the visualization of reactive oxygen species and the level of DNA damage. The authors found that PAW generated by glow discharge could have a positive effect on most of the measured parameters. They concluded that the concentrations of stable reactive oxygen–nitrogen species in PAW, favorable for accelerating the transition to aerobic metabolism in the pea, may not be suitable for different plant species, such as barley [17]. The authors suggest that the response to treatment depends not only on PAW but also on the plant species.

Reviews (3). Three reviews included in the Special Issue focus on the effects of seed/plant treatment with NTP [29–31]. An overview of the existing knowledge on changes in both biochemical and physiological processes induced by seed treatment with NTP was published by Mildaziene et al. [29]. The effects of seed treatment with NTP on DNA methylation, wide-scale changes in gene and protein expression, and enzyme activities in the affected seeds and growing plants are considered in the context of the observed effects on plant growth and yield. Particular attention is paid to the importance of seed dormancy, the role of reactive oxygen and nitrogen species and phytohormones, the mobilization of secondary metabolism, increased adaptability to stress, and the effects on the plant-associated microbiome. This review also outlines possible future research directions.

Other authors focused their review [30] on a group of legumes (*Fabaceae* family) whose representatives (19 species) have frequently been used in experiments examining the effects of NTP or PAW on plants. The paper summarizes and critically evaluates the current results on seed germination and initial seedling growth, surface microbial decontamination, and the induced changes in seed wettability and metabolic activity.

Starič et al. [31] reviewed the main concepts and underlying principles of NTP treatment techniques as well as the various aspects of NTP's interaction with seeds. Different plasma-generation methods and setups are described, and the impact of NTP treatments on DNA damage, gene expression, enzymatic activities, morphological and chemical traits, seed germination and plant resistance to stress is considered. Important parameters of the NTP and the interactions of plasma species with the seed surface are presented and discussed.

Conclusions. Recent breakthroughs in research on the effects of NTP seed treatments on plants are strongly linked to discoveries in plant physiology and biochemistry and are related to plant plasticity, adaptability, stress responses and communication. Short NTP treatments of plant materials can induce various changes in plant development and metabolism that persist for a long time. We are only beginning to understand how to use very complex molecular mechanisms for the mobilization of plant resources and for the improvement of agricultural plant performance. It is likely that investigations of plasma-induced changes in plant physiological and biochemical processes may reveal new facts of both fundamental and applied importance.

Author Contributions: Conceptualization, V.M. and B.S.; writing—original draft preparation, B.S. and V.M. All authors have read and agreed to the published version of the manuscript.

Funding: This research received no external funding.

Institutional Review Board Statement: Not applicable.

Informed Consent Statement: Not applicable.

Data Availability Statement: All the data included in the main text.

Acknowledgments: We would like to thank all the colleagues who contributed to this Special Issue and the editorial office for their helpful support during its compilation.

Conflicts of Interest: The authors declare no conflict of interest.

References

1. Piel, A. *Plasma Physics. An Introduction to Laboratory, Space, and Fusion Plasmas*; Univ Kiel: Kiel, Germany, 2010.
2. Shintani, H.; Sakudo, A. (Eds.) *Gas Plasma Sterilization in Microbiology: Theory, Applications, Pitfalls and New Perspectives*; Caister Academic Press: Poole, UK, 2016. [CrossRef]
3. Tendero, C.; Tixier, C.; Tristant, P.; Desmaison, J.; Leprince, P. Atmospheric Pressure Plasmas: A Review. *Spectroc. Acta B* **2006**, *61*, 2. [CrossRef]
4. Akishev, Y.; Grushin, M.; Karalnik, V.; Kochetov, I.; Napartovich, A.; Trushkin, N. Genaration of atmospheric pressure non-thermal plasma by diffusive and constricted discharges in air and nitrogen at the rest and flow. *J. Phys. Conf. Ser.* **2010**, *257*, 012014. [CrossRef]
5. Fridman, G. Non-equilibrium Plasmas in Biology and Medicine. In *Biological and Environmental Application of Gas Discharge Plasmas*; Brelles, M.G., Ed.; California State Polytechnic University, Pomona: Claremont, CA, USA, 2012; pp. 95–184.
6. Graves, D.B. The Emerging Role of Reactive Oxygen and Nitrogen Species in Redox Biology and Some Implications for Plasma Applications to Medicine and Biology. *J. Phys. D Appl. Phys.* **2012**, *45*, 263001. [CrossRef]
7. Attri, P.; Ishikawa, K.; Okumura, T.; Koga, K.; Masaharu, S.M. Plasma Agriculture from Laboratory to Farm: A Review. *Processes* **2020**, *8*, 1002. [CrossRef]
8. Pankaj, S.K.; Wan, Z.; Keener, K.M. Effects of Cold Plasma on Food Quality: A Review. *Foods* **2018**, *7*, 4. [CrossRef] [PubMed]
9. Cheruthazhekatt, S.; Cernak, M.; Slavicek, P.; Havel, J. Gas Plasmas and Plasma Modified Materials in Medicine. *J. Appl. Biomed.* **2021**, *8*, 55. [CrossRef]
10. Domonkos, M.; Tichá, P.; Trejbal, J.; Demo, P. Applications of Cold Atmospheric Pressure Plasma Technology in Medicine, Agriculture and Food Industry. *Appl. Sci.* **2021**, *11*, 4809. [CrossRef]
11. Paužaitė, G.; Malakauskienė, A.; Naučienė, Z.; Žūkienė, R.; Filatova, I.; Lyushkevich, V.; Azarko, I.; Mildažienė, V. Changes in Norway Spruce Germination and Growth Induced by Pre-sowing Seed Treatment with Cold Plasma and Electromagnetic Field: Short-Term versus Long-Term Effects. *Plasma Process. Polym.* **2018**, *15*, 1700068. [CrossRef]
12. Metelmann, H.R.; von Woedtke, T.; Weltmann, K.D. (Eds.) *Comprehensive Clinical Plasma Medicine. Cold Physical Plasma for Medical Application*; Springer: Cham, Switzerland, 2018; ISBN 331967627X.
13. Ten Bosch, L.; Pfohl, K.; Avramidis, G.; Wieneke, S.; Wolfgang, V.W.; Karlovsky, P. Plasma-Based Degradation of Mycotoxins Produced by *Fusarium*, *Aspergillus* and *Alternaria* Species. *Toxins* **2017**, *9*, 97. [CrossRef]
14. Scholtz, V.; Šerá, B.; Khun, J.; Šerý, M.; Julák, J. Effects of Nonthermal Plasma on Wheat Grains and Products. *J. Food Qual.* **2019**, *2019*, 7917825. [CrossRef]
15. Scholtz, V.; Jirešová, J.; Šerá, B.; Julák, J. A Review of Microbicidal Decontamination and Disinfection of Cereals by Non-thermal Plasma. *Food* **2021**, *10*, 2927. [CrossRef]
16. Šerá, B.; Vanková, R.; Roháček, K.; Šerý, M. Gliding Arc Plasma Treatment of Maize (*Zea mays* L.) Grains Promotes Seed Germination and Early Growth, Affecting Hormone Pools, but Not Significantly Photosynthetic Parameters. *Agronomy* **2021**, *11*, 2066. [CrossRef]
17. Kostoláni, D.; Ndiffo Yemeli, G.B.; Švubová, R.; Kyzek, S.; Machala, Z. Physiological Responses of Young Pea and Barley Seedlings to Plasma-Activated Water. *Plants* **2021**, *10*, 1750. [CrossRef]
18. Kučerová, K.; Henselová, M.; Slováková, Ľ.; Hensel, K. Effects of Plasma Activated Water on Wheat: Germination, Growth Parameters, Photosynthetic Pigments, Soluble Protein Content, and Antioxidant Enzymes Activity. *Plasma Process. Polym.* **2019**, *16*, 1800131. [CrossRef]

19. Ndiffo Yemeli, G.B.; Švubová, R.; Kostolani, D.; Kyzek, S.; Machala, Z. The Effect of Water Activated by Nonthermal Air Plasma on the Growth of Farm Plants: Case of Maize and Barley. *Plasma Process. Polym.* **2021**, *18*, 2000205. [CrossRef]
20. Judickaitė, A.; Lyushkevich, V.; Filatova, I.; Mildažienė, V.; Žūkienė, R. The Potential of Cold Plasma and Electromagnetic Field as Stimulators of Natural Sweeteners Biosynthesis in *Stevia rebaudiana* Bertoni. *Plants* **2022**, *11*, 611. [CrossRef]
21. Starič, P.; Grobelnik Mlakar, S.; Junkar, I. Response of Two Different Wheat Varieties to Glow and Afterglow Oxygen Plasma. *Plants* **2021**, *10*, 1728. [CrossRef]
22. Sera, B.; Sery, M.; Gavril, B.; Gajdova, I. Seed Germination and Early Growth Responses to Seed Pre-Treatment by Non-Thermal Plasma in Hemp Cultivars (*Cannabis sativa* L.). *Plasma Chem. Plasma Process.* **2017**, *37*, 207–221. [CrossRef]
23. Mildaziene, V.; Pauzaite, G.; Nauciene, Z.; Zukiene, R.; Malakauskiene, A.; Norkeviciene, E.; Stukonis, V.; Slepetiene, A.; Olsauskaite, V.; Padarauskas, A.; et al. Effect of Seed Treatment with Cold Plasma and Electromagnetic Field on Red Clover Germination, Growth and Content of Major Isoflavones. *J. Phys. D Appl. Phys.* **2020**, *53*, 26. [CrossRef]
24. Ivankov, A.; Naučienė, Z.; Degutytė-Fomins, L.; Žūkienė, R.; Januškaitienė, I.; Malakauskienė, A.; Jakstas, V.; Ivanauskas, L.; Romanovskaja, D.; Slepetiene, A.; et al. Changes in Agricultural Performance of Common Buckwheat Induced by Seed Treatment with Cold Plasma and Electromagnetic Field. *Appl. Sci.* **2021**, *11*, 4391. [CrossRef]
25. Cannazzaro, S.; Traversari, S.; Cacini, S.; Di Lonardo, S.; Pane, C.; Burchi, G.; Massa, D. Non-Thermal Plasma Treatment Influences Shoot Biomass, Flower Production and Nutrition of Gerbera Plants Depending on Substrate Composition and Fertigation Level. *Plants* **2021**, *10*, 689. [CrossRef]
26. Cortese, E.; Settimi, A.G.; Pettenuzzo, S.; Cappellin, L.; Galenda, A.; Famengo, A.; Dabalà, M.; Antoni, V.; Navazio, L. Plasma-Activated Water Triggers Rapid and Sustained Cytosolic Ca^{2+} Elevations in *Arabidopsis thaliana*. *Plants* **2021**, *10*, 2516. [CrossRef]
27. Danilejko, Y.K.; Belov, S.V.; Egorov, A.B.; Lukanin, V.I.; Sidorov, V.A.; Apasheva, L.M.; Dushkov, V.Y.; Budnik, M.I.; Belyakov, A.M.; Kulik, K.N.; et al. Increase of Productivity and Neutralization of Pathological Processes in Plants of Grain and Fruit Crops with the Help of Aqueous Solutions Activated by Plasma of High-Frequency Glow Discharge. *Plants* **2021**, *10*, 2161. [CrossRef]
28. Lukacova, Z.; Svubova, R.; Selvekova, P.; Hensel, K. The Effect of Plasma Activated Water on Maize (*Zea mays* L.) under Arsenic Stress. *Plants* **2021**, *10*, 1899. [CrossRef] [PubMed]
29. Mildaziene, V.; Ivankov, A.; Sera, B.; Baniulis, D. Biochemical and Physiological Plant Processes Affected by Seed Treatment with Non-Thermal Plasma. *Plants* **2022**, *11*, 856. [CrossRef]
30. Šerá, B.; Scholtz, V.; Jirešová, J.; Khun, J.; Julák, J.; Šerý, M. Effects of Non-Thermal Plasma Treatment on Seed Germination and Early Growth of Leguminous Plants—A Review. *Plants* **2021**, *10*, 1616. [CrossRef] [PubMed]
31. Starič, P.; Vogel-Mikuš, K.; Mozetič, M.; Junkar, I. Effects of Nonthermal Plasma on Morphology, Genetics and Physiology of Seeds: A Review. *Plants* **2020**, *9*, 1736. [CrossRef] [PubMed]

Article

The Potential of Cold Plasma and Electromagnetic Field as Stimulators of Natural Sweeteners Biosynthesis in *Stevia rebaudiana* Bertoni

Augustė Judickaitė [1], Veronika Lyushkevich [2], Irina Filatova [2], Vida Mildažienė [1] and Rasa Žūkienė [1,*]

[1] Department of Biochemistry, Faculty of Natural Sciences, Vytautas Magnus University, Vileikos Str. 8, LT-44404 Kaunas, Lithuania; auguste.judickaite@vdu.lt (A.J.); vida.mildaziene@vdu.lt (V.M.)
[2] B. I. Stepanov Institute of Physics, National Academy of Sciences of Belarus, 68 Prospekt Nezavisimosti, BY-220072 Minsk, Belarus; v.lyushkevich@ifanbel.bas-net.by (V.L.); filatova@presidium.bas-net.by (I.F.)
* Correspondence: rasa.zukiene@vdu.lt

Abstract: Stevioside (Stev) and rebaudioside A (RebA) are the most abundant steviol glycosides (SGs) responsible for the sweetness of *Stevia rabaudiana* Bertoni. As compared to Stev, RebA has a higher sweetening potency, better taste and therefore is the most preferred component of the stevia leaf extracts. The aim of this study was to determine the effect of pre-sowing seed treatment with abiotic stressors cold plasma (CP) and electromagnetic field (EMF) on the amount and ratio of RebA and Stev in the leaves of stevia. Additionally, the effect on total phenolic content, flavonoid content and antioxidant activity was investigated. Seeds were treated 5 and 7 min with cold plasma (CP5 and CP7 groups) and 10 min with electromagnetic field (EMF10 group) six days before sowing. The germination tests in vitro demonstrated that all treatments slightly increased germination rate and percentage. HPLC analysis revealed that CP and EMF had strong stimulating effect on SGs accumulation. All treatments increased RebA concentration approximately 1.6-fold; however, the ratio of RebA/Stev decreased from 8.5 in the control to 1.9, 2.5 and 1.1 in CP5, CP7 and EMF10 groups respectively, since the concentration of Stev increased more than RebA, 7.1, 4.6 and 11.0-fold, respectively, compared to control. However, treatments had opposite effect on total phenolic content, flavonoid content, and antioxidant activity. We have demonstrated for the first time that short time pre-sowing treatment of stevia seeds with CP and EMF can be a powerful tool for the enhancement of biosynthesis of RebA and Stev, however it can have negative impact on the content of other secondary metabolites.

Keywords: *Stevia rebaudiana* Bertoni; cold plasma; electromagnetic field; steviol glycosides

Citation: Judickaitė, A.; Lyushkevich, V.; Filatova, I.; Mildažienė, V.; Žūkienė, R. The Potential of Cold Plasma and Electromagnetic Field as Stimulators of Natural Sweeteners Biosynthesis in *Stevia rebaudiana* Bertoni. *Plants* **2022**, *11*, 611. https://doi.org/10.3390/plants11050611

Academic Editor: Corina Danciu

Received: 31 January 2022
Accepted: 22 February 2022
Published: 24 February 2022

Publisher's Note: MDPI stays neutral with regard to jurisdictional claims in published maps and institutional affiliations.

1. Introduction

Stevia rebaudiana Bert. (Bertoni) is a perennial shrub indigenous to Paraguay, South America, and nowadays it is cultivated abundantly in many countries as economically important source of natural low-calorie sweeteners, steviol glycosides (SGs) [1]. Stevia-based sweeteners have increased in market usage due to growing consumer demand for natural products with low or no added sugars. The global stevia market was valued at USD 650 Million in 2020 and is projected to reach USD 1.28 billion by 2028, growing at a compound annual growth (CAGR) of 8.95% [2].

There are at least 38 steviol glycosides identified in stevia to date [3]. Rebaudioside A (RebA) and stevioside (Stev) are the most abundant steviol glycosides (SGs) responsible for the sweetness of stevia (Figure 1). As compared to Stev, RebA has an additional glucose monomer that gives it a higher sweetening potency and therefore RebA is the most preferred component of the stevia leaf extracts. In dried leaves, Stev and RebA account for more than 90% of the total SGs found in stevia leaves. Stev is 110–270 times sweeter than conventional sugar (sucrose) [4]. For comparison, RebA is estimated to be 140–400 times sweeter than sucrose [5]. RebA also lacks liquorice off-taste and lingering sweet aftertaste characteristic to Stev. Therefore, in stevia product industry the stimulation of RebA biosynthesis rather than Stev and higher RebA/Stev ratio is preferable. The aftertaste is eliminated if RebA and Stev are present at least in equal quantities [6].

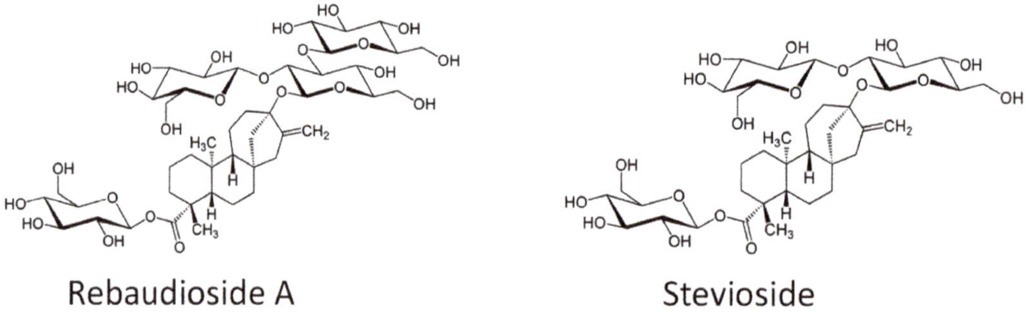

Rebaudioside A

Stevioside

Figure 1. Chemical formulas of rebaudioside A and stevioside.

Results of many studies and clinical trials confirmed that beside the sweet taste, stevia extract and SGs may offer a multitude of beneficial effects on health. Next to diterpenes SGs, non-sweetener fraction is rich of phenolic compounds, giving additional health benefits to the leaf material adding extra value to the product. Stevia extracts and SGs are associated with anti-hypertensive, anti-hyperglycemic, antioxidant, anti-inflammatory, antifungal, anti-microbial activities, and anti-cariogenic action (reviewed in [7]). Due to these various beneficial attributes and absence of side effects in long term use, sweeteners produced from stevia plants are gaining popularity.

Increased demand for stevia products forces the search of new methods for enhancement of economical and ecological production. Different methods and their combinations to increase SG yield are used: selection and breeding of cultivars with high concentrations of SGs, micropropagation [6], optimization of cultivation conditions (photoperiod length, moisture, temperature) [8], fertilization including biofertilizers [9], treatment with nanoparticles [10,11], and modifying after-harvest procedures such as drying, extraction conditions, and enzymatic conversion of Stev to RebA [12–14]. The main disadvantages of these methods often are time-consuming, expensive, or polluting, growing conditions that are climate-dependent and therefore, are not easily modified. Stevia propagation is also problematic: seed germination is poor (15–50%), and vegetative propagation results in low yields. Seed treatment with a physical stressor to obtain higher quality is regarded as clean and cheap technology compared to the use of chemicals. Seed treatment with cold plasma (CP) and electromagnetic field (EMF) is an environmentally friendly method used to improve various plant properties and stimulate synthesis of secondary metabolites [15]. Non-thermal plasma or cold plasma is a non-equilibrium gas discharge plasma, consisting of charged particles, such as ions, free electrons, and neutral particles, including gas molecules, free radicals and UV photons. In cold plasma, the particles are not in thermodynamic equilibrium: electron temperature of a plasma can be several orders of magnitude higher than the temperature of the neutral species or of the ions, which is near room temperature [16]. Radiofrequency (RF) EMF is non-ionizing radiation in which the energy and momentum are carried by alternating magnetic and electric fields, and numerous effects of exposure to EMF has been documented for plants and their seeds [17]. CP and EMF applications in agriculture and food production are gaining increasing attention in recent decades' research. Stimulating effects of these stressors on seed germination, morphometric parameters and biomass production of various plants are well described (see recent reviews [17–22]); however, much less is known about CP and EMF-induced changes in secondary metabolite biosynthesis and underlying mechanisms. We have demonstrated the potential of CP and EMF to increase the amount of vitamin C, caffeic acid derivatives and radical scavenging capacity in purple coneflower (*Echinacea purpurea*) [23], non-psychotropic canabinoids in industrial hemp (*Cannabis sativa*) [24], isoflavones in leaves [25] and root exudates [26] of red clover (*Trifolium pratense*), different secondary metabolites in common buckwheat (*Fagopyrum esculentum*) [27]. Additionally, the germination rate or yield was increased in most of the mentioned studies.

On the other hand, although CP and EMF treatment effects were assessed for numerous plant species and on various morphometric and biochemical plant traits [17–27], it was never applied for treatment of *Stevia rebaudiana*.

The aim of this study was to evaluate the potential of pre-sowing seed treatment with low pressure capacitively coupled CP or RF EMF for the stimulation of main steviol glycosides (RebA and Stev) biosynthesis in stevia plants. Additionally, we have investigated the impact of these treatments on seed germination and content of other secondary metabolites—phenolic compounds, flavonoids, and the resulting effects on antioxidant activity. We have reported for the first time that short time pre-sowing treatment of stevia seeds with CP and EMF can be a powerful tool for the enhancement of biosynthesis/accumulation of RebA and Stev; however, CP and EMF treatments decreased RebA/Stev ratio, the content of phenolics, flavonoids and antioxidant activity.

2. Results

2.1. Effects on Germination In Vitro

Based on some our previous studies on different plant seeds [25,26,28,29], the chosen durations for seed treatments were 5 and 7 min for cold plasma (these treatments are further abbreviated as CP5 and CP7, respectively), and 10 min for EMF treatment (this treatment is abbreviated as EMF10). The results of in vitro germination test performed 6 days after CP and EMF treatments for the control and treated seeds of *S. rebaudiana* are presented in Figure 2a. The germination curve for control group indicated slower germination rate and lower final germination percentage compared to the curves of all treated groups. Richards plots were used to determine the main indices of germination kinetics for quantitation of differences and data are presented in Table 1. All treatments decreased median germination time (Me) by ~1 day, indicating increase in the germination rate. Both cold plasma treatments (CP5 and CP7) slightly increase quartile deviation (Qu) indicating bigger dispersion of germination time (less uniform germination). The most positive response in germination percentage (Vi) was achieved in CP5 treatment group where Vi was 29% higher compared to control. CP7 and EMF treatments increased Vi 17% and 13%, respectively.

Stevia seed treatment with CP and EMF not only ameliorated germination rate and percentage but also had positive effect on seedling root development. Typical 5-day old seedlings are shown in Figure 2b. Control group had more seedlings with less developed roots (1 mm or less) compared to CP and EMF groups, and EMF group had much longer roots in comparison to all other groups.

Table 1. Effect of CP and EMF on *Stevia rebaudiana* germination kinetics indices.

Group	Vi, %	Me, Days	Qu, Days
Control	27.03 ± 2.13	5.06 ± 0.12	0.68 ± 0.10
CP5	35.00 ± 1.01 *	4.39 ± 0.18 *	1.14 ± 0.08 *
CP7	31.58 ± 3.90 *	4.59 ± 0.09 *	0.98 ± 0.17 *
EMF10	30.51 ± 0.26 *	4.36 ± 0.19 *	0.79 ± 0.09

Vi, the final germination percentage; Me, the median germination time; Qu, the quartile deviation; Mean ± SEM ($n = 4$); * statistically significant difference compared to control ($p < 0.05$).

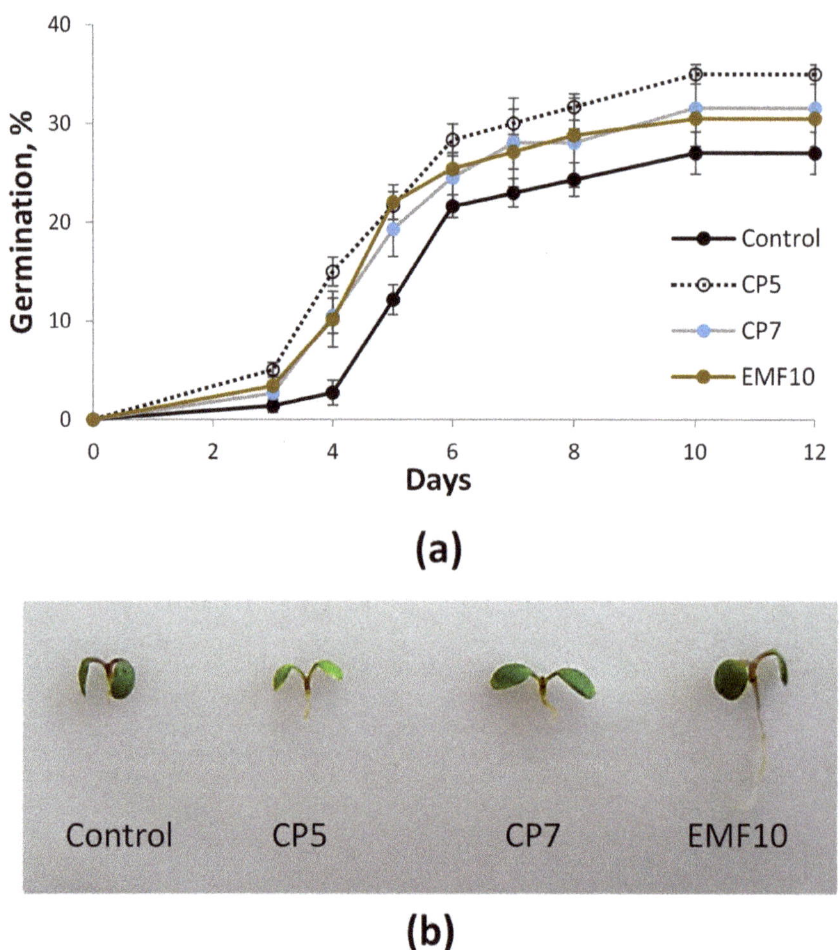

Figure 2. Germination kinetics of *Stevia rebaudiana* seeds (Mean ± SEM, *n* = 4) (**a**) and typical 5-day old seedlings of *Stevia rebaudiana* (**b**).

2.2. Effects on Concentrations of Steviol Glycosides

All seed treatments considerably increased Stev and RebA concentration in leaves as compared to the control (Table 2). This increase in folds in respect to the control is shown in Figure 3. RebA concentration increased approximately 1.6-fold; however, the ratio of RebA/Stev decreased from 8.4 in the control group to 1.9, 2.5 and 1.1 in CP5, CP7 and EMF10 groups, respectively. RebA/Stev ratio inversely correlated with RebA (the linear correlation coefficient was $r^2 = 0.96$), total amount of RebA and Stev ($r^2 = 0.91$), but less with Stev ($r^2 = 0.77$). RebA/Stev ratio was mainly affected by the concentration of Stev that increased more than RebA, 7.1, 4.6 and 11.0-fold, respectively, compared to the control (Figure 3).

Table 2. Effect of CP and EMF on *Stevia rebaudiana* leaf steviol glycoside content (mg·g^{-1} of DW) and ratio (Mean $\pm$ SEM, n = 4).

	RebA	Stev	RebA+Stev	RebA/Stev	RebA/(RebA+Stev)	Stev/(RebA+Stev)
Control	36.71 $\pm$ 3.10	5.27 $\pm$ 1.63	41.98 $\pm$ 4.71	8.35 $\pm$ 1.62	0.88 $\pm$ 0.02	0.12 $\pm$ 0.02
CP5	56.63 $\pm$ 9.07 *	37.35 $\pm$ 8.83 *	93.99 $\pm$ 17.89 *	1.86 $\pm$ 0.24 *	0.64 $\pm$ 0.03 *	0.36 $\pm$ 0.03 *
CP7	59.58 $\pm$ 9.12 *	24.35 $\pm$ 4.14 *	83.93 $\pm$ 13.25 *	2.50 $\pm$ 0.07 *	0.71 $\pm$ 0.01 *	0.29 $\pm$ 0.01 *
EMF10	60.77 $\pm$ 0.33 *	58.15 $\pm$ 0.15 *	118.93 $\pm$ 0.18 *	1.05 $\pm$ 0.01 *	0.51 $\pm$ 0.00 *	0.49 $\pm$ 0.00 *

* statistically significant difference compared to control ($p < 0.05$).

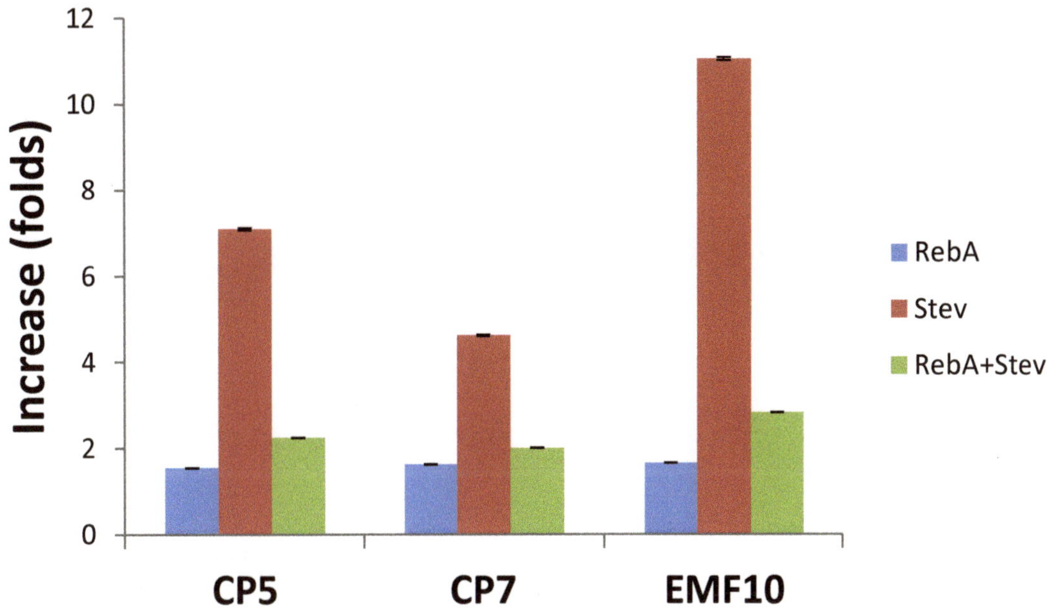

Figure 3. CP and EMF-induced increase in concentration of RebA, Stev and total concentration of RebA and Stev in stevia leaves compared to control. Mean $\pm$ SEM (n = 4), statistically significant differences compared to control ($p \leq 0.05$) were obtained in all treatment groups.

2.3. Effects on Total Phenolic Content, Flavonoid Content, and Antioxidant Activity

In contrast to CP and EMF-induced SGs production stimulation, these treatments had negative impact on the content of total phenolics (TPC), flavonoids (TFC) and antioxidant activity (Figure 4).

CP5 decreased TPC by 13% (Figure 4a). CP7 and EMF10 decreased the concentration much more, 2.2- and 2.1-fold, respectively. TFC was only slightly decreased by CP and EMF treatment (Figure 4b). CP7 treatment has the strongest effect (−25%).

The effect of CP and EMF treatment on antioxidant activity in stevia leaves was evaluated by measuring the scavenging of the stable 2,2-diphenyl-1-picrylhydrazyl free radical (DPPH) and the results are shown in Figure 4c. The pattern of the effect is following the pattern of changes in TPC with the correlation of r^2 = 0.9965.

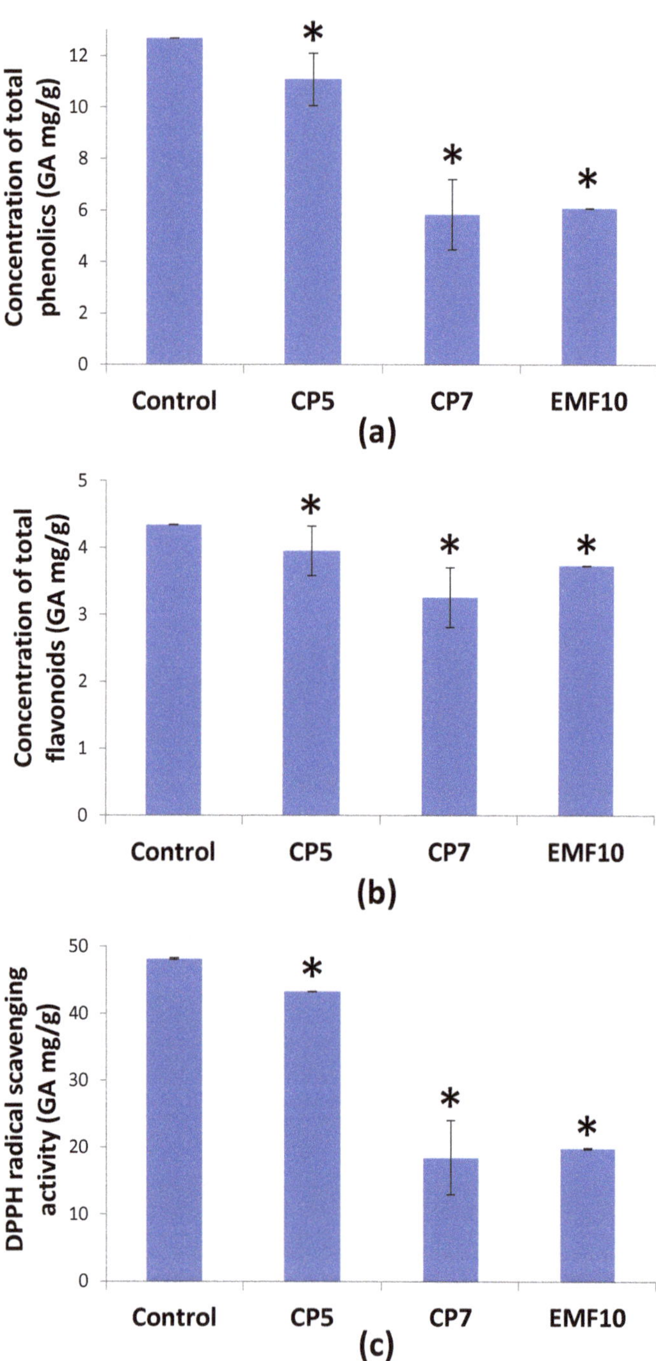

Figure 4. Concentration of total phenolics (**a**), flavonoids (**b**) and antioxidant activity (**c**) in stevia leaves. Mean ± SEM ($n = 4$), * statistically significant differences compared to control ($p \leq 0.05$).

3. Discussion

The purpose of this study was to evaluate the potential of pre-sowing seed treatment with CP and EMF to stimulate the production of main steviol glycosides—RebA and Stev—in *Stevia rebaudiana* plant. These effects of stevia seed treatments with these physical stressors have never been studied before.

The early effect of CP or EMF manifested in stimulation of germination. Pre-sowing seed treatments with CP and EMF are known to ameliorate the germination yield and rate for seeds of numerous plant species [17–22]. Seeds of stevia are characterized by physiological dormancy but often are in a non-dormant state [30]. Hence, low germination yields of stevia seeds are not related to dormancy but are mostly caused by sterility. Therefore, a remarkable increase in germination yield was not expected in this study. Infertile achenes are formed due to sporophytic self-incompatibility. Fertile seeds have fully developed embryos and water-permeable seed coat. Stevia plants are conventionally propagated through cuttings, but this traditional method cannot produce a large number of plants and raising seedlings by sexual plant reproduction is limited [31]. Seed treatment with physical stressors increased germination yield by 13–29%. The molecular mechanism of CP- and EMF-induced germination improvement is still unclear. Some factors involved in seed response to such treatments were, however, elucidated recently. The eustress response to CP and EMF treatment is related to the increased ratio between gibberellins and abscisic acid contents [29,32]. Another well documented mechanism is CP-induced seed surface modification (mainly oxidation due to active particles present in plasma) and consequent increase in hydrophilicity and improved water uptake [33,34]. EMF does not induce such changes on seed surface, because seeds do not interact with aggressive CP-generated particles upon exposure to EMF [35].

The effects of seed treatment with two different stressors, EMF and CP, were compared in this study. CP is complex stressor consisting of the electrical component (discharge), numerous charged particles (free electrons, ions) and neutral active species, including gas molecules, free radicals, metastable particles, and generated photons (including UV) [36]. RF EMF has relatively low quantum energy and does not ionize atoms and molecules; however, EMF may activate molecules (water, components of membranes) by causing electronic excitation and increasing the frequency of collisions while penetrating through the seed tissues [17,37]. Numerous studies (reviewed in [17]) reported that effects of seed treatment with EMF can induce substantial positive effects on germination and seedling growth. Comparison of CP and EMF effects on the same plant species (including effects on the seed electron paramagnetic resonance (EPR) signal [38], seed phytohormones and protein expression in seedlings [39], content of secondary metabolites [23–27]) showed that, for certain plant species, EMF treatment is not less effective tool for seed priming than CP, although the reactive species are not involved in seed interaction with EMF.

Our results show that effects of seed treatment with two different stressors (CP and EMF) on the performance of stevia follow similar trends: positive effects of germination and synthesis of SGs is associated with negative effect on TPC and flavonoid amount as well as decrease in antioxidant activity. Compared to CP, EMF had stronger stimulating effects on the growth of seedling roots and induced the strongest increase in the concentration of Stev (Table 2, Figure 3). The effect of EMF on antioxidant activity and content of such antioxidants as TPC or flavonoids did not differ from the effect of CP7 and was stronger compared to effect of CP5.

Accumulation of steviol glycosides is known as a complex and dynamic process, reflecting an underlying physiological and biochemical processes that are currently intensively studied but not yet fully understood [8]. As our results show, the picture is complicated even further by the unknown action mechanisms of abiotic stressors, such as CP and EMF on the accumulation of SGs. We have demonstrated for the first time that pre-sowing seed treatment with CP and EMF can considerably increase Stev and RebA concentration in leaves (in times in the respect to the control) (Table 2) and at different extent for each SG. *Stevia rebaudiana* cultivar Criolla used in this study is a breeding product characterized by high content of RebA compared to Stev, but even in such a cultivar RebA content can be increased by a very short seed treatment. Unfortunately, the RebA /Stev ratio was not increased in this study. The result "quantity over quality" was obtained since the total SGs content increased mainly due to stronger increase in Stev compared to RebA concentration: RebA concentration increased only 1.6-fold, while the concentration of Stev increased 4.6-11.0 -fold depending on treatment (Figure 3). In this study we have determined that EMF treatment for 10 min is the most effective treatment concerning SGs production stimulation in stevia from all treatments used. Nevertheless, current data do not permit to presuppose that CP treatment is less effective and cannot be a promising method. CP5 treatment results in higher Stev and total SGs concentrations compared to CP7 (Table 2) meaning that 7-min treatment with CP can be already too long and the optimal duration to be determined could be less than 5 min and give similar results to EMF treatment. In addition, to further determine the yield of SGs in industrial applications, the dynamics of SGs concentration and consequential crop performance and adaptivity to environmental conditions must be evaluated, i.e., effects on plant growth, morphology, biomass per plant or per soil area, leaf-to-stem ratio, etc. We did not observe CP- or EMF-induced changes in plant biomass in investigated 8-week-old plants; however, the optimal harvest time is at later vegetative stages were the dynamics of the induced changes on a longer time scale should be evaluated.

SGs biosynthesis is complex and poorly understood process. Therefore, further investigations by applying combined approaches of transcriptomics, RNomics, proteomics, metabolomics and fluxomics are required to extend understanding of the mechanism of CP- and EMF-induced SGs accumulation. Lucho et al. [40] demonstrated the lack of correlation between stevioside content and the transcription of the corresponding biosynthetic genes in their study, what, according to Kumar and others [41], may also be due to the fact that the up-regulated genes (or their gene products) are nonlimiting/nonregulatory. Parallel to this, Saifi and others [42] identified two miRNAs that may up-regulate, and nine miRNAs that may down-regulate their target genes of the steviol glycosides biosynthetic pathway in *S. rebaudiana*. Overall, there is scarce information about the regulation of the SG biosynthesis pathways and master switches for this regulation.

In contrast to CP and EMF-induced SGs production stimulation, these treatments had negative impact on the content of total phenolics (TPC), flavonoids (TFC) and antioxidant activity. Various abiotic physical and chemical stressors usually simultaneously increase production of SGs and phenolic compounds. Such effects were demonstrated for PEG 6000-stimulated drought stress [43]. The large amount of secondary metabolites in *S. rebaudiana* provoke considerations about the role of SGs in plant adaptivity and overlapping of SGs functions with those of phenolic compounds, especially given the significant metabolic cost.

SGs play important role in the adaptation of plants to stress environments via alleviating stress associated effects [44,45]. Libik-Konieczny et al. [46] presented a hypothesis that steviol glycosides might function in the protection of photosynthetic apparatus against adverse environmental conditions. A standard literature states that the primary function of polyphenols is also to act as UV protection agents in plants [47,48]. However, this theory was criticized by Kuhnert and Karaköse [49], since they could not correlate the sunshine hours to phenolics nor a single flavonoid or a single chlorogenic acid, except for cis-caffeoyl derivatives. Ceunen and Geuns [8] presented several hypotheses about SGs function in a plant: steviol, the aglycone of SGs, could act as a gibberellin precursor; SGs synthesis is a defense mechanism against insects; SGs might serve as a long-term energy reserve and might play a role in the cellular antioxidant network since they have the capacity to act as potent scavengers of reactive oxygen species; however, all these hypothesis remains inconclusive.

The opposing effects of stressors on SGs and TPC we demonstrated in stevia (increase in SGs content and decrease in TPC) were not observed by other authors. Nevertheless, the tendency found in this study was reproducible in similar study using other type of cold plasma (dielectric barrier discharge plasma) treatment and different varieties of stevia (unpublished data). Libik-Konieczny et al. [46], however, demonstrated the opposite reaction of stevia to adverse environmental conditions, i.e., decrease in SGs amount, but similar tendency of inversely proportional changes in SGs and TPC—the climatic stress resulted in lower SGs production but higher TPC. There is little information on the crosstalk of the SG biosynthesis pathways with biosynthetic pathways of phenolic compounds and flavonoids. Plastid-derived terpenoids SGs are synthesized by the methyl-erythritol phosphate (MEP) pathway (Figure 5), whereas phenolic compounds and flavonoids—via the shikimate/phenylpropanoid pathway. The terpenoid and flavonoid biosynthetic pathways are supplied with carbon skeletons generated in a primary metabolism but, otherwise, are thought to operate independently of one another in most plant tissues [50] (Figure 5). One of the possible explanations for the opposite effects on biosynthesis of SGs and phenolic or flavonoid compounds (as well as antioxidant activity, which is determined by the phenylpropanoids) observed in this study is competition between MEP and phenylpropanoid pathways for the common precursor phosphoenolpyruvate. The last decade's studies, however, are beginning to uncover metabolic and regulatory connections between these two major branches of the specialized plant metabolism. Some metabolites from MEP pathway such as DMAPP can be used for the synthesis of both flavonoids and terpenoids [51–54]. Moreover, certain transcription factors coordinate metabolic activities between the flavonoid and terpenoid biosynthetic pathways [55,56]. Clearly, much more research must be done to uncover all levels of regulation of metabolic pathways in order to explain fundamentals and targets of CP and EMF action and gain knowledge for treatment protocol development and effect prediction.

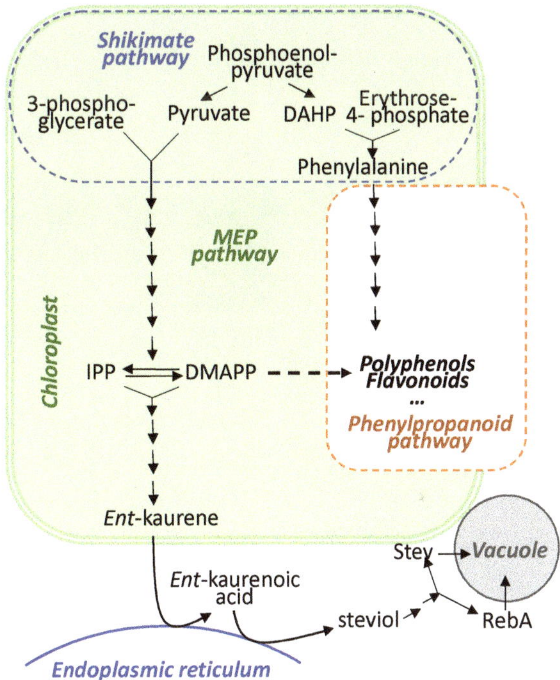

Figure 5. Schematic overview of terpenoid and flavonoid/ polyphenols biosynthetic pathways. The names of compounds are as follows: DAHP, 3-deoxy-D-arabino-heptulosonate 7-phosphate; DMAPP, dimethylallyl diphosphate; MEP, methyl-erythritol phosphate. Lines of arrows indicate multistep reactions, dashed arrow—putative reaction direction.

Whatever molecular mechanisms are involved in the observed effects, our findings provide a new promising tool for manipulation in SGs' production in *Stevia rebaudiana*. Future research could be directed to find out the optimal treatment dose/duration, to investigate the persistence of the effects in the time course of treated seed storage, the dynamics of the induced changes during different vegetation stages and optimal harvest time, the trait transfer possibility to vegetatively or sexually propagated plants, and the reproducibility of the effect by applying different plasma sources on different cultivars of *Stevia rebaudiana*.

4. Materials and Methods

4.1. Chemicals and Reagents

The standard of rutin, galic acid, Folin-Ciocalteu's phenol reagent, HPLC grade methanol, 2,2-diphenyl-1-picrylhydrazyl, ethanol were obtained from Sigma Aldrich (St.Louis, MO, USA), stevioside and rebaudioside A were from TransMIT (Geiben, Germany), HPLC-grade acetonitrile, sodium acetate were from Sharlau Chemie S. A. (Sentmenat, Spain), HCl, sodium carbonate, acetic acid were from Carl Roth (Karlsruhe, Germany), hexamethylenetetramine, aluminum chloride were from Thermo Fisher Scientific (Lankashire, UK). All solutions were prepared with ultrapure 18.2 MΩ water from a Watek ultrapure water purification system (Watek Ltd., Ledeč nad Sázavou, Czech Republic).

4.2. Plant Material

Seeds of *Stevia rebaudiana* Bertoni cultivar Criolla were received from Academy of Agricultural Sciences, Vytautas Magnus University (Kaunas, Lithuania). Seed quality was checked using a stereomicroscope, and pale seeds were removed as sterile, thus, dark seeds only were used for experiments.

4.3. Seed Treatment with CP and EMF

A schematic diagram of the experimental setup for seed treatment with CP and RF EMF is presented in Figure 6.

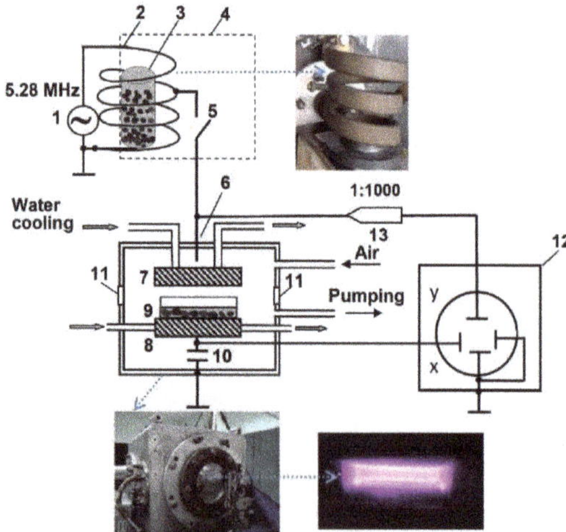

Figure 6. Schematic diagram of the experimental setup: 1—RF generator, 2—inductor (the view right), 3—dielectric container with seeds, 4—screen, 5—commutator, 6—vacuum chamber (the view is shown below), 7—powered electrode, 8—lower electrode, 9—Petri dish with seeds, 10—measuring capacitor, 11—window (the view of ignited plasma is shown below right), 12—oscilloscope, 13—voltage probe.

Low pressure capacitively coupled plasma was produced in a planar geometry reactor consisting of two water-cooled copper electrodes placed at 20 mm from each other in a stainless-steel hermetic chamber. RF voltage was applied to the upper electrode by the commutator 5 (Figure 6). Plasma diagnostic methods including optical emission spectroscopy (OES) based on nitrogen emissions and the discharge characteristic measurement is used to control plasma parameters during the treatments. The effective electron temperature T_e was determined using the technique based on the measurement of the ratio I^{391}/I^{394} of the peak intensities of emission of the ionic N_2^+ (λ = 391.4 nm) and molecular N_2 (λ = 394.3 nm) nitrogen bands [57]. Emission spectra were recorded in the range from 220 to 950 nm by a spectrometer SL100 (SOL Instruments Ltd.) equipped with a CCD area image sensor S10141 (Hamamatsu Photonics Norden AB, Sweden). Typical emission spectra of air plasma can be found elsewhere [24,58]. The effective electron density n_e was estimated from the expression: $j = n_e \cdot e \cdot v_d$, where j—current density, v_d—drift velocity of the electrons, e—electron charge. The current density j was calculated as $j = I/S$, where I is the discharge current, S is the area of the electrode. Experimental setup for measurement the discharge characteristics is presented in [59]. Open sterile glass Petri dish with evenly dispersed 100 seeds was placed on a grounded electrode. Before igniting the discharge (parameters are given in Table 3), the air was pumped from the chamber for about 7 min to reach the working pressure.

Seed treatment with RF EMF was carried out by placing the dielectric container with seeds in three-turn water-cooled coil of the RF generator. The characteristics of EMF are shown in Table 3, the electric and magnetic strength components in the axial zone of the coil were as described previously [35]. The treatment was performed in ambient air at atmospheric pressure and room temperature. Based on some our previous studies on different plant seeds [25,26,28,29], the chosen duration for seed treatments was 5 and 7 min for cold plasma (this treatment is further abbreviated as CP5 and CP7, respectively), and 10 min for EMF treatment (this treatment is abbreviated as EMF10).

Control, CP- and EMF-treated seeds were stored at room temperature (19–22 °C) for 6 days until sowing in vitro.

Table 3. Values of different parameters of CP and EMF applied in *Stevia rebaudiana* seed treatment.

Parameter	Value
CP treatment	
Discharge frequency	5.28 MHz
Pressure	200 Pa
Input power	~8.4 W
Effective electron temperature (T_e)	~2.3 eV
Effective electron density (n_e)	~5 × 10^8 cm^{-3}
Electrode diameter	120 mm
Distance between electrodes	20 mm
EMF treatment	
Frequency	5.28 MHz
Pressure	Atmospheric
Amplitude value of the magnetic component	835 A/m ($B \approx$ 1 mT)
Amplitude value of the electric component	17.96 kV/m

4.4. Seed Germination Test

The untreated (control) seeds and seeds exposed to EMF and CP were evenly distributed on two layers of filter paper in 90-mm-diameter plastic Petri dishes (four replicates of 25 seeds each) and watered with 5 mL distilled water. Petri dishes with seeds were placed in a climatic chamber (Pol-Eko-Aparatura KK 750, Poland) with automatic control of relative humidity (60%), light, and temperature. Alternating light regimes were maintained in the chamber (16 h light, 8 h dark) and constant temperature of 25 ± 1 °C. Seeds were provided additional water, if necessary, to prevent drying. A seed was considered germinated when the initial emergence of the radicle was observed. Germinated seeds were counted daily until their number stopped increasing.

The effects of stressors on germination were estimated by the induced changes in parameters of germination kinetics, derived using application of Richards' function [60] for the analysis of germinating seed population [61]: Vi (%)— the final germination percentage indicating seed viability, Me (days)—the median germination time ($t_{50\%}$) indicating the germination halftime of a seed lot or germination rate, and Qu (days)—the quartile deviation indicating the dispersion of germination time in a seed lot (half of seeds with an average growth time germinate in the range Me $\pm$ Qu).

4.5. Plant Cultivation

After germination, seedlings were carefully transferred from Petri dishes to plastic growth containers filled with substrate BioSoil (SIA "Green-PIK LAT", reg. No. K0.02-1386-16, Latvia), consisting of high-quality vermicompost, moss peat and sand. Characteristics: nitrogen (N)—0.3%, phosphorus (P_2O_5)—0.2%, potassium (K_2O)—0.3%, organic substances—min 14.6%, humidity—max 30%, pH 6–7. Seedlings were planted in 9 $\times$ 9 $\times$ 10 cm containers and grown under greenhouse conditions, with long day photoperiod (16 h light, 8 h dark), relative humidity of 60% and constant temperature of 25 $\pm$ 1 °C.

4.6. Extract Preparation

The leaves were collected from 8-week-old plants and dried at 40 °C for 24 h. Dried leaves were powdered using a batch mill with disposable grinding chamber (Tube-Mill control, IKA, Staufen, Germany), and 1 g of powder was mixed with 50 mL of 70% ethanol. The extraction was carried out in triplicate by sonication for 60 min at 25 °C. The mixture was centrifuged at 16,000$\times$ g for 10 min, the supernatant was collected and kept at -20 °C until analysis.

4.7. HPLC Analysis of Steviol Glycosides

Steviol glycosides rebaudioside A (RebA) and stevioside (Stev) were separated and quantified using high-performance liquid chromatography (HPLC) [62]. An Agilent 1200 series HPLC system (Agilent Technologies Inc., Santa Clara, CA, USA) with a diode array detector was used. Samples were filtered through a syringe filter with a PVDF membrane (pore diameter 0.22 μm) and separated on a reversed phase column (Purospher STAR RP-18e 5 μm Hibar 2 $\times$ 250 mm, Merck, Germany) with a precolumn. Injection volume was 10 μL at 70 °C column temperature. Isocratic elution at a flow rate of 0.25 mL min^{-1} with a mobile phase consisting of 70% deionized water acidified with HCl to pH 2.75 and 30% acetonitrile was used for separation with an additional washing step with 50% acetonitrile. RebA and Stev were detected at the wavelength of 210 nm. Calibration was done by plotting the peak area responses against the concentration values in the concentration range from 1 to 1000 μg mL^{-1} with linear dependence for both analytes. Each analysis was repeated three times, and the mean value was used.

4.8. Determination of Total Phenolic Content

Total phenolic content was determined using the modified Folin–Ciocalteu method [62]. An amount of 0.2 mL of stevia extract was mixed with 1 mL of 0.2 N Folin–Ciocalteu reagent and 0.8 mL 7.5% sodium carbonate solution. After 60 min of incubation in the dark at room temperature, absorbance was measured at 760 nm. Gallic acid was used as a standard, and results were expressed by mg of gallic acid equivalent (GAE) mg g^{-1} of dry weight (DW).

4.9. Determination of Total Flavonoid Content

Total flavonoid content was analyzed by a colorimetric method based on the complexation of phenolic compounds with Al (III) [62]. An amount of 80 μL of stevia extract was mixed with 1920 μL of a reagent containing 40% ethanol, 0.7% acetic acid, 0.4% hexamethylenetetramine, and 0.6% aluminum chloride. After 30 min of incubation in the dark at 4 °C, absorbance was measured at 407 nm. Rutin was used as a standard, and results were expressed by mg of rutin equivalent (RUE) mg g^{-1} of DW.

4.10. Determination of Antioxidant Activity

Antioxidant activity was measured based on the scavenging of the stable 2,2-diphenyl-1-picrylhydrazyl free radical (DPPH) as described [62]. Accordingly, 50 µL of stevia extract were mixed with 1950 µL of a DPPH solution (0.025 mg mL, prepared in acetonitrile:methanol:sodium acetate buffer (100 mM, pH 5.5) (1:1:2)). After 15 min of incubation in the dark at room temperature, absorbance was measured at 515 nm. Rutin was used as a standard, and antioxidant activity was expressed by mg of rutin equivalent (RUE) mg g^{-1} of DW.

4.11. Statistical Analysis

Statistical analysis of the results was performed using Statistica 10 software (issued by IBM Lietuva, Vilnius, Lithuania to Vytautas Magnus University). Data are presented as means $\pm$ SEM ($n = 4$). The number of measured plants in the control and treatment groups varied from 24 to 33. Statistical significance of CP and EMF effects was evaluated using Student's t-test (unpaired). The differences were assumed to be statistically significant when $p < 0.05$.

5. Patents

Patent application LT2020 560 "The method for the increase of steviol glycosides amount in stevia plants by seed treatment with cold plasma before sowing" (3 December 2020) at The State Patent Bureau of the Republic of Lithuania resulted from the work reported in this manuscript (inventors: Rasa Žūkienė, Vida Mildažienė).

Author Contributions: Conceptualization, R.Ž. and V.M; methodology, A.J., I.F., V.L. and R.Ž.; validation, R.Ž., A.J. and V.M.; formal analysis, V.M.; investigation, R.Ž. and A.J.; data curation, R.Ž.; writing—original draft preparation, R.Ž.; writing—review and editing, V.M.; visualization, R.Ž.; supervision, R.Ž. All authors have read and agreed to the published version of the manuscript.

Funding: This research received no external funding.

Institutional Review Board Statement: Not applicable.

Informed Consent Statement: Not applicable.

Data Availability Statement: Not applicable.

Conflicts of Interest: The authors declare no conflict of interest.

References

1. Gantait, S.; Das, A.; Banerjee, J. Geographical Distribution, Botanical Description and Self-Incompatibility Mechanism of Genus Stevia. *Sugar Tech* **2018**, *20*, 1–10. [CrossRef]
2. Verified Market Research. *Global Stevia Market Size By Form, By Application, By Geographic Scope and Forecast*; Verified Market Research: New York, NY, USA, 2021; p. 202. Available online: https://www.verifiedmarketresearch.com (accessed on 31 January 2022).
3. Libik-Konieczny, M.; Capecka, E.; Tuleja, M.; Konieczny, R. Synthesis and production of steviol glycosides: Recent research trends and perspectives. *Appl. Microbiol. Biotechnol.* **2021**, *105*, 3883–3900. [CrossRef] [PubMed]
4. Tavarini, S.; Angelini, L.G. Stevia rebaudiana Bertoni as a source of bioactive compounds: The effect of harvest time, experimental site and crop age on Steviol glycoside content and antioxidant properties. *J. Sci. Food Agric.* **2012**, *93*, 2121–2129. [CrossRef] [PubMed]
5. Lemus-Mondaca, R.; Vega-Galvez, A.; Zura-Bravo, L.; Ah-Hen, K. Stevia rebaudiana Bertoni, source of a high-potency natural sweetener: A comprehensive review on the biochemical, nutritional and functional aspects. *Food Chem.* **2012**, *132*, 1121–1132. [CrossRef]
6. Yadav, A.K.; Singh, S.; Dhyani, D.; Ahuja, P. A review on the improvement of stevia [Stevia rebaudiana (Bertoni)]. *Can. J. Plant Sci.* **2011**, *91*, 1–27. [CrossRef]

7. Testai, L.; Calderone, V. Stevia rebaudiana Bertoni: Beyond Its Use as a Sweetener. Pharmacological and Toxicological Profile of Steviol Glycosides of Stevia rebaudiana Bertoni. In *Steviol Glycosides: Cultivation, Processing, Analysis and Applications in Food*, 2nd ed.; Wölwer-Rieck, U., Ed.; Royal Society of Chemistry: Cambridge, UK, 2019; pp. 148–161. [CrossRef]

8. Ceunen, S.; Geuns, J.M.C. Steviol Glycosides: Chemical Diversity, Metabolism, and Function. *J. Nat. Prod.* **2013**, *76*, 1201–1228. [CrossRef]

9. Tavarini, S.; Clemente, C.; Bender, C.; Angelini, L.G. Health-promoting compounds in stevia: The effect of mycorrhizal symbiosis, phosphorus supply and harvest time. *Molecules* **2020**, *25*, 5399. [CrossRef]

10. Javed, R.; Usman, M.; Yücesan, B.; Zia, M.; Gürel, E. Effect of zinc oxide (ZnO) nanoparticles on physiology and steviol glycosides production in micropropagated shoots of Stevia rebaudiana Bertoni. *Plant Physiol. Biochem.* **2017**, *110*, 94–99. [CrossRef]

11. Majlesi, Z.; Ramezani, M.; Gerami, M. Investigation on some main glycosides content of Stevia rebaudiana B. under different concentrations of commercial and synthesized silver nanoparticles. *Pharm. Biomed. Res.* **2018**, *4*, 8–14. [CrossRef]

12. Lemus-Mondaca, R.; Ah-Hen, K.; Vega-Gálvez, A.; Honores, C.; Moraga, N.O. Stevia rebaudiana leaves: Effect of drying process temperature on bioactive components, antioxidant capacity and natural sweeteners. *Plant Foods Hum. Nutr.* **2016**, *71*, 49–56. [CrossRef]

13. Bursać Kovačević, D.; Maras, M.; Barba, F.J.; Granato, D.; Roohinejad, S.; Mallikarjunan, K.; Montesano, D.; Lorenzo, J.M.; Putnik, P. Innovative technologies for the recovery of phytochemicals from Stevia rebaudiana Bertoni leaves: A review. *Food Chem.* **2018**, *268*, 513–521. [CrossRef] [PubMed]

14. Chen, L.; Cai, R.; Weng, J.; Li, Y.; Jia, H.; Chen, K.; Yan, M.; Ouyang, P. Production of rebaudioside D from stevioside using a UGTSL2 Asn358Phe mutant in a multi-enzyme system. *Microb. Biotechnol.* **2020**, *13*, 974–983. [CrossRef] [PubMed]

15. Waskow, A.; Howling, A.; Furno, I. Mechanisms of plasma-seed treatments as a potential seed processing technology. *Front. Phys.* **2021**, *9*, 617345. [CrossRef]

16. Misra, N.; Schlutter, O.; Cullen, P. Plasma in Food and agriculture. In *Cold Plasma in Food and Agriculture: Fundamentals and Applications*, 3rd ed.; Patrick, J., Ed.; Elsevier: Amsterdam, The Netherlands, 2016; pp. 1–16.

17. Kaur, S.; Vian, A.; Chandel, S.; Singh, H.P.; Batish, D.R.; Kohli, R.K. Sensitivity of plants to high frequency electromagnetic radiation: Cellular mechanisms and morphological changes. *Rev. Environ. Sci. Biotechnol.* **2021**, *20*, 55–74. [CrossRef]

18. Attri, P.; Ishikawa, K.; Okumura, T.; Koga, K.; Shiratani, M. Plasma agriculture from laboratory to farm: A review. *Processes* **2020**, *8*, 1002. [CrossRef]

19. Staric, P.; Vogel-Mikuš, K.; Mozetic, M.; Junkar, I. Effects of nonthermal plasma on morphology, genetics and physiology of seeds: A review. *Plants* **2020**, *9*, 1736. [CrossRef]

20. Adhikari, B.; Adhikari, M.; Park, G. The effects of plasma on plant growth, development, and sustainability. *Appl. Sci.* **2020**, *10*, 6045. [CrossRef]

21. Song, J.-S.; Kim, S.B.; Ryu, S.; Oh, J.; Kim, D.-S. Emerging plasma technology that alleviates crop stress during the early growth stages of plants: A Review. *Front. Plant Sci.* **2020**, *11*, 988. [CrossRef]

22. Holubová, L.; Kyzek, S.; Durovcová, I.; Fabová, J.; Horváthová, E.; Ševcovicová, A.; Gálová, E. Non-thermal plasma—a new green priming agent for plants? *Int. J. Mol. Sci.* **2020**, *21*, 9466. [CrossRef]

23. Mildaziene, V.; Pauzaite, G.; Nauciené, Z.; Malakauskiene, A.; Zukiene, R.; Januskaitiene, I.; Jakstas, V.; Ivanauskas, L.; Filatova, I.; Lyushkevich, V. Pre-sowing Seed Treatment with Cold Plasma and Electromagnetic Field Increases Secondary Metabolite Content in Purple Coneflower (*Echinacea purpurea*) Leaves. *Plasma Process. Polym.* **2018**, *15*, 1700059. [CrossRef]

24. Ivankov, A.; Nauciene, Z.; Zukiene, R.; Degutyte-Fomins, L.; Malakauskiene, A.; Kraujalis, P.; Venskutonis, P.R.; Filatova, I.; Lyushkevich, V.; Mildaziene, V. Changes in Growth and Production of Non-Psychotropic Cannabinoids Induced by Pre-Sowing Treatment of Hemp Seeds with Cold Plasma, Vacuum and Electromagnetic Field. *Appl. Sci.* **2020**, *10*, 8519. [CrossRef]

25. Mildaziene, V.; Paužaitė, G.; Naučienė, Z.; Zukiene, R.; Malakauskienė, A.; Norkeviciene, E.; Slepetiene, A.; Stukonis, V.; Olauskaite, V.; Padarauskas, A.; et al. Effect of Seed Treatment with Cold Plasma and Electromagnetic Field on Red Clover Germination, Growth and Content of Major Isoflavones. *J. Phys. D* **2020**, *53*, 264001. [CrossRef]

26. Mildaziene, V.; Ivankov, A.; Pauzaite, G.; Naučienė, Z.; Zukiene, R.; Degutyte-Fomins, L.; Pukalskas, A.; Venskutonis, P.R.; Filatova, I.; Lyushkevich, V. Seed Treatment with Cold Plasma and Electromagnetic Field Induces Changes in Red Clover Root Growth Dynamics, Flavonoid Exudation, and Activates Nodulation. *Plasma Process. Polym.* **2020**, *18*, 2000160. [CrossRef]

27. Ivankov, A.; Naučienė, Z.; Degutytė-Fomins, L.; Žūkienė, R.; Januškaitienė, I.; Malakauskienė, A.; Jakštas, V.; Ivanauskas, L.; Romanovskaja, D.; Šlepetienė, A.; et al. Changes in Agricultural Performance of Common Buckwheat Induced by Seed Treatment with Cold Plasma and Electromagnetic Field. *Appl. Sci.* **2021**, *11*, 4391. [CrossRef]

28. Ivankov, A.; Zukiene, R.; Nauciene, Z.; Degutyte-Fomins, L.; Filatova, I.; Lyushkevich, V.; Mildaziene, V. The effects of red clover seed treatment with cold plasma and electromagnetic field on germination and seedling growth are dependent on seed color. *Appl. Sci.* **2021**, *11*, 4676. [CrossRef]

29. Degutytė-Fomins, L.; Paužaitė, G.; Žūkienė, R.; Mildažienė, V.; Koga, K.; Shiratani, M. Relationship between cold plasma treatment-induced changes in radish seed germination and phytohormone balance. *Jpn. J. Appl. Phys.* **2020**, *59*, SH1001. [CrossRef]

30. Baskin, C.C.; Baskin, J.M. Types of Seeds and Kinds of Seed Dormancy. In *Seeds*, 2nd ed.; Baskin, C.C., Baskin, J.M., Eds.; Academic Press: Waltham, MA, USA, 2014; pp. 37–77. [CrossRef]

31. Aman, N.; Hadi, F.; Khalil, S.A.; Zamir, R.; Ahmad, N. Efficient regeneration for enhanced steviol glycosides production in Stevia rebaudiana (Bertoni). *Comptes Rendus Biol.* **2013**, *336*, 486–492. [CrossRef]

32. Zukiene, R.; Nauciene, Z.; Januskaitiene, I.; Pauzaite, G.; Mildaziene, V.; Koga, K.; Shiratani, M. Dielectric barrier discharge plasma treatment induced changes in sunflower seed germination, phytohormone balance, and seedling growth. *Appl. Phys. Express* **2019**, *12*, 126003. [CrossRef]

33. Varnagiris, S.; Vilimaite, S.; Mikelionyte, I.; Urbonavicius, M.; Tuckute, S.; Milcius, D. The Combination of Simultaneous Plasma Treatment with Mg Nanoparticles Deposition Technique for Better Mung Bean Seeds Germination. *Processes* **2020**, *8*, 1575. [CrossRef]

34. Holc, M.; Mozetič, M.; Recek, N.; Primc, G.; Vesel, A.; Zaplotnik, R.; Gselman, P. Wettability Increase in Plasma-Treated Agricultural Seeds and Its Relation to Germination Improvement. *Agronomy* **2021**, *11*, 1467. [CrossRef]

35. Mildaziene, V.; Pauzaite, G.; Malakauskiene, A.; Zukiene, R.; Nauciene, Z.; Filatova, I.; Azharonok, V.; Lyushkevich, V. Response of perennial woody plants to seed treatment by electromagnetic field and low-temperature plasma. *Bioelectromagnetics* **2016**, *37*, 536–548. [CrossRef] [PubMed]

36. Piel, A. *Plasma Physics. An Introduction to Laboratory, Space, and Fusion Plasmas*, 2nd ed.; Springer International Publishing AG: Cham, Switzerland, 2017.

37. Bera, K.; Dutta, P.; Sadhukhan, S. Seed priming with non-ionizing physical agents: Plant responses and underlying physiological mechanisms. *Plant Cell Rep.* **2022**, *41*, 53–73. [CrossRef] [PubMed]

38. Paužaitė, G.; Malakauskienė, A.; Naučienė, Z.; Žūkienė, R.; Filatova, I.; Lyushkevich, V.; Azarko, I.; Mildažienė, V. Changes in Norway spruce germination and growth induced by pre-sowing seed treatment with cold plasma and electromagnetic field: Short-term versus long-term effects. *Plasma Process. Polym.* **2018**, *15*, 1700068. [CrossRef]

39. Mildažienė, V.; Aleknavičiūtė, V.; Žūkienė, R.; Paužaitė, G.; Naučienė, Z.; Filatova, I.; Lyushkevich, V.; Haimi, P.; Tamošiūnė, I.; Baniulis, D. Treatment of Common sunflower (*Helianthus annus* L.) seeds with radio-frequency electromagnetic field and cold plasma induces changes in seed phytohormone balance, seedling development and leaf protein expression. *Sci. Rep.* **2019**, *9*, 6437. [CrossRef]

40. Lucho, S.R.; do Amaral, M.N.; Milech, C.; Ferrer, M.Á.; Calderón, A.A.; Bianchi, V.J.; Braga, E.J.B. Elicitor-Induced Transcriptional Changes of Genes of the Steviol Glycoside Biosynthesis Pathway in Stevia rebaudiana Bertoni. *J. Plant Growth Regul.* **2018**, *37*, 971–985. [CrossRef]

41. Kumar, H.; Kaul, K.; Bajpai-Gupta, S.; Kaul, V.K.; Kumar, S. A comprehensive analysis of fifteen genes of steviol glycosides biosynthesis pathway in Stevia rebaudiana (Bertoni). *Gene* **2012**, *492*, 276–284. [CrossRef]

42. Saifi, M.; Nasrullah, N.; Ahmad, M.M.; Ali, A.; Khan, J.A.; Abdin, M.Z. In silico analysis and expression profiling of miRNAs targeting genes of steviol glycosides biosynthetic pathway and their relationship with steviol glycosides content in different tissues of Stevia rebaudiana. *Plant Physiol. Biochem.* **2015**, *94*, 57–64. [CrossRef]

43. Ahmad, M.A.; Javed, R.; Adeel, M.; Rizwan, M.; Yang, Y. PEG 6000-Stimulated Drought Stress Improves the Attributes of In Vitro Growth, Steviol Glycosides Production, and Antioxidant Activities in Stevia rebaudiana Bertoni. *Plants* **2020**, *9*, 1552. [CrossRef]

44. Hill, C.B.; Roessner, U. Advances in high-throughput untargeted LC-MS analysis for plant metabolomics. In *Advanced LC-MS Applications for Metabolomics*, 2nd ed.; De Vooght-Johnson, R., Ed.; Future Medicine Ltd.: London, UK, 2015; pp. 58–71. [CrossRef]

45. Ramakrishna, A.; Ravishankar, G.A. Influence of abiotic stress signals on secondary metabolites in plants. *Plant Signal. Behav.* **2011**, *6*, 1720–1731. [CrossRef]

46. Libik-Konieczny, M.; Capecka, E.; Kąkol, E.; Dziurka, M.; Grabowska-Joachimiak, A.; Sliwinska, E.; Pistellie, L. Growth, development and steviol glycosides content in the relation to the photosynthetic activity of several Stevia rebaudiana Bertoni strains cultivated under temperate climate conditions. *Sci. Hortic.* **2018**, *234*, 10–18. [CrossRef]

47. Haslam, E. Plant polyphenols (syn. vegetable tannins) and chemical defense—A reappraisal. *J. Chem. Ecol.* **1988**, *14*, 1789–1805. [CrossRef] [PubMed]

48. Haslam, E. Natural Polyphenols (Vegetable Tannins) as Drugs: Possible Modes of Action. *J. Nat. Prod.* **1996**, *59*, 205–215. [CrossRef] [PubMed]

49. Kuhnert, N.; Karaköse, H. Presentation and Analysis of Other Constituents in the Leaves: Polyphenolics in Stevia rebaudiana Leaves. In *Steviol Glycosides: Cultivation, Processing, Analysis and Applications in Food*, 2nd ed.; Wölwer-Rieck, U., Ed.; Royal Society of Chemistry: Cambridge, UK, 2019; pp. 113–124. [CrossRef]

50. Kang, J.H.; McRoberts, J.; Shi, F.; Moreno, J.E.; Jones, A.D.; Howe, G.A. The flavonoid biosynthetic enzyme chalcone isomerase modulates terpenoid production in glandular trichomes of tomato. *Plant Physiol.* **2014**, *164*, 1161–1174. [CrossRef]

51. Nagel, J.; Culley, L.K.; Lu, Y.P.; Liu, E.W.; Matthews, P.D.; Stevens, J.F.; Page, J.E. EST analysis of hop glandular trichomes identifies an O-methyltransferase that catalyzes the biosynthesis of xanthohumol. *Plant Cell* **2008**, *20*, 186–200. [CrossRef]

52. Wang, G.D.; Tian, L.; Aziz, N.; Broun, P.; Dai, X.B.; He, J.; King, A.; Zhao, P.X.; Dixon, R.A. Terpene biosynthesis in glandular trichomes of hop. *Plant Physiol.* **2008**, *148*, 1254–1266. [CrossRef] [PubMed]

53. Shen, G.A.; Huhman, D.; Lei, Z.T.; Snyder, J.; Sumner, L.W.; Dixon, R.A. Characterization of an isoflavonoid-specific prenyltransferase from Lupinus albus. *Plant Physiol.* **2012**, *159*, 70–80. [CrossRef] [PubMed]

54. Voo, S.S.; Grimes, H.D.; Lange, B.M. Assessing the biosynthetic capabilities of secretory glands in Citrus peel. *Plant Physiol.* **2012**, *159*, 81–94. [CrossRef] [PubMed]

55. Bedon, F.; Bomal, C.; Caron, S.; Levasseur, C.; Boyle, B.; Mansfield, S.D.; Schmidt, A.; Gershenzon, J.; Grima-Pettenati, J.; Séguin, A.; et al. Subgroup 4 R2R3-MYBs in conifer trees: Gene family expansion and contribution to the isoprenoid- and flavonoid-oriented responses. *J. Exp. Bot.* **2010**, *61*, 3847–3864. [CrossRef]
56. Ben Zvi, M.M.; Shklarman, E.; Masci, T.; Kalev, H.; Debener, T.; Shafir, S.; Ovadis, M.; Vainstein, A. PAP1 transcription factor enhances production of phenylpropanoid and terpenoid scent compounds in rose flowers. *New Phytol.* **2012**, *195*, 335–345. [CrossRef]
57. Britun, N.; Gaillard, M.; Ricard, A.; Kim, Y.M.; Kim, K.S.; Han, J.G. Determination of the vibrational, rotational and electron temperatures in N_2 and Ar–N_2 rf discharge. *J. Phys. D Appl. Phys.* **2007**, *40*, 1022. [CrossRef]
58. Filatova, I.I.; Azharonok, V.; Lushkevich, V.; Zhukovsky, A.; Gadzhieva, G.; Spaisić, K.; Spasić, K.; Živković, S.; Puać, N.; Lasović, S.; et al. Plasma seeds treatment as a promising technique for seed germination improvement. In Proceedings of the 31st International Conference on Phenomena in Ionized Gases, ICPIG (International Conference on Phenomena in Ionized Gases), Granada, Spain, 14–19 July 2013; pp. 4–7.
59. Filatova, I.; Lyushkevich, V.; Goncharik, S.; Zhukovsky, A.; Krupenko, N.; Kalatskaja, J. The effect of low-pressure plasma treatment of seeds on the plant resistance to pathogens and crop yields. *J. Phys. D Appl. Phys.* **2020**, *53*, 244001. [CrossRef]
60. Richards, F.J. A flexible growth function for empirical use. *J. Exp. Bot.* **1959**, *10*, 290–300. [CrossRef]
61. Hara, Y. Calculation of population parameters using Richards function and application of indices of growth and seed vigor to rice plants. *Plant Prod. Sci.* **1999**, *2*, 129–135. [CrossRef]
62. Blinstrubienė, A.; Burbulis, N.; Juškevičiūtė, N.; Vaitkevičienė, N.; Žūkienė, R. Effect of growth regulators on *Stevia rebaudiana* Bertoni callus genesis and influence of auxin and proline to steviol glycosides, phenols, flavonoids accumulation, and antioxidant activity in vitro. *Molecules* **2020**, *25*, 2759. [CrossRef] [PubMed]

Article

Plasma-Activated Water Triggers Rapid and Sustained Cytosolic Ca^{2+} Elevations in *Arabidopsis thaliana*

Enrico Cortese [1], Alessio G. Settimi [2], Silvia Pettenuzzo [3,4,5], Luca Cappellin [5], Alessandro Galenda [6], Alessia Famengo [6], Manuele Dabalà [2], Vanni Antoni [7] and Lorella Navazio [1,8,*]

[1] Department of Biology, University of Padova, Via U. Bassi 58/B, 35131 Padova, Italy; enrico.cortese@unipd.it
[2] Department of Industrial Engineering, University of Padova, Via F. Marzolo 9, 35131 Padova, Italy; alessiogiorgio.settimi@unipd.it (A.G.S.); manuele.dabala@unipd.it (M.D.)
[3] Center Agriculture Food Environment (C3A), University of Trento, Via E. Mach 1, 38010 San Michele all'Adige, Italy; silvia.pettenuzzo-1@unitn.it
[4] Research and Innovation Centre, Edmund Mach Foundation, Via E. Mach 1, 38010 San Michele all'Adige, Italy
[5] Department of Chemical Sciences, University of Padova, Via F. Marzolo 1, 35131 Padova, Italy; luca.cappellin@unipd.it
[6] CNR Institute of Condensed Matter Chemistry and Technologies for Energy (ICMATE), Corso Stati Uniti 4, 35127 Padova, Italy; alessandro.galenda@cnr.it (A.G.); alessia.famengo@cnr.it (A.F.)
[7] Consorzio RFX, Corso Stati Uniti 4, 35127 Padova, Italy; vanni.antoni@igi.cnr.it
[8] Botanical Garden, University of Padova, Via Orto Botanico 15, 35123 Padova, Italy
* Correspondence: lorella.navazio@unipd.it

Citation: Cortese, E.; Settimi, A.G.; Pettenuzzo, S.; Cappellin, L.; Galenda, A.; Famengo, A.; Dabalà, M.; Antoni, V.; Navazio, L. Plasma-Activated Water Triggers Rapid and Sustained Cytosolic Ca^{2+} Elevations in *Arabidopsis thaliana*. *Plants* **2021**, *10*, 2516. https://doi.org/10.3390/plants10112516

Academic Editors: Vida Mildažienė and Božena Šerá

Received: 29 October 2021
Accepted: 16 November 2021
Published: 19 November 2021

Publisher's Note: MDPI stays neutral with regard to jurisdictional claims in published maps and institutional affiliations.

Abstract: Increasing evidence indicates that water activated by plasma discharge, termed as plasma-activated water (PAW), can promote plant growth and enhance plant defence responses. Nevertheless, the signalling pathways activated in plants in response to PAW are still largely unknown. In this work, we analysed the potential involvement of calcium as an intracellular messenger in the transduction of PAW by plants. To this aim, *Arabidopsis thaliana* (Arabidopsis) seedlings stably expressing the bioluminescent Ca^{2+} reporter aequorin in the cytosol were challenged with PAW generated by a plasma torch. Ca^{2+} measurement assays demonstrated the induction by PAW of rapid and sustained cytosolic Ca^{2+} elevations in Arabidopsis seedlings. The dynamics of the recorded Ca^{2+} signals were found to depend upon different parameters, such as the operational conditions of the torch, PAW storage, and dilution. The separate administration of nitrate, nitrite, and hydrogen peroxide at the same doses as those measured in the PAW did not trigger any detectable Ca^{2+} changes, suggesting that the unique mixture of different reactive chemical species contained in the PAW is responsible for the specific Ca^{2+} signatures. Unveiling the signalling mechanisms underlying plant perception of PAW may allow to finely tune its generation for applications in agriculture, with potential advantages in the perspective of a more sustainable agriculture.

Keywords: aequorin; *Arabidopsis thaliana*; calcium signalling; cytosolic Ca^{2+} changes; plasma-activated water; plasma torch; reactive oxygen species; reactive nitrogen species

1. Introduction

Cold atmospheric plasmas are weakly ionized gases that can be generated in ambient air. At a relatively low consumption of energy, they constitute a unique delivery system of a rich family of short- and long-lived chemicals, such as reactive oxygen (ROS) and nitrogen (RNS) species, called RONS when grouped together. Cold plasmas have already proven effective in medicine applications such as in regenerative medicine for blood coagulation and dental treatment, as well as in sanitizing surfaces and medical tools, and further applications, such as anti-cancer treatments, are under investigation [1]. When interacting with a liquid, cold plasmas can generate further new chemical species, as in the case of the so-called plasma-activated water (PAW). The nature and concentration of the RONS generated depend on the sources and gases used for plasma generation, on the chemical environment, and can be modulated by varying parameters such as voltage, distance between the liquid and the plasma, exposure time, and type of electrodes used. In the generated PAW short-living species such as hydroxyl- ($^{\bullet}$OH), superoxide- ($O_2^{\bullet -}$), nitric oxide- ($^{\bullet}$NO) radicals, and ozone (O_3) are formed and further react, yielding nitric oxide (NO), nitrate (NO_3^-) and nitrite (NO_2^-), peroxynitrite ($ONOO^-$), and hydrogen peroxide (H_2O_2) [2].

The complex chemistry occurring during the PAW generation has recently attracted a great deal of interest due to a variety of applications in agriculture and in the food sector [2–5]. In plant biology, cold plasma and PAW have been shown to increase the seed germination rate, even under osmotic and saline stresses, as well as to promote plant growth [6–11]. Moreover, PAW irrigation of tomato plants has been reported to induce defence gene expression [12,13] and accumulation of the defence hormones salicylic acid and jasmonic acid [12,14]. A differential expression of genes involved in the main plant defence pathways was also confirmed in PAW-treated periwinkle and grapevine plants [15]. These studies suggest that PAW can play beneficial roles in agriculture by promoting plant growth and pre-alerting plant defence prior to a potential subsequent attack by pathogens, a phenomenon defined as "priming" [8,16]. PAW may, therefore, represent an attractive eco-friendly alternative to pesticides, whose administration in bulk quantities represent a matter of growing concern for their impact on the environment. Nevertheless, studies addressing the signalling pathways activated in plants in response to PAW have been lacking so far.

In this work, we investigated the signalling mechanisms underpinning the effects played by PAW on plants. In particular, we evaluated the potential involvement of calcium in the plant perception and transduction of the mixture of molecules contained in PAW. Calcium is a universal signalling element involved in a wide range of physiological processes in all living organisms [17]. In plants, Ca^{2+} serves as an intracellular messenger of primary importance in many different signal transduction pathways [18]. A plethora of abiotic stimuli, such as touch/wind [19,20], salinity, drought [21–23], oxidative stress [24], and cold/heat stress [25–31], as well as biotic stimuli in pathogenic and beneficial plant-microbe interactions [32,33], have been shown to evoke in plants specific spatio-temporal Ca^{2+} signals, which are further transduced by Ca^{2+} sensor proteins into transcriptional and metabolic responses [34–36]. Notably, Ca^{2+}-based signalling circuits are well conserved along the green lineages, from algae to embryophytes [30,37].

To test the effect of PAW on the induction of transient changes in the cytosolic concentration of the ion ($[Ca^{2+}]_{cyt}$), we used as an experimental system a transgenic line of the model plant *Arabidopsis thaliana* (Arabidopsis) stably expressing the genetically encoded Ca^{2+} indicator aequorin in the cytosol. The obtained results showed that Arabidopsis perception of PAW generated by two different plasma torches is mediated by rapid and sustained cytosolic Ca^{2+} elevations. Further studies are needed in the future to address conserved and unique features of Ca^{2+}-mediated sensing mechanisms of PAW in phylogenetically distant plant species, as well as in plants of economic interest. A better understanding of the biochemical and molecular bases of plant perception of PAW may allow to finely tune the chemical composition of PAW for an optimal application in agriculture.

2. Results

2.1. Generation of PAW by Plasma Torch

During the course of this work two plasma torches (torch #1 and torch #2) were used, both consisting of a non-transferred arc generated through a narrow nozzle less than 1 cm wide, so that a relatively high-power plasma could be generated and concentrated in a relatively narrow surface. To produce PAW, samples of 50 mL deionized H_2O were treated for different time intervals, ranging from 1 to 10 min, with the torch at a distance of 1 to 10 cm from the H_2O surface, in a cooling bath of ice and salt. Operational parameters of the plasma torch were set to operate in a range of power 450–1800 W and with a pressure from 1 to 3 bar. Upon generation, PAW was quickly fractionated in small single-use aliquots and immediately frozen in liquid N_2.

2.2. PAW Triggers a Cytosolic Ca^{2+} Increase in Aequorin-Expressing Arabidopsis thaliana Seedlings

To evaluate the potential involvement of Ca^{2+} signalling in the perception of PAW by plants, an Arabidopsis line stably expressing the bioluminescent Ca^{2+} indicator aequorin in the cytosol was used [38–40]. Ca^{2+} measurement assays were carried out in 7-day-old transgenic Arabidopsis intact seedlings, which were challenged with PAW generated by the above-described plasma torch settings. Plant treatment with PAW induced rapid and sustained cytosolic Ca^{2+} increases (Figure 1a), whose magnitude was found to correlate with the duration of the exposure of H_2O to plasma (Figure 1b). No $[Ca^{2+}]_{cyt}$ changes were observed in control samples, in which deionized H_2O (without activation by plasma) was applied to seedlings (Figure 1a). These data demonstrate that PAW sensing by plants occurs through intracellular Ca^{2+} changes, characterized by a specific signature.

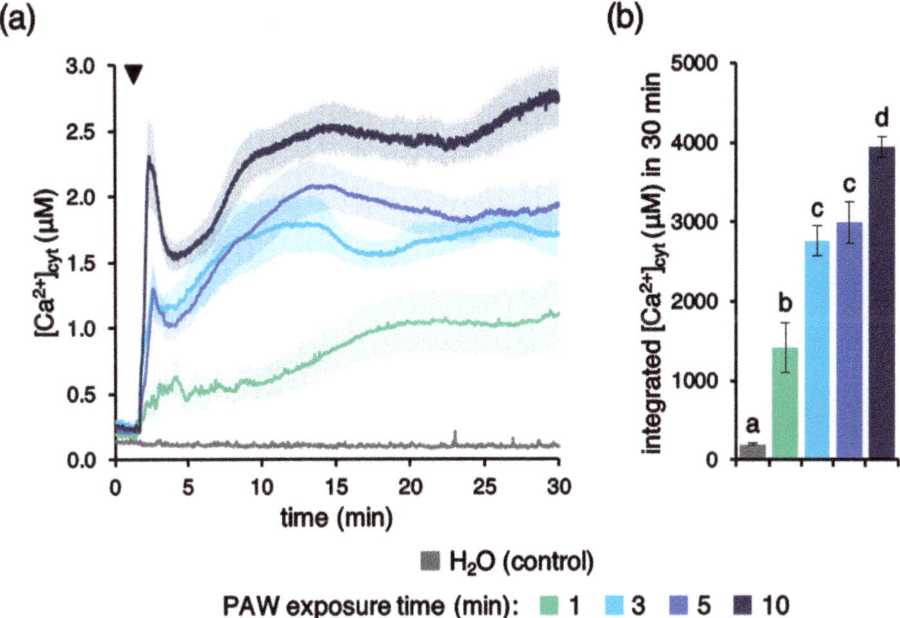

Figure 1. Monitoring of changes in cytosolic Ca^{2+} concentration ($[Ca^{2+}]_{cyt}$) induced by plasma-activated water (PAW) in *Arabidopsis thaliana* (Arabidopsis). Ca^{2+} measurement assays were conducted in Arabidopsis seedlings stably expressing aequorin in the cytosol. Seven-day-old intact seedlings were challenged with 1:2 dilutions of different PAWs, obtained by exposing deionized H_2O to cold plasma (generated at 900 W power, torch #1) for different time intervals: 1 min (green), 3 min (light blue), 5 min (blue), or 10 min (dark blue). Untreated deionized H_2O was administered to control samples (grey). (a) Data are the means (solid lines) ± SE (shading) of six seedlings derived from three independent growth replicates. The arrowhead indicates the time of stimulus application (at 100 s); (b) statistical analyses of integrated $[Ca^{2+}]_{cyt}$ dynamics over 30 min. Bars labelled with different letters differ significantly ($p < 0.05$, Student's t test).

2.3. The Dynamics of the Elicited Ca^{2+} Signals Depend on PAW Features

We next investigated the potential dependence of PAW-induced Ca^{2+} signals on the characteristics of PAW. It is known that the production of plasma-induced chemistry and, in particular, RONS generated within the aqueous medium depends on the plasma source and on several parameters, among which is power [41]. Operating the plasma torch at different power regimes, ranging from 450 to 1800 W, was found to affect the amplitude of the recorded Ca^{2+} changes (Figure 2a,b), confirming a key role played by the modulation of the energy transferred to the pressurized gas during the generation of the cold plasma.

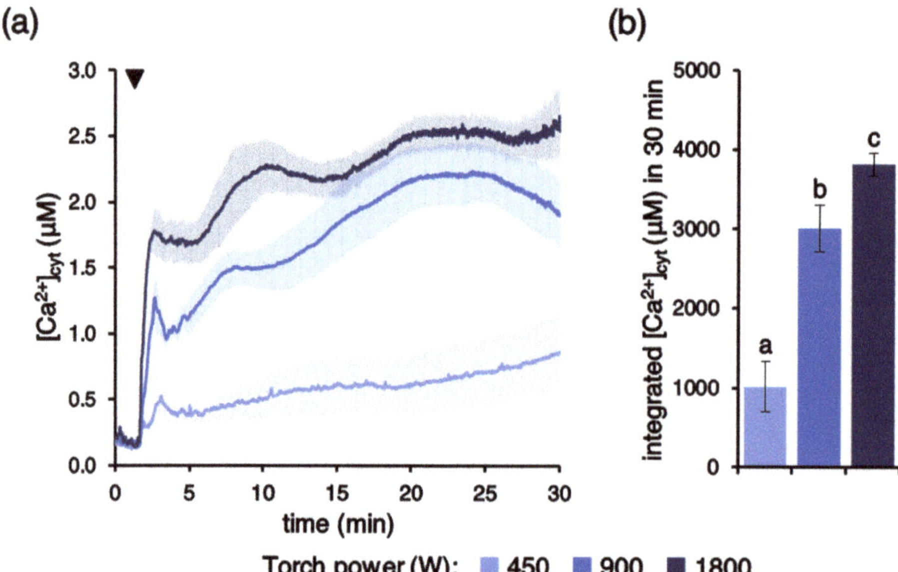

Figure 2. Dependence of the PAW-induced $[Ca^{2+}]_{cyt}$ elevation dynamics on the power of the plasma torch used to generate PAW. Ca^{2+} measurement assays were conducted in aequorin-expressing Arabidopsis seedlings. Seedlings were challenged with 1:2 dilutions of different PAWs, obtained after 5 min exposure of deionized H_2O to cold plasma generated under various torch #1 power conditions: 450 W (pale blue), 900 W (blue), and 1800 W (dark blue). (**a**) Data are the means (solid lines) $\pm$ SE (shading) of six seedlings derived from three independent growth replicates. The arrowhead indicates the time of stimulus application (at 100 s); (**b**) statistical analyses of integrated $[Ca^{2+}]_{cyt}$ dynamics over 30 min. Bars labelled with different letters differ significantly ($p < 0.05$, Student's *t* test).

Ca^{2+} signals with progressively reduced magnitude were triggered by increasing ratios of PAW dilution (Figure 3a,b), demonstrating a dose-dependent effect in the PAW-induced intracellular Ca^{2+} changes triggered in Arabidopsis seedlings.

2.4. Effects of Different Temperature and Time Intervals of PAW Storage on Cytosolic Ca^{2+} Changes

An additional factor that was taken into consideration was the effect of different temperature and time intervals of PAW storage. Upon production, PAW was kept at different temperatures ($-80\ °C$, $4\ °C$, and room temperature [RT]) for increasing time intervals, ranging from the immediate use up to 3 months. Figure 4 shows that PAW stored at $-80\ °C$ retains unvaried inducing activity on $[Ca^{2+}]_{cyt}$ elevations for at least 3 months. On the other hand, storage at either $4\ °C$ or RT was found to severely affect PAW properties, resulting in greatly reduced Ca^{2+} signals already after 1 day from PAW production (Figure 4).

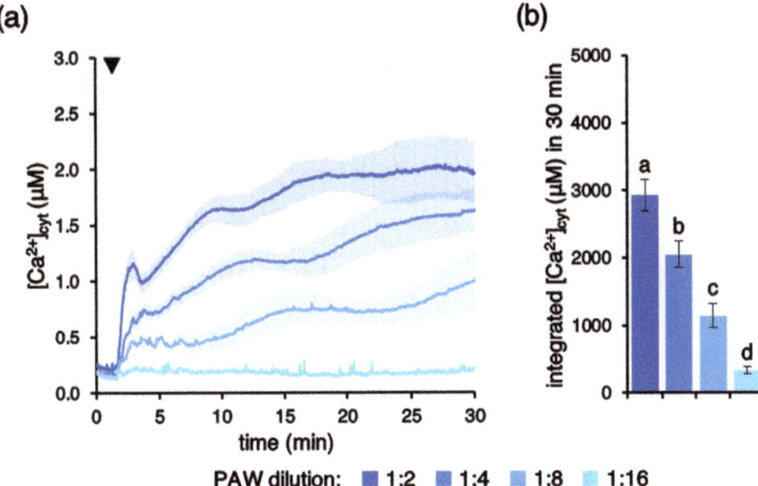

Figure 3. Concentration dependence of PAW-induced $[Ca^{2+}]_{cyt}$ increases. Ca^{2+} measurement assays were conducted in aequorin-expressing Arabidopsis seedlings. Seedlings were challenged with progressive dilutions of PAW (lighter colours indicate more diluted PAWs) generated by exposing deionized H_2O to cold plasma for 5 min at 900 W torch #1 power. (**a**) Data are the means (solid lines) ± SE (shading) of six seedlings derived from three independent growth replicates. The arrowhead indicates the time of stimulus application (at 100 s); (**b**) statistical analyses of integrated $[Ca^{2+}]_{cyt}$ dynamics over 30 min are shown. Bars labelled with different letters differ significantly ($p < 0.05$, Student's *t* test).

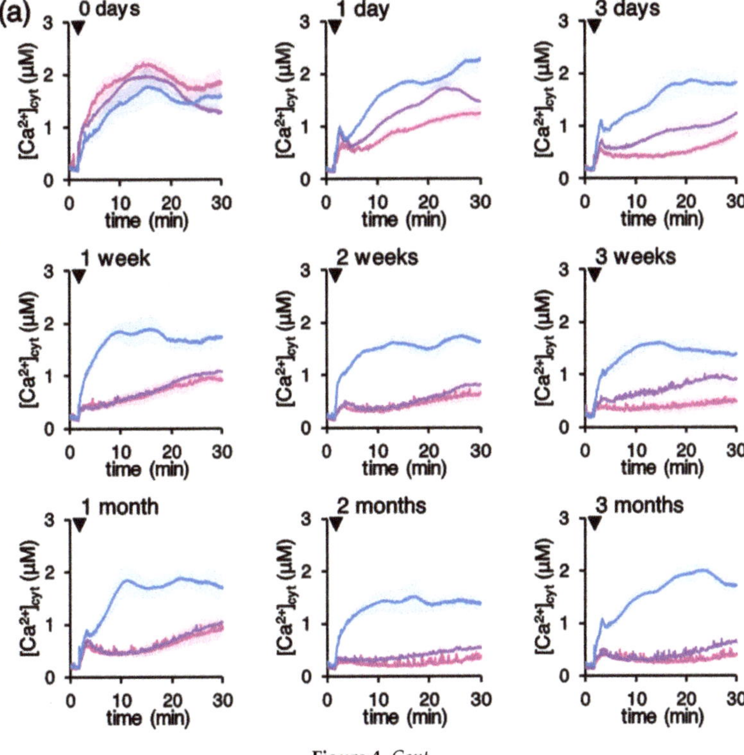

Figure 4. *Cont.*

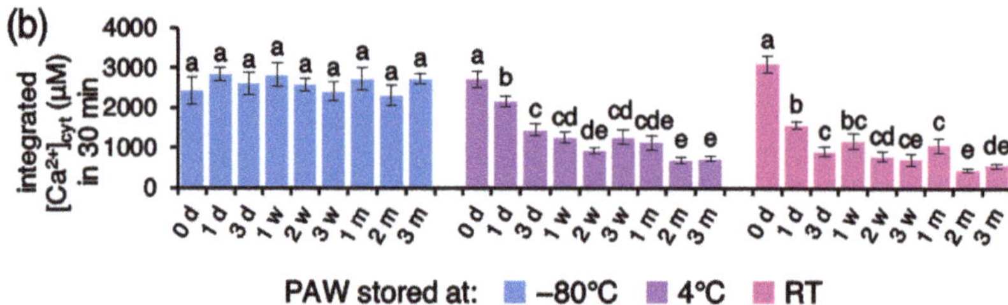

Figure 4. (**a**) Effects of different temperatures and time intervals of PAW storage on PAW-induced $[Ca^{2+}]_{cyt}$ increases. Ca^{2+} measurement assays were conducted in aequorin-expressing Arabidopsis seedlings. Seedlings were challenged with 1:4 dilutions of PAW generated by exposing deionized H_2O to cold plasma for 5 min at 900 W torch #1 power. Upon production, PAW was stored at different temperatures prior to plant treatment: $-80\,^{\circ}C$ (blue), $4\,^{\circ}C$ (purple), and room temperature (RT) (pink) for various time intervals (ranging from 0 days up to 3 months), as indicated on top of the panels. Data are the means (solid lines) $\pm$ SE (shading) of six seedlings derived from three independent growth replicates. Arrowheads indicate the time of stimulus application (at 100 s); (**b**) statistical analyses of integrated $[Ca^{2+}]_{cyt}$ dynamics over 30 min are shown. Key: d, days; w, weeks; and m, months. Bars labelled with different letters differ significantly ($p < 0.05$, Student's t test).

2.5. PAW Induces a Long-Lasting Cytosolic Ca^{2+} Elevation, but Not Cell Death

Ca^{2+} measurement assays demonstrated that the $[Ca^{2+}]_{cyt}$ elevation evoked by PAW appeared long-lasting, with sustained Ca^{2+} levels as high as ~1 µM after 1 h (Figure 5a). Nevertheless, viability assays carried out in Arabidopsis suspension-cultured cells [39] by the Evans blue test demonstrated the lack of cytotoxic effects by the PAW treatment. Indeed, no significant increase in cell death was found either at 1 h or even 48 h after PAW administration, in comparison with control samples (Figure 5b).

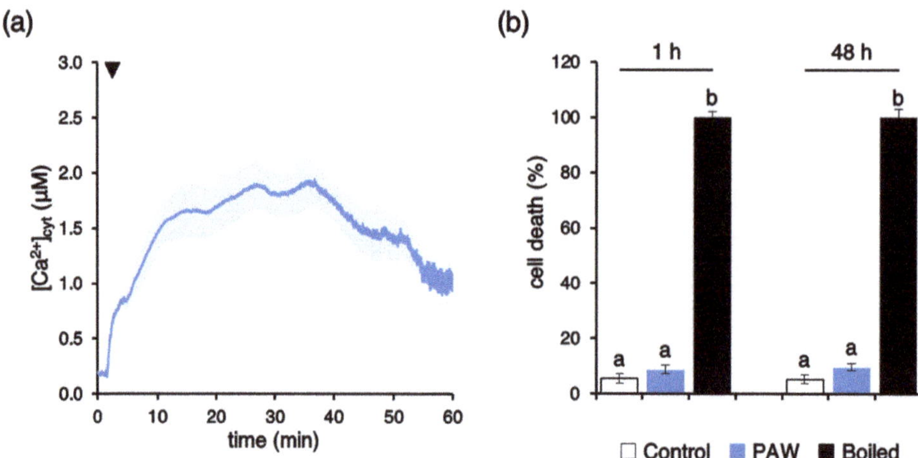

Figure 5. The long-lasting $[Ca^{2+}]_{cyt}$ elevation induced by PAW is not cytotoxic. (**a**) Ca^{2+} measurement assays were conducted in aequorin-expressing Arabidopsis seedlings. At 100 s (arrowhead) seedlings were challenged with a 1:4 dilution of PAW generated by exposing deionized H_2O to cold plasma for 5 min at 900 W torch #1 power. Changes in $[Ca^{2+}]_{cyt}$ were continuously recorded for 1 h. Data are the means (solid lines) $\pm$ SE (shading) of six different seedlings derived from three independent growth replicates; (**b**) viability of Arabidopsis cell suspension cultures treated with PAW (1:4 diluted) for either 1 h or 48 h (blue bars). Control cells were incubated with cell culture medium only (white bars). The 100% value corresponds to cells treated for 10 min at 100 °C (black bars). Bars labelled with different letters differ significantly ($p < 0.05$, Student's t test).

2.6. The Peculiar Chemical Environment Generated by Activation of H_2O by Plasma Discharge Accounts for the Specific PAW-Induced Ca^{2+} Signature

Chemical analyses performed by ion chromatography allowed the detection and quantification of nitrate (NO_3^-) and nitrite/nitrous acid (NO_2^-/HNO_2) in the PAW (Figure 6a,b). The content in ammonium (NH_4^+), measured by ion chromatography, as well as the content of hydrogen peroxide (H_2O_2), measured through spectrophotometric analysis of Ti^{IV}/H_2O_2 adduct, resulted under the detection limit of the assays [42]. Additional analyses performed by the ferrous oxidation in xylenol orange (FOX1) method [43] provided a quantification of H_2O_2 content at 0.59 ± 0.02 mg/L (Figure 6b).

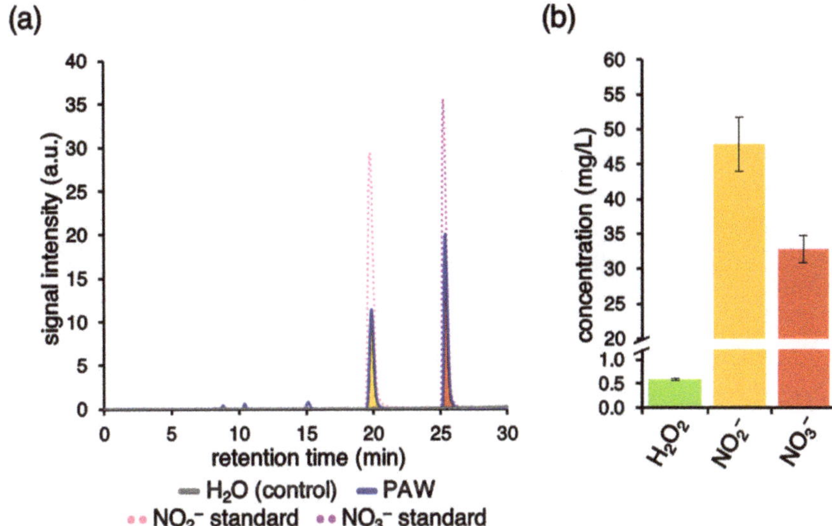

Figure 6. Chemical analyses for the detection of reactive oxygen and nitrogen species (RONS) in PAW. (**a**) Undiluted PAW (blue) generated by exposing deionized H_2O to cold plasma for 5 min at 900 W torch #1 power was analysed by ion chromatography. Deionized H_2O (grey) was used as a control. NO_2^- (pink) and NO_3^- (purple) solutions were used as standards. Representative traces are shown. (**b**) Determination of H_2O_2 (green), NO_2^- (yellow), and NO_3^- (red) concentrations in PAW generated in six different batches.

Notably, the separate administration of the single chemical components (H_2O_2, NO_2^-, and NO_3^-), at the same concentrations as those measured in the PAW, did not trigger detectable Ca^{2+} response in transgenic Arabidopsis seedlings (Figure 7). No differences were observed when NO_2^- and NO_3^- were provided as either K^+ salts or Na^+ salts (Figure 7). These results indicate that the unique mixture of the different chemical species found in the PAW is responsible for the induced Ca^{2+} signature.

Measurements of pH and conductivity of PAW demonstrated that the exposure of deionized H_2O to cold plasma resulted in remarkable changes of its chemical properties. Figure S1 shows a decrease in pH from about 5.5 in deionized H_2O to about 3.0 (in PAW generated by 5 min exposure to plasma torch #1 operating at 900 W). The extent of the pH drop was found to depend on the time interval of plasma discharge (Figure S1a), torch power (Figure S1c), and PAW dilution (Figure S1e). Likewise, the conductivity of deionized H_2O was found to be affected by the plasma discharge, with significantly higher values in PAW obtained through increasing time intervals of exposure to plasma (Figure S1b), increasing torch powers (Figure S1d), and PAW concentration (Figure S1f).

To check if the low pH could be responsible for the observed PAW-induced Ca^{2+} elevations, aequorin-expressing Arabidopsis seedlings were challenged with deionized H_2O that had been previously acidified to the same pH as PAW at 1:4 dilution (from pH 5.5 to pH 3.5). The pH change alone was found to only slightly perturb the resting level of cytosolic Ca^{2+} (Figure 8a); therefore, the low pH-induced integrated $[Ca^{2+}]_{cyt}$ change could not account for the remarkably higher Ca^{2+} signal induced by PAW (Figure 8b). This result further confirms that the peculiar chemical environment generated in the PAW by plasma discharge is the ultimate responsible factor for the specific PAW-induced Ca^{2+} signature. In agreement with these data, no significant changes in the PAW pH were found as a consequence of PAW storage at different temperatures for different time intervals (Figure S2).

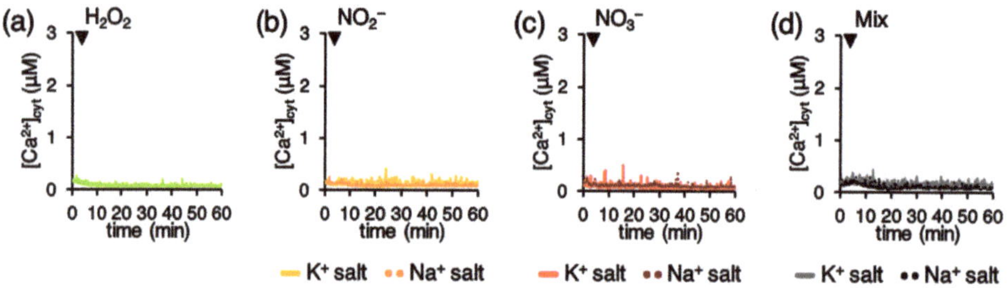

Figure 7. Monitoring of $[Ca^{2+}]_{cyt}$ in response to hydrogen peroxide, nitrite, and nitrate at the same doses as those measured in the PAW at 1:2 dilution. Ca^{2+} measurement assays were conducted in aequorin-expressing Arabidopsis seedlings. At 100 s (arrowheads) seedlings were challenged separately with: (**a**) 8.7 μM H_2O_2 (green trace); (**b**) 520.1 μM NO_2^- (provided as K^+ salt, yellow trace; provided as Na^+ salt, orange trace); (**c**) 264.2 μM NO_3^- (provided as K^+ salt, red trace; provided as Na^+ salt, brown trace); and (**d**) a solution (mix) containing all the above chemicals (with NO_2^- and NO_3^- provided as K^+ salt, grey trace; with NO_2^-_and NO_3^- provided as Na^+ salt, black trace). Data are the means (solid lines) $\pm$ SE (shading) of six different seedlings derived from three independent growth replicates.

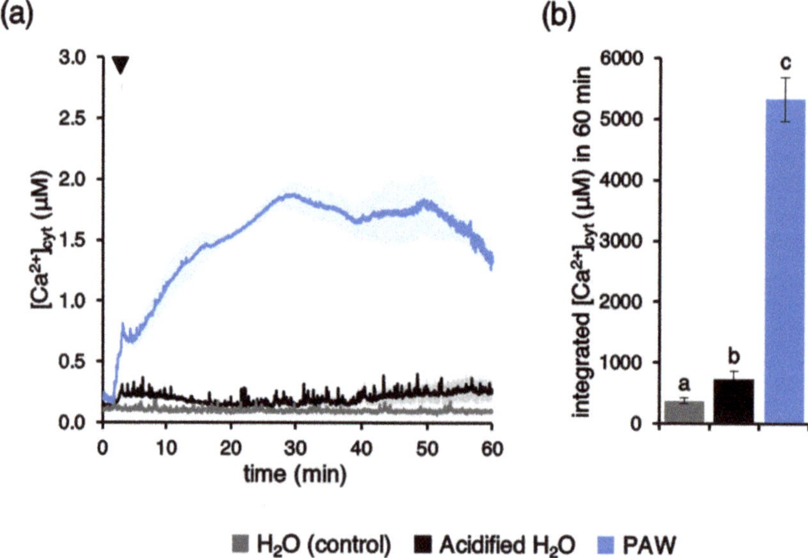

Figure 8. Comparison between the $[Ca^{2+}]_{cyt}$ responses induced by PAW and by an acidified H_2O. Ca^{2+} assays were conducted in aequorin-expressing Arabidopsis seedlings. Seedlings were challenged with 1:4 dilution of PAW generated by exposing deionized H_2O to cold plasma for 5 min at 900 W torch #1 power (blue) or deionized H_2O acidified to the same pH as PAW at 1:4 dilution (pH 3.5) (black). Untreated deionized H_2O (pH 5.5) was administered to control samples (grey). (**a**) Data are the means (solid lines) $\pm$ SE (shading) of six different seedlings derived from three independent growth replicates. The arrowhead indicates the time of stimulus application (at 100 s). (**b**) Statistical analyses of integrated $[Ca^{2+}]_{cyt}$ dynamics over 60 min are shown. Bars labelled with different letters differ significantly ($p < 0.05$, Student's *t* test).

2.7. The Plant Ca^{2+} Response to PAW Depends on the Total Energy Transferred to the H$_2$O during Plasma Discharge

Further experiments were carried out with an additional plasma torch (torch #2) operating in a different range of power and gas pressure (see Materials and Methods). Administration of the PAW generated by this plasma source to Arabidopsis seedlings resulted in the induction of a cytosolic Ca^{2+} increase that closely mirrored the one generated by the first plasma torch (Figure 9a). Spectrophotometric analyses also confirmed, in this case, the presence of nitrate and nitrite in the PAW. Figure 9b shows the UV-Vis spectrum of the obtained PAW. The main absorption in the range 190–250 nm is ascribed to nitrite, nitrate, and hydrogen peroxide. In the 270–420 nm interval, the structured band typical of acidic NO$_2$$^-$ is present at 354 nm, while the maximum of the nitrate absorption reported at 300 nm is not visible due to the lower concentration of NO$_3$$^-$ species [44]. It must be noted that with this alternative torch, an exposure of the H$_2$O to plasma lasting only 90 s was found to be sufficient to induce a Ca^{2+} increase in comparatively similar intensity to the one generated by 5 min exposure with the first torch (Figure 1). Indeed, the magnitude of the evoked plant Ca^{2+} response was found to depend on the overall power of the torch, which is the critical factor determining the energy transferred to the H$_2$O during plasma discharge.

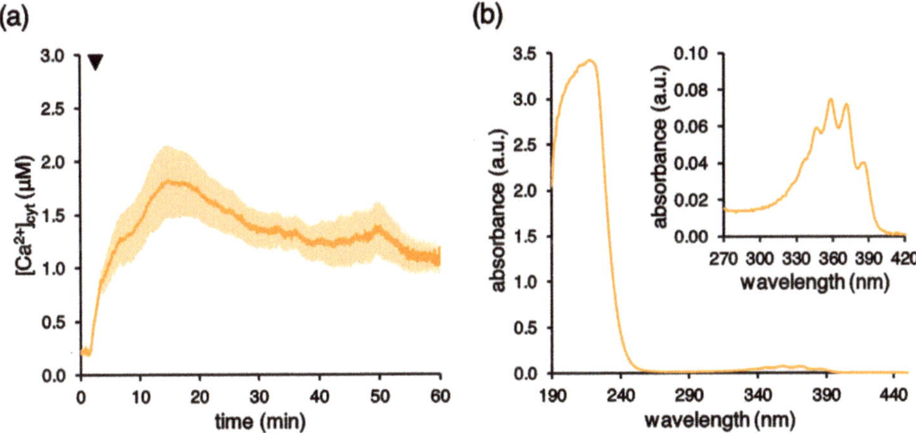

Figure 9. Monitoring of [Ca^{2+}]$_{cyt}$ in Arabidopsis seedlings challenged with PAW generated by an alternative plasma torch. (**a**) Ca^{2+} measurement assays were conducted in aequorin-expressing Arabidopsis seedlings. At 100 s (arrowhead) seedlings were challenged with 1:2 dilution of a PAW, obtained after 90 s exposure of deionized H$_2$O to cold plasma generated by a different plasma torch (torch #2). Data are the means (solid lines) ± SE (shading) of three independent experiments. (**b**) Spectrophotometric UV-Vis analyses of the generated PAW, showing the characteristic spectrum of absorbance of nitrite, nitrate, and hydrogen peroxide. In the inset, a magnification of the region between 270 and 420 nm, highlighting the five-fingers band of NO$_2$$^-$ around 354 nm, is shown.

3. Discussion

In the last few years there has been a surge of papers in the field of cold plasma technology. Several studies have addressed the effects of the administration of PAW to plants, concerning the promotion of growth and development [5,8,16], as well as the induction of defence responses (see [5,9,14,16] for reviews).

In this work we focused our attention on the elucidation of the mechanisms underlying PAW perception by plants. By using an Arabidopsis line stably expressing the bioluminescent Ca^{2+} reporter aequorin in the cytosol, we demonstrated that PAW evokes rapid and sustained cytosolic Ca^{2+} elevations, characterized by specific signatures, that were found to depend upon several parameters, such as: (a) operational conditions of the torches used to generate PAW; (b) time interval of H$_2$O exposure to plasma; (c) dose of PAW administered to plants; and (d) temperature and time interval of PAW storage. In particular, the magnitude of the recorded Ca^{2+} signals, measured as integral of the overall PAW-induced Ca^{2+} increases, was found to depend on the torch power, determining the energy transferred to the water during PAW generation. This was also confirmed by the use of a different torch as plasma generating device. The consistency of the results using two torches operating at different operational parameters and with different nozzle geometries has, therefore, confirmed the key role of the transferred energy to determine the Ca^{2+} signal evolution.

The involvement of Ca^{2+} signalling in the transduction mechanisms triggered by PAW in plants could somehow be anticipated, because PAW is known to contain a mixture of ROS and RNS. In the literature, a tight link between ROS and Ca^{2+} has been firmly highlighted [45,46]. Moreover, the involvement of Ca^{2+} also in nitrate sensing is increasingly emerging [47,48]. However, it must be noted that the treatment of Arabidopsis seedlings with the same doses of H_2O_2, nitrate, and nitrite as those measured in the PAW did not result in detectable Ca^{2+} changes. Indeed, the concentrations of H_2O_2 and nitrate commonly reported as capable of inducing cytosolic Ca^{2+} elevations are much higher, i.e., in the millimolar range [24,47]. These data suggest that the induction of the PAW-induced Ca^{2+} signature may be attributable to a complex "cocktail" of different reactive chemical species contained in the PAW, rather than to a single component (such as H_2O_2 or nitrate).

The PAW-induced cytosolic Ca^{2+} signature is characterized by a fast, immediate increase in $[Ca^{2+}]_{cyt}$ up to 10–25 times the resting level of the ion; notably, the Ca^{2+} change was found to reach, after about 10 min, a plateau that is maintained for additional 20 min, before slowly decreasing. Continuous monitoring of Ca^{2+} showed that even after 1 h the Ca^{2+} elevation did not dissipate completely. However, cell viability assays demonstrated the lack of cytotoxic effects of PAW in our experimental set up. An intriguing possibility is that the unique dynamics of the PAW-activated Ca^{2+} signals, i.e., a sustained, long-lasting Ca^{2+} elevation, may be a crucial determinant of the induced "priming" condition, consisting in the activation of defence gene expression and antioxidant activities that bolster plant resistance to subsequent pathogen attacks. The lack of return of $[Ca^{2+}]_{cyt}$ back to pre-stimulus resting values (about 100 nM) may lead to a state of plant alert in which the plant gets more prepared to face subsequent battles [49]. Future research will be directed at unravelling whether PAW-induced Ca^{2+} signalling underlies only plant self-defence responses or also plant growth promoting effects.

Ca^{2+} measurement assays performed by challenging transgenic Arabidopsis seedlings with PAWs stored for different time intervals at different temperatures demonstrate that only PAW quickly frozen in liquid N_2 upon production and then stored at $-80\,^{\circ}C$ retained the ability to induce a Ca^{2+}-mediated response in plants, whereas the Ca^{2+}-inducing activity of PAW stored at $4\,^{\circ}C$ or room temperature rapidly decreased over time. It has previously been reported that only PAW stored at $-80\,^{\circ}C$ maintained an unvaried bactericidal activity against *S. aureus*, providing important cues for optimal applications in disinfection and food preservation [50]. Our results confirm those observations about the dependence of the physico-chemical properties of PAW on temperature storage, by extending the range of PAW activities to Ca^{2+}-mediated responses elicited in plants. The obtained data highlight the necessity for either cryopreservation or generation of the PAW ready-to-use and in situ, in order to allow an effective and quick administration to plants upon production.

In summary, in this work we provided evidence that Ca^{2+} acts as an intracellular messenger in the signalling pathway triggered by PAW in Arabidopsis. Our data indicate the possibility to use aequorin-based Ca^{2+} measurements as a rapid and reliable assay to rapidly monitor early plant responses to PAW. Establishing a sound scientific ground will provide the key elements to develop tools and treatments aimed to improve plant growth and resistance to pathogens, in order to increase crop yield in a sustainable and eco-friendly way.

4. Materials and Methods

4.1. Plant Material and Growth Conditions

An Arabidopsis *thaliana* ecotype Columbia (Col-0) transgenic line (Cyt_YA) stably expressing in the cytosol the bioluminescent Ca^{2+} reporter aequorin fused to yellow fluorescent protein (YFP) [38–40] was used in this study. Seeds were surface-sterilized and sown on half-strength Murashige and Skoog medium ($\frac{1}{2}$ MS) (Duchefa Biochemie, Haarlem, The Netherlands) supplemented with 1.5% (w/v) sucrose, 0.8% (w/v) agar. Seedlings were grown for 7 days under a 16/8 h light/dark photoperiod at 21 °C. In some experiments, cell suspension cultures derived from the Cyt-YA line [39] were used. They were maintained and subcultured weekly in MS medium containing 0.5% (w/v) sucrose, 0.5 μg/mL 2,4-dichlorophenoxiacetic acid (2,4-D), and 0.25 μg/mL 6-benzylaminopurine (BAP) (Merck, Darmstadt, Germany), supplemented with 10 μg/mL kanamycin as selective agent, as recently described [51].

4.2. Generation of PAW

PAW was generated by exposing deionized H_2O at room temperature to the cold plasma obtained by two different plasma torches operating in a range of power 450–2700 W and with pressure from 1 to 3 bar. Torch #1 was a single rotating FLUME Jet RD1004 with an FG 1001 plasma generator (Plasmatreat, Elgin, IL, USA), that worked with an excitation frequency between 16 and 20 kHz and generated a plasma with a maximum power of 2.7 kW (Voltage = 230 V, Current = 12 A). Torch #2 was an AcXys ULS series atmospheric pressure cold plasma (AcXys Technologies, Saint-Martin-le-Vinoux, France) fed with purified air; plasma nozzle ϕ = 5 mm. Initial tests were performed varying the H_2O amount, the distance of the nozzle from the H_2O surface (from 1 to 10 cm), and the treatment time (from 1 to 10 min). H_2O was kept in beakers immersed in an ice and salt cooling bath, in order to keep the H_2O temperature increase within 40 °C for the longer treatments and the maximum power. Taking the pH as a target reference (i.e., 3.0), it has been found that the best compromise between temperature rise and pH change was, for torch #1, with the nozzle at 1.5 cm from the H_2O surface and with treatment duration between 3 and 5 min. Concerning torch #2, the target pH was achieved with the nozzle at 10 cm from the H_2O surface and 90 s of treatment time. Therefore, most of the results reported in this paper refer to 50 mL deionized H_2O with its surface exposed at 1.5 cm from the torch nozzle with the standard protocol parameters set at about 900 W for 5 min concerning torch #1 and 10 cm from torch to H_2O surface at 800 W for 90 s concerning torch #2. After generation, PAW was divided in single-use aliquots (1 mL for Ca^{2+} measurement purposes, 5 mL for pH and conductivity measurements or 25 mL for chemical analyses), immediately cryogenically frozen through immersion in liquid N_2, and then stored at −80 °C. For long-term storage tests, some PAW aliquots were kept also at 4 °C or room temperature.

4.3. Chemical Analyses of PAW

Nitrite and nitrate in PAW were quantified by Dionex ICS-6000 SP ion chromatography on a Dionex IonPac ASIP-4 μm column 2 × 250 mm (Thermo Fisher Scientific, Waltham, MA, USA). Ammonium concentration was determined by Dionex Easion ion chromatography equipped with a Dionex IonPac CS12A RFIC column 4 × 250 mm (Thermo Fisher Scientific). PAW conductivity and pH were measured with electrode-based instruments, Cond7+ (XS Instruments, Carpi, Italy) and pH METER BasiC 20 (Crison, Alella, Spain), respectively.

H_2O_2 content in PAW was measured by spectrophotometric analysis of Ti^{IV}/H_2O_2 adduct as described by [52]. Briefly, 0.5 mL of titanium (IV) oxysulfate solution was added to 1 mL of the sample and diluted with 8.5 mL deionized H_2O. Spectrophotometric analysis of the peroxidic complex $[Ti(O_2)OH(H_2O)_3]^+_{aq}$ was then carried out measuring the absorption at 409 nm with a double beam spectrometer Varian Cary 100 Bio (Varian, Palo Alto, CA, USA), using a solution containing 0.5 mL of $TiOSO_4$ and 9.5 mL of deionized H_2O

as reference. H_2O_2 content in PAW was also determined through the FOX1 method [43]. The assay is based on a colorimetric reaction caused by the peroxide-mediated oxidation of Fe^{2+} followed by the reaction of Fe^{3+} with xylenol orange. A total of 50 µL of diluted PAW samples (consisting in 10 µL of PAW and 40 µL of H_2O) were added to 950 µL of assay solution (0.25 mM ammonium ferrous sulfate, 25 mM H_2SO_4, 0.1 mM xylenol orange, and 100 mM sorbitol) and the absorbance at 560 nm was detected after 1 h incubation.

UV-Vis absorbance was measured immediately after the production of PAW by torch #2 using a double-beam UV-Vis spectrophotometer UV-2600 (Shimadzu, Kyoto, Japan) on quartz cuvettes with standard optical path of 10 mm. The spectra were recorded from 190 to 450 nm with a spectral resolution of 1 nm and a scan speed of 480 nm/min.

4.4. Aequorin-Based Ca^{2+} Measurement Assays

Transgenic Arabidopsis seedlings (7-day-old) were incubated overnight in the dark with 5 µM coelenterazine (Prolume, Pinetop, AZ, USA) to reconstitute the functional aequorin probe. Prior to the start of the experiment, each seedling was gently rinsed and placed in 250 µL deionized H_2O inside the chamber of a custom-made luminometer (ET Enterprises Ltd., Uxbridge, UK), in close proximity to a low-noise photomultiplier, with a built-in amplifier discriminator. The output was captured using a photon-counting board. After 100 s, 250 µL of either PAW (tested at various dilutions) or deionized H_2O (that has not been exposed to plasma; used as control) were injected. In some experiments, Arabidopsis seedlings were challenged with H_2O_2, NO_2^-, and NO_3^- at the same concentrations as those measured in the PAW. NO_2^- and NO_3^- were administered as either Na^+ or K^+ salts (Merck). At the end of the experiment, 500 µL of a solution containing 1 M $CaCl_2$, 30% (v/v) ethanol was added to completely discharge the remaining Ca^{2+} probe, allowing for the conversion of the collected light signal into $[Ca^{2+}]_{cyt}$ by means of a built-in algorithm based on the calibration curve of aequorin [53]. Integrated $[Ca^{2+}]_{cyt}$ values were obtained as the sum of each instantaneous $[Ca^{2+}]_{cyt}$ value for the entire duration of the experiment.

4.5. Cell Viability Assay

Cell viability was determined by the Evans blue method [54]. Briefly, Arabidopsis cell suspension cultures obtained from the transgenic line [39] were kept in control conditions or treated at mid-exponential phase (4 days) with PAW (diluted 1:4) for either 1 h or 48 h. After 15 min incubation with 0.05% (w/v) Evans blue (Merck), excess and unbound dye was removed by extensive washing with H_2O. The dye bound to dead cells was solubilized in 1% (w/v) SDS, 50% (v/v) methanol for 30 min at 55 °C. The percentage of cell death was assessed by measuring the absorbance at 600 nm. As positive control (100% cell death), cells were incubated for 10 min at 100 °C.

Supplementary Materials: The following are available online at https://www.mdpi.com/article/10.3390/plants10112516/s1: Figure S1, effects of different plasma exposure time intervals, torch power and PAW dilution on pH and conductivity of PAWs; Figure S2, effect of time interval and temperature of PAW storage on pH.

Author Contributions: Conceptualization, L.N. and V.A.; methodology, L.N., E.C., V.A., M.D., L.C., A.G. and A.F.; software, E.C.; validation, E.C., A.G.S., S.P. and L.C.; formal analysis, E.C.; investigation, E.C., A.G.S., S.P., L.C., A.G. and A.F.; resources, V.A., M.D., L.C., A.F. and L.N.; data curation, E.C., L.C., A.G. and V.A.; writing—original draft preparation, L.N. and E.C.; writing—review and editing, L.N., E.C., V.A., L.C, A.G., A.F. and M.D.; visualization, E.C.; supervision, L.N. and V.A.; project administration, L.N.; funding acquisition, L.N. All authors have read and agreed to the published version of the manuscript.

Funding: This research was funded by the University of Padova, Italy, PRID 2018 (prot. BIRD180317) and DOR 2018–2021 to L.N.; L.C. acknowledges funding from P-DiSC#02BIRD2019-UNIPD. E.C. is the recipient of a post-doctoral grant from the Department of Biology, University of Padova, Italy (MIUR Excellence Department Project 2018).

Data Availability Statement: The data presented in this study are available on request from the corresponding author.

Acknowledgments: We thank U. Vothknecht (Bonn, Germany) for kindly providing *A. thaliana* transgenic seeds (Cyt-YA) expressing cytosolic aequorin. We also acknowledge the Plant Genome Editing Facility of the Department of Biology, University of Padova, for support with in vitro and in vivo plant growth.

Conflicts of Interest: The authors declare no conflict of interest.

References

1. Kaushik, N.K.; Ghimire, B.; Li, Y.; Adhikari, M.; Veerana, M.; Kaushik, N.; Jha, N.; Adhikari, B.; Lee, S.J.; Masur, K.; et al. Biological and medical applications of plasma-activated media, water and solutions. *Biol. Chem.* **2018**, *400*, 39–62. [CrossRef] [PubMed]
2. Thirumdas, R.; Kothakota, A.; Annapure, U.; Siliveru, K.; Blundell, R.; Gatt, R.; Valdramidis, V.P. Plasma activated water (PAW): Chemistry, physico-chemical properties, applications in food and agriculture. *Trends Food Sci. Technol.* **2018**, *77*, 21–31. [CrossRef]
3. Bourke, P.; Ziuzina, D.; Boehm, D.; Cullen, P.J.; Keener, K. The potential of cold plasma for safe and sustainable food production. *Trends Biotechnol.* **2018**, *36*, 615–626. [CrossRef] [PubMed]
4. Herianto, S.; Hou, C.Y.; Lin, C.M.; Chen, H.L. Nonthermal plasma-activated water: A comprehensive review of this new tool for enhanced food safety and quality. *Compr. Rev. Food Sci. Food Saf.* **2021**, *20*, 583–626. [CrossRef]
5. Guo, D.; Liu, H.; Zhou, L.; Xie, J.; He, C. Plasma-activated water production and its application in agriculture. *J. Sci. Food Agric.* **2021**, *101*, 4891–4899. [CrossRef] [PubMed]
6. Bafoil, M.; Jemmat, A.; Martinez, Y.; Merbahi, N.; Eichwald, O.; Dunand, C.; Yousfi, M. Effects of low temperature plasmas and plasma activated waters on *Arabidopsis thaliana* germination and growth. *PLoS ONE* **2018**, *13*, e0195512. [CrossRef] [PubMed]
7. Bafoil, M.; Le Ru, A.; Merbahi, N.; Eichwald, O.; Dunand, C.; Yousfi, M. New insights of low-temperature plasma effects on germination of three genotypes of *Arabidopsis thaliana* seeds under osmotic and saline stresses. *Sci. Rep.* **2019**, *9*, 8649. [CrossRef] [PubMed]
8. Adhikari, B.; Adhikari, M.; Park, G. The effects of plasma on plant growth, development, and sustainability. *Appl. Sci.* **2020**, *10*, 6045. [CrossRef]
9. Song, J.S.; Kim, S.B.; Ryu, S.; Oh, J.; Kim, D.S. Emerging plasma technology that alleviates crop stress during the early growth stages of plants: A review. *Front. Plant Sci.* **2020**, *11*, 988. [CrossRef] [PubMed]
10. Starič, P.; Vogel-Mikuš, K.; Mozetič, M.; Junkar, I. Effects of nonthermal plasma on morphology, genetics and physiology of seeds: A review. *Plants* **2020**, *9*, 1736. [CrossRef]
11. Šerá, B.; Scholtz, V.; Jirešová, J.; Khun, J.; Julák, J.; Šerý, M. Effects of non-thermal plasma treatment on seed germination and early growth of leguminous plants—A review. *Plants* **2021**, *10*, 1616. [CrossRef]
12. Adhikari, B.; Adhikari, M.; Ghimire, B.; Park, G.; Choi, E.H. Cold atmospheric plasma-activated water irrigation induces defense hormone and gene expression in tomato seedlings. *Sci. Rep.* **2019**, *9*, 16080. [CrossRef] [PubMed]
13. Perez, S.M.; Biondi, E.; Laurita, R.; Proto, M.; Sarti, F.; Gherardi, M.; Bertaccini, A.; Colombo, V. Plasma activated water as resistance inducer against bacterial leaf spot of tomato. *PLoS ONE* **2019**, *14*, e0217788. [CrossRef] [PubMed]
14. Adhikari, B.; Pangomm, K.; Veerana, M.; Mitra, S.; Park, G. Plant disease control by non-thermal atmospheric-pressure plasma. *Front. Plant Sci.* **2020**, *11*, 77. [CrossRef] [PubMed]
15. Zambon, Y.; Contaldo, N.; Laurita, R.; Várallyay, E.; Canel, A.; Gherardi, M.; Colombo, V.; Bertaccini, A. Plasma activated water triggers plant defence responses. *Sci. Rep.* **2020**, *10*, 19211. [CrossRef] [PubMed]
16. Holubová, L.; Kyzek, S.; Ďurovcová, I.; Fabová, J.; Horváthová, E.; Ševčovičová, A.; Gálová, E. Non-thermal plasma—A new green priming agent for plants? *Int. J. Mol. Sci.* **2020**, *21*, 9466. [CrossRef] [PubMed]
17. Berridge, M.J.; Lipp, P.; Bootman, M.D. The versatility and universality of calcium signalling. *Nat. Rev. Mol. Cell. Biol.* **2000**, *1*, 11–21. [CrossRef]
18. Sanders, D.; Pelloux, J.; Brownlee, C.; Harper, J.F. Calcium at the crossroads of signaling. *Plant Cell* **2002**, *14* (Suppl. 1), S401–S417. [CrossRef]
19. Knight, M.R.; Campbell, A.K.; Smith, S.M.; Trewavas, A.J. Transgenic plant aequorin reports the effects of touch and cold-shock and elicitors on cytoplasmic calcium. *Nature* **1991**, *352*, 524–526. [CrossRef] [PubMed]
20. Knight, M.R.; Smith, S.M.; Trewavas, A.J. Wind-induced plant motion immediately increases cytosolic calcium. *Proc. Natl. Acad. Sci. USA* **1992**, *89*, 4967–4971. [CrossRef]
21. Knight, H.; Trewavas, A.J.; Knight, M.R. Calcium signalling in *Arabidopsis thaliana* responding to drought and salinity. *Plant J.* **1997**, *5*, 1067–1078. [CrossRef] [PubMed]
22. Sun, J.; Wang, M.J.; Ding, M.Q.; Deng, S.R.; Liu, M.Q.; Lu, C.F.; Zhou, X.Y.; Shen, X.; Zheng, X.J.; Zhang, Z.K.; et al. H_2O_2 and cytosolic Ca^{2+} signals triggered by the PM H^+-coupled transport system mediate K^+/Na^+ homeostasis in NaCl-stressed *Populus euphratica* cells. *Plant Cell Environ.* **2010**, *33*, 943–958. [CrossRef] [PubMed]
23. Chen, S.; Wang, X.; Jia, H.; Li, F.; Ma, Y.; Liesche, J.; Liao, M.; Ding, X.; Liu, C.; Chen, Y.; et al. Persulfidation-induced structural change in SnRK2.6 establishes intramolecular interaction between phosphorylation and persulfidation. *Mol. Plant* **2021**, *14*, 1814–1830. [CrossRef]

24. Rentel, M.C.; Knight, M.R. Oxidative stress-induced calcium signaling in Arabidopsis. *Plant Physiol.* **2004**, *135*, 1471–1479. [CrossRef] [PubMed]

25. Knight, H.; Trewavas, A.J.; Knight, M.R. Cold calcium signaling in Arabidopsis involves two cellular pools and a change in calcium signature after acclimation. *Plant Cell* **1996**, *8*, 489–503. [CrossRef] [PubMed]

26. Gong, M.; van der Luit, A.H.; Knight, M.R.; Trewavas, A.J. Heat-shock-induced changes in intracellular Ca^{2+} level in tobacco seedlings in relation to thermotolerance. *Plant Physiol.* **1998**, *116*, 429–437. [CrossRef]

27. Saidi, Y.; Finka, A.; Muriset, M.; Bromberg, Z.; Weiss, Y.G.; Maathuis, F.J.; Goloubinoff, P. The heat shock response in moss plants is regulated by specific calcium-permeable channels in the plasma membrane. *Plant Cell* **2009**, *21*, 2829–2843. [CrossRef] [PubMed]

28. Finka, A.; Cuendet, A.F.; Maathuis, F.J.; Saidi, Y.; Goloubinoff, P. Plasma membrane cyclic nucleotide gated calcium channels control land plant thermal sensing and acquired thermotolerance. *Plant Cell* **2012**, *24*, 3333–3348. [CrossRef]

29. Lenzoni, G.; Knight, M.R. Increases in absolute temperature stimulate free calcium concentration elevations in the chloroplast. *Plant Cell Physiol.* **2019**, *60*, 538–548. [CrossRef] [PubMed]

30. de Vries, J.; de Vries, S.; Curtis, B.A.; Zhou, H.; Penny, S.; Feussner, K.; Pinto, D.M.; Steinert, M.; Cohen, A.M.; von Schwartzenberg, K.; et al. Heat stress response in the closest algal relatives of land plants reveals conserved stress signaling circuits. *Plant J.* **2020**, *103*, 1025–1048. [CrossRef]

31. Liu, Q.; Ding, Y.; Shi, Y.; Ma, L.; Wang, Y.; Song, C.; Wilkins, K.A.; Davies, J.M.; Knight, H.; Knight, M.R.; et al. The calcium transporter ANNEXIN1 mediates cold-induced calcium signaling and freezing tolerance in plants. *EMBO J.* **2021**, *40*, e104559. [CrossRef]

32. Zipfel, C.; Oldroyd, G. Plant signalling in symbiosis and immunity. *Nature* **2017**, *543*, 328–336. [CrossRef]

33. Aldon, D.; Mbengue, M.; Mazars, C.; Galaud, J. Calcium signalling in plant biotic interactions. *Int. J. Mol. Sci.* **2018**, *19*, 665. [CrossRef]

34. Dodd, A.N.; Kudla, J.; Sanders, D. The language of calcium signaling. *Annu. Rev. Plant Biol.* **2010**, *61*, 593–620. [CrossRef]

35. Costa, A.; Navazio, L.; Szabo, I. The contribution of organelles to plant intracellular calcium signalling. *J. Exp. Bot.* **2018**, *69*, 4175–4193. [CrossRef] [PubMed]

36. Kudla, J.; Becker, D.; Grill, E.; Hedrich, R.; Hippler, M.; Kummer, U.; Parniske, M.; Romeis, T.; Schumacher, K. Advances and current challenges in calcium signaling. *New Phytol.* **2018**, *218*, 414–431. [CrossRef]

37. Edel, K.H.; Marchadier, E.; Brownlee, C.; Kudla, J.; Hetherington, A.M. The evolution of calcium-based signalling in plants. *Curr. Biol.* **2017**, *27*, R667–R679. [CrossRef] [PubMed]

38. Mehlmer, N.; Parvin, N.; Hurst, C.H.; Knight, M.R.; Teige, M.; Vothknecht, U.C. A toolset of aequorin expression vectors for in planta studies of subcellular calcium concentrations in *Arabidopsis thaliana*. *J. Exp. Bot.* **2012**, *63*, 1751–1761. [CrossRef] [PubMed]

39. Sello, S.; Perotto, J.; Carraretto, L.; Szabò, I.; Vothknecht, U.C.; Navazio, L. Dissecting stimulus-specific Ca^{2+} signals in amyloplasts and chloroplasts of *Arabidopsis thaliana* cell suspension cultures. *J. Exp. Bot.* **2016**, *67*, 3965–3974. [CrossRef] [PubMed]

40. Sello, S.; Moscatiello, R.; Mehlmer, N.; Leonardelli, M.; Carraretto, L.; Cortese, E.; Zanella, F.G.; Baldan, B.; Szabò, I.; Vothknecht, U.C.; et al. Chloroplast Ca^{2+} fluxes into and across thylakoids revealed by thylakoid-targeted aequorin probes. *Plant Physiol.* **2018**, *177*, 38–51. [CrossRef] [PubMed]

41. Zhou, R.; Zhou, R.; Wang, P.; Xian, Y.; Mai-Prochnow, A.; Lu, X.; Cullen, P.J.; Ostrikov, K.; Bazaka, K. Plasma activated water (PAW): Generation, origin of reactive species and biological applications. *J. Phys. D Appl. Phys.* **2020**, *53*, 303001. [CrossRef]

42. Judée, F.; Simon, S.; Bailly, C.; Dufour, T. Plasma-activation of tap water using DBD for agronomy applications: Identification and quantification of long lifetime chemical species and production/consumption mechanisms. *Water Res.* **2018**, *133*, 47–59. [CrossRef] [PubMed]

43. Wolff, S.P. [18] Ferrous ion oxidation in presence of ferric ion indicator xylenol orange for measurement of hydroperoxides. *Method Enzymol.* **1994**, *233*, 182–189. [CrossRef]

44. Brisset, J.L.; Pawlat, J. Chemical effects of air plasma species on aqueous solutes in direct and delayed exposure modes: Discharge, post-discharge and plasma activated water. *Plasma Chem. Plasma Process.* **2016**, *36*, 355–381. [CrossRef]

45. Gilroy, S.; Suzuki, N.; Miller, G.; Choi, W.G.; Toyota, M.; Devireddy, A.R.; Mittler, R. A tidal wave of signals: Calcium and ROS at the forefront of rapid systemic signaling. *Trends Plant Sci.* **2014**, *19*, 623–630. [CrossRef] [PubMed]

46. Marcec, M.J.; Gilroy, S.; Poovaiah, B.W.; Tanaka, K. Mutual interplay of Ca^{2+} and ROS signaling in plant immune response. *Plant Sci.* **2019**, *283*, 343–354. [CrossRef] [PubMed]

47. Riveras, E.; Alvarez, J.M.; Vidal, E.A.; Oses, C.; Vega, A.; Gutiérrez, R.A. The calcium ion is a second messenger in the nitrate signaling pathway of Arabidopsis. *Plant Physiol.* **2015**, *169*, 1397–1404. [CrossRef] [PubMed]

48. Wang, X.; Feng, C.; Tian, L.; Hou, C.; Tian, W.; Hu, B.; Zhang, Q.; Ren, Z.; Niu, Q.; Song, J.; et al. A transceptor–channel complex couples nitrate sensing to calcium signaling in *Arabidopsis*. *Mol. Plant.* **2021**, *14*, 774–786. [CrossRef] [PubMed]

49. Prime-A-Plant Group; Conrath, U.; Beckers, G.J.; Flors, V.; García-Agustín, P.; Jakab, G.; Mauch, F.; Newman, M.A.; Pieterse, C.M.; Poinssot, B.; et al. Priming: Getting ready for battle. *Mol. Plant Microbe Interact.* **2006**, *19*, 1062–1071. [CrossRef]

50. Shen, J.; Tian, Y.; Li, Y.; Ma, R.; Zhang, Q.; Zhang, J.; Fang, J. Bactericidal effects against *S. aureus* and physicochemical properties of plasma activated water stored at different temperatures. *Sci. Rep.* **2016**, *6*, 28505. [CrossRef]

51. Cortese, E.; Carraretto, L.; Baldan, B.; Navazio, L. Arabidopsis photosynthetic and heterotrophic cell suspension cultures. *Methods Mol. Biol.* **2021**, *2200*, 167–185. [CrossRef] [PubMed]

52. Sandri, F.; Danieli, M.; Zecca, M.; Centomo, P. Comparing catalysts of the direct synthesis of hydrogen peroxide in organic solvent: Is the measure of the product an issue? *Chem. Cat. Chem.* **2021**, *13*, 2653–2663. [CrossRef]

53. Brini, M.; Marsault, R.; Bastianutto, C.; Alvarez, J.; Pozzan, T.; Rizzuto, R. Transfected aequorin in the measurement of cytosolic Ca^{2+} concentration ($[Ca^{2+}]_c$). A critical evaluation. *J. Biol. Chem.* **1995**, *270*, 9896–9903. [CrossRef] [PubMed]

54. Baker, C.J.; Mock, N.M. An improved method for monitoring cell-death in cell-suspension and leaf disc assays using evans blue. *Plant Cell Tissue Organ Cult.* **1994**, *39*, 7–12. [CrossRef]

Article

Increase of Productivity and Neutralization of Pathological Processes in Plants of Grain and Fruit Crops with the Help of Aqueous Solutions Activated by Plasma of High-Frequency Glow Discharge

Yuri K. Danilejko [1], Sergey V. Belov [1], Alexey B. Egorov [1], Vladimir I. Lukanin [1], Vladimir A. Sidorov [1], Lyubov M. Apasheva [2], Vladimir Y. Dushkov [2], Mikhail I. Budnik [2], Alexander M. Belyakov [3], Konstantin N. Kulik [3], Shamil Validov [4], Denis V. Yanykin [1], Maxim E. Astashev [1], Ruslan M. Sarimov [1], Valery P. Kalinichenko [5], Alexey P. Glinushkin [5] and Sergey V. Gudkov [1,*]

[1] Prokhorov General Physics Institute of the Russian Academy of Sciences, 119991 Moscow, Russia; dyuk42@list.ru (Y.K.D.); ser79841825@yandex.ru (S.V.B.); a261@rambler.ru (A.B.E.); vladimirlukanin@yandex.ru (V.I.L.); 2017vovas@gmail.com (V.A.S.); ya-d-ozh@rambler.ru (D.V.Y.); astashev.max@gmail.com (M.E.A.); rusa@kapella.gpi.ru (R.M.S.)

[2] Semenov Institute of Chemical Physics of Russian Academy of Sciences, 119991 Moscow, Russia; apasheva@rambler.ru (L.M.A.); vdushkov@yandex.ru (V.Y.D.); ziraf@mail.ru (M.I.B.)

[3] Federal Scientific Center for Agroecology, Integrated Land Reclamation and Protective Afforestation of the Russian Academy of Sciences, 400062 Volgograd, Russia; dokbam49@mail.ru (A.M.B.); kulikn@yandex.ru (K.N.K.)

[4] Federal Research Center Kazan Scientific Center of Russian Academy of Sciences, 420008 Kazan, Russia; sh.validov@knc.ru

[5] All-Russian Phytopathology Research Institute, 143050 Big Vyazyomy, Russia; kalinitch@mail.ru (V.P.K.); glinale@mail.ru (A.P.G.)

* Correspondence: S_makariy@rambler.ru

Citation: Danilejko, Y.K.; Belov, S.V.; Egorov, A.B.; Lukanin, V.I.; Sidorov, V.A.; Apasheva, L.M.; Dushkov, V.Y.; Budnik, M.I.; Belyakov, A.M.; Kulik, K.N.; et al. Increase of Productivity and Neutralization of Pathological Processes in Plants of Grain and Fruit Crops with the Help of Aqueous Solutions Activated by Plasma of High-Frequency Glow Discharge. *Plants* 2021, 10, 2161. https://doi.org/10.3390/plants10102161

Academic Editors: Vida Mildažienė and Božena Šerá

Received: 25 September 2021
Accepted: 8 October 2021
Published: 12 October 2021

Publisher's Note: MDPI stays neutral with regard to jurisdictional claims in published maps and institutional affiliations.

Abstract: In this work, we, for the first time, manufactured a plasma-chemical reactor operating at a frequency of 0.11 MHz. The reactor allows for the activation of large volumes of liquids in a short time. The physicochemical properties of activated liquids (concentration of hydrogen peroxide, nitrate anions, redox potential, electrical conductivity, pH, concentration of dissolved gases) are characterized in detail. Antifungal activity of aqueous solutions activated by a glow discharge has been investigated. It was shown that aqueous solutions activated by a glow discharge significantly reduce the degree of presence of phytopathogens and their effect on the germination of such seeds. Seeds of cereals (sorghum and barley) and fruit (strawberries) crops were studied. The greatest positive effect was found in the treatment of sorghum seeds. Moreover, laboratory tests have shown a significant increase in sorghum drought tolerance. The effectiveness of the use of glow-discharge-activated aqueous solutions was shown during a field experiment, which was set up in the saline semi-desert of the Northern Caspian region. Thus, the technology developed by us makes it possible to carry out the activation of aqueous solutions on an industrial scale. Water activated by a glow discharge exhibits antifungicidal activity and significantly accelerates the development of the grain and fruit crops we studied. In the case of sorghum culture, glow-discharge-activated water significantly increases drought resistance.

Keywords: high-frequency glow discharge; low temperature plasma; fusarium; PAW; germination of seeds; plant resistance

1. Introduction

Increasing the yield and improving the quality of agricultural products, reducing the costs of its production is impossible without the introduction of economically profitable and environmentally friendly technologies that ensure an increase in the productivity of agricultural plants. One of the promising directions in the creation of such technologies is the use of biologically activated water [1–7]. Usually, to obtain activated aqueous solutions, devices are used, the principle of operation of which is based on electrochemical methods using electrolyzers of various designs, including the separation of water into the constituents "anolyte" and "catholyte" [7,8]. In recent years, a method for activating water, as well as an aqueous solution of various electrolytes, has become widespread by exposing its surface to a low-temperature non-equilibrium plasma generated by a high-voltage electric discharge in an atmosphere of various gases [9,10], a microwave discharge in an argon atmosphere [11,12], high-voltage discharge in the volume of water under conditions of intense cavitation [13,14].

An alternative approach is the activation of solutions using a high-frequency glow discharge in water vapor [15–17]. Glow discharge is a type of stationary self-sustaining electric discharge, which is formed at low pressure and low current. Generally, a glow discharge is produced in gases at low pressure. The main characteristics of a glow discharge remain relatively stable over time [18]. In a glow discharge, the gas conducts electricity well due to its strong ionization. Gas ionization is achieved through the emission of electrons from the cathode under the action of a strong electric field (or high temperature). In addition, the secondary electron emission of electrons from the cathode, caused by the bombardment of the cathode with positively charged ions of the medium. In general, a glow discharge can be maintained at a voltage significantly lower than the dielectric breakdown voltage of the medium [18]. In recent years, methods have been developed for obtaining a glow discharge in liquids. In this case, the decomposition of water vapor occurs due to the transfer of energy from an electron gas with a temperature not higher than 5 eV to water molecules. Water decomposition can proceed through two channels: a channel through stepwise excitation of vibrational degrees of freedom of water molecules and a channel through the dissociative attachment of electrons to water molecules [19]. In both cases, they form reactive oxygen and nitrogen species with significant biological activity. Activation of water using a glow discharge is the most economically viable approach, this approach is easily scalable and allows you to handle large volumes of liquids [20].

The main goal of this work was to create a plasma-chemical reactor operating at frequencies of several tens of kilohertz. We assumed that such a reactor could process aqueous solutions more efficiently, which ultimately would allow obtaining activated water at a lower cost. We also assume that in a high-frequency plasma-chemical reactor, we can obtain equilibrium peroxide concentrations of more than 1 mM. Solutions with similar concentrations of biologically active substances can be used to combat phytopathogens. In addition, one of the objectives of the study was the use of the obtained activated aqueous solutions in agricultural practice.

In this work, we, for the first time, manufactured a plasma-chemical reactor operating at a frequency of 0.11 MHz. The reactor allows for the activation of large volumes of liquids in a short time. The physicochemical properties of activated liquids (concentration of hydrogen peroxide, nitrate anions, redox potential, electrical conductivity, pH, concentration of dissolved gases) are characterized in detail. Antifungal activity of aqueous solutions activated by a glow discharge was investigated. It was shown that aqueous solutions activated by a glow discharge significantly reduced the degree of presence of phytopathogens and their effect on germination of such seeds. Seeds of cereals (sorghum and barley) and fruit (strawberries) crops were studied. The greatest positive effect was found in the treatment of sorghum seeds. Moreover, laboratory tests showed a significant increase in sorghum drought tolerance. The effectiveness of the use of glow-discharge-activated aqueous solutions was shown during a field experiment, which was set up in the saline semi-desert of the Northern Caspian region. Thus, the technology developed by us makes it possible to carry out the activation of aqueous solutions on an industrial scale. Water activated by a glow discharge exhibits antifungicidal activity and significantly accelerates the development of the grain and fruit crops we studied. In the case of sorghum culture, glow-discharge-activated water significantly increases drought resistance.

2. Results

2.1. Physicochemical Properties of Aqueous Solutions Activated by Glow Discharge Plasma

The influence of the plasma of a glow discharge on the physicochemical parameters of water was investigated (Table 1). It was found that the change in the physicochemical parameters of water occurs within 40–50 min of treatment. With further processing, the rate of change of the parameters significantly decreases or, as in the case of hydrogen peroxide, it ceases to change, reaching a stationary value of 7×10^{-3} M.

Table 1. Changes in the physicochemical parameters of aqueous solutions after exposure to glow discharge plasma for 20 or 40 min.

Exposure Time	Measured Parameters					
	EC **, mS/cm	[O$_2$], µM	pH	Redox, mV	NO$_3^-$, mM	H$_2$O$_2$, mM
0 min	7.3 ± 0.5	273 ± 5	6.7 ± 0.1	303 ± 7	<0.01	<0.01
20 min	16.4 ± 1.0 *	264 ± 7	7.8 ± 0.1 *	510 ± 32 *	10.98 ± 0.61 *	3.48 ± 0.43 *
40 min	24.9 ± 1.2 *	261 ± 8	8.3 ± 0.2 *	598 ± 26 *	22.05 ± 0.98 *	7.12 ± 0.68 *

* Statistical differences relative to control ($p < 0.05$); ** EC—electrical conductivity.

During processing, the conductivity of the solution increased almost linearly. The rate of increase in the specific conductivity of the solution was approximately 450 µS/cm/min. In this case, the concentration of molecular oxygen dissolved in water tends to decrease. At the same time, the decrease in the concentration of molecular oxygen during the first 20 min of treatment was much more intense than during the next 20 min. In addition, during the activation process, the pH of the solution changes. During the first 20 min of treatment, the pH value increased by almost one. With further activation within 20 min, the pH value changed only by 0.5 units. Similar changes were observed in the values of the redox potential. In the first 20 min of processing, the potential increased by almost 200 mV. With subsequent processing, an increase of less than 100 mV was observed. During activation, nitrate anions accumulate in aqueous solutions. The increase in the concentration of nitrate anions linearly depends on the processing time. It should be noted that, in contrast to other studied parameters, the concentration of nitrate anions continues to increase after 40 min of exposure. Accumulation of nitrate anions was not observed when experiments were carried out in an argon atmosphere. The increase in the concentration of hydrogen peroxide was linearly dependent on the processing time. The above-described patterns were applicable at processing times of 40–50 min. With further activation of aqueous solutions, the concentration of hydrogen peroxide ceased to change significantly. In some experiments, a decrease in the concentration of hydrogen peroxide was observed. We carried out a number of experiments using electrolytes based on salts of phosphoric, nitric, and a number of organic acids. It was shown that the generation of hydrogen peroxide substantially depends on the composition and concentration of the starting compounds in an aqueous solution. In addition, depending on the composition and concentration of salts, the duration of the activity of the resulting solution depends. Basically, to assess the preservation of the activity of the plasma-activated solution, we used the concentration of hydrogen peroxide. It was shown that under different conditions, the concentration of hydrogen peroxide decreased by two times within 1–3 weeks. Based on the results obtained, in further studies, we used activated solutions containing the maximum amount of active substances (treatment time 40 min).

2.2. Fungicidal Properties of Aqueous Solutions Activated by Glow Discharge Plasma

The fungicidal properties of aqueous solutions activated by glow discharge plasma were studied using healthy (−) and fusarium-infected (+) seeds of *Sorghum bicolor* (L.) *Moench*, wheat *Triticum aestivum*, and strawberry *Fragaria* L. The proportion of affected seeds according to microscopy data was 98, 92, and 76% for wheat, sorghum, and strawberry, respectively. Fusarium blight was identified by the presence of the pathogens *Fusarium* sp.

by microscopy (Table 2). It was shown that soaking seeds in deionized water did not significantly reduce the presence of fungus on the surface of the grains. Soaking the seeds for 5 h in the activated solution led to a significant decrease in the degree of seed infestation. In some cases, using microscopy, it was difficult to distinguish the degree of seed infestation, as well as to reliably identify the type of phytopathogen. Moreover, the microscopy we used did not allow us to distinguish between viable and non-viable forms of the fungus.

Table 2. Influence of aqueous solutions activated by glow discharge plasma on seed infestation with fusarium.

	Microscopy, Seed Contamination Level, %					
	Sorghum		Wheat		Strawberry	
	+	−	+	−	+	−
Control	98	0	92	0	76	0
Deionized water	97	0	90	0	76	0
Activated solution	35	0	61	0	52	0
	RT-PCR, Seed Infection Rate, Ct					
Control	10	>40	17	>40	29	>40
Deionized water	16	>40	18	>40	28	>40
Activated solution	>40	>40	27	>40	35	>40

The infection was verified by RT-PCR. In samples of sorghum (+) and wheat (+) cereals, the DNA of *F. avenaceum* (Ct ~10) and *F. graminearum* (Ct ~17) was identified. In samples of all cereals (-), DNA of both species was not detected. *F. oxysporum* DNA (Ct ~29) was identified in strawberries by real-time PCR. Soaking seeds in deionized water did not significantly affect the concentration of pathogen DNA. After soaking in a solution activated by glow discharge plasma, the concentration of the original DNA of the pathogen significantly decreased. The best results have been achieved with sorghum seeds. Treatment of strawberry seeds was slightly less effective. The least effective treatment was the wheat seed. The presence or absence of pathogen DNA cannot answer the question of the viability of the pathogen. In this regard, seed germination was carried out. It was shown that among all the studied crops, the largest percentage of Fusarium-infected seeds of the treated activated solutions germinates in sorghum; strawberry treatment was slightly less effective. Wheat seed treatment yielded the least significant results.

The results obtained do not allow determining the reasons for the better germination and germination of seeds treated with activated water. On the one hand, of course, activated water has a fungicidal effect on fusarium. On the other hand, activated water can contribute to the development of increased resistance of crops to fungal and bacterial diseases. To answer this question, a series of experiments with healthy seeds was carried out. Since the treatment of sorghum seeds turned out to be the most effective, we carried out further tests only on sorghum seeds.

2.3. Study of the Effectiveness of the Action of Activated Water on the Yield of Sorghum Crops (Laboratory Tests)

The main goal of the experimental studies was to assess the effect of the activated solution at all stages of plant growth and development. To solve this problem, research was carried out in laboratory and field conditions. In the first treatment variant, sorghum seeds were soaked in the activated solution without dilution. To find the optimal processing time, sorghum seeds in the amount of 100 pcs. were soaked in the test solution for a certain time, then washed in distilled water and germinated at 25 °C in Petri dishes on filter paper moistened with distilled water. After 48 h from the start of the treatment, the number of hatching seeds was measured. The seed was considered to have hatched if the root length at the time of measurement was more than 5 mm. The experiment was carried out in triplicate. The average value of seed germination energy depending on the treatment time is presented in Table 3. It is shown that the optimal seed treatment time was in the range of 4–7 h, with the best indicator in the region of 5 h. Soaking seeds in a solution for more than 11 h leads to a decrease in germination. In the future, we would use sorghum seeds, pre-moistened for 5 h with activated water without dilution (the best option, Table 4). Germination of treated infected seeds in Petri dishes on filter paper moistened with distilled water in all replicates did not lead to mold formation, even on dead, not germinated seeds. In control experiments with distilled water, non-germinated seeds were covered with mold after 36 h.

Table 3. Germination of sorghum seeds after short-term soaking in activated water.

	Soaking Time, h						
	1	**3**	**5**	**7**	**9**	**11**	**24**
Germination, %	80	85	93	91	81	69	0

Table 4. Main parameters of young sorghum sprouts 120 h after treatment.

	Germination, %	Root Length, mm	Sprout Weight, g
Control	80 ± 3	8.2 ± 1.2	6.1 ± 0.5
Activated solution	91 ± 2 *	9.7 ± 0.7	7.6 ± 0.2 *

* Statistical differences relative to control ($p < 0.05$).

In the second variant, laboratory tests were carried out in soil on a Klasmann peat mixture, recipe "883". This soil was optimal for the cultivation of young greenery due to the low concentration of minerals and the presence of perlite in its composition. Sorghum seeds in the amount of 100 pcs. were soaked for 5 h in an activated solution without dilution, then washed in distilled water and sown in a container with soil with a volume of 0.7 L to a depth of 5 cm. After 120 h from the moment the seeds were soaked, the young sprouts were carefully washed from the soil, after which the average full germination of seeds, the average length of the main root, and the average weight of the sprout, taking into account the hypocotyl, were measured. The experiment was carried out in triplicate, and the averaged data are shown in Table 4.

When seeds were treated with water activated by glow discharge plasma, seedlings appeared 12 h earlier than control. Such a temporary reserve allowed plants to start the process of photosynthesis earlier, which means that they quickly begin to build up the root system and vegetative mass. Thus, the length of the root in the group of plants treated with activated water was 20% longer than in the control. The biomass of one plant in the group of plants treated with activated water was 25% higher than in the control group. It should be noted that in the group of plants treated with activated water, the standard errors of the mean were less than in the control group.

Figure 1 shows examples of sorghum sprouts on the 5th day from the moment of seed treatment in control and experiment. In the experiment, a more developed leaf surface of young shoots was clearly observed. If we express the leaf surface area in numbers, then in the group of plants treated with activated water, the leaf area will be 2.5 times larger than in the control.

Figure 2 shows the results of an experiment in soil under conditions of provoking root rot (thickened sowing—40 seeds per pot with a volume of 0.7 L), excessive soil moisture and watering with cold water, low air temperature. Sorghum seeds were pre-moistened for 5 h in activated water. Distilled water was used as a control. After 10 days from the moment of seed treatment, root rot was recorded only in the control samples. The source of phytopathogens, apparently, are the seeds themselves, and this factor, among other things, negatively affects the varietal and sowing qualities of seeds.

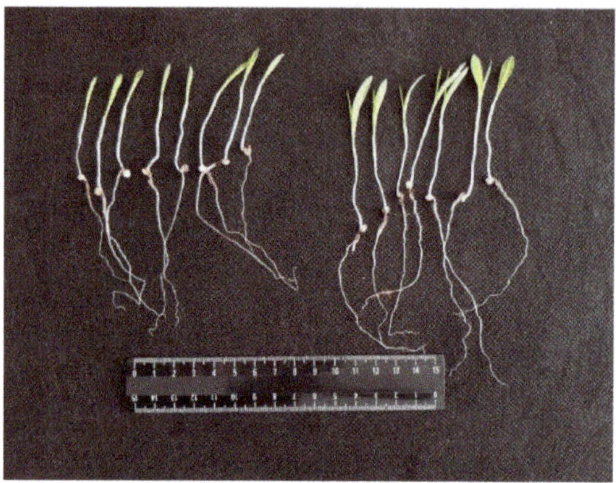

Figure 1. Photographs of randomly selected young shoots of sorghum on the 5th day after seed treatment (**left**—control, **right**—experiment (seeds were soaked in water with activated glow discharge plasma for 5 h)).

(A)   (B)   (C)

Figure 2. General view of sorghum plants on the 10th day after seed treatment (**A,C**) (pot in the center)—treated with activated water, (**B,C**) (left and right pots)—control. An area of infection and spread of root rot is highlighted around.

In the third variant, the moistening of seeds with an activated solution was carried out with different dilutions with deionized water and an increase in the time of seed treatment. In this experiment, 100 sorghum seeds (in 3 replicates) were laid out on filter paper moistened with the test solution with a supposed promising dilution and germinated in Petri dishes at a temperature of 25 °C. After 10 days, the main parameters of the seedlings were measured. Seed energy was assessed by the length of the main root and germination (Table 5).

As can be seen from Table 5, the optimal dilution ratio of the activated solution for sorghum seeds when soaked for 1 or 2 days was 1:500. With this dilution, even on the first day of germination, a more developed main root and higher germination energy were visually recorded. This effect was most pronounced 24 h after seed treatment (Figure 3).

Table 5. The main parameters of sorghum seedlings after 120 h from the moment of soaking in activated water and its solutions for 24 and 48 h.

Dilution Rate	Soaking Time, h	Germination, %	Root Length, mm
Control	24	78 ± 3	8 ± 2
	48	81 ± 3	13 ± 4
Without dilution	24	0	0
	48	0	0
1:250	24	54 ± 5 *	6 ± 2
	48	67 ± 4 *	14 ± 3
1:500	24	84 ± 2 *	12 ± 2 *
	48	87 ± 2 *	22 ± 5 *

* Statistical differences relative to control ($p < 0.05$).

<table>
<tr><td>Time After Soaking</td><td>Without Dilution</td><td>1:250</td><td>1:500</td><td>Control</td></tr>
</table>

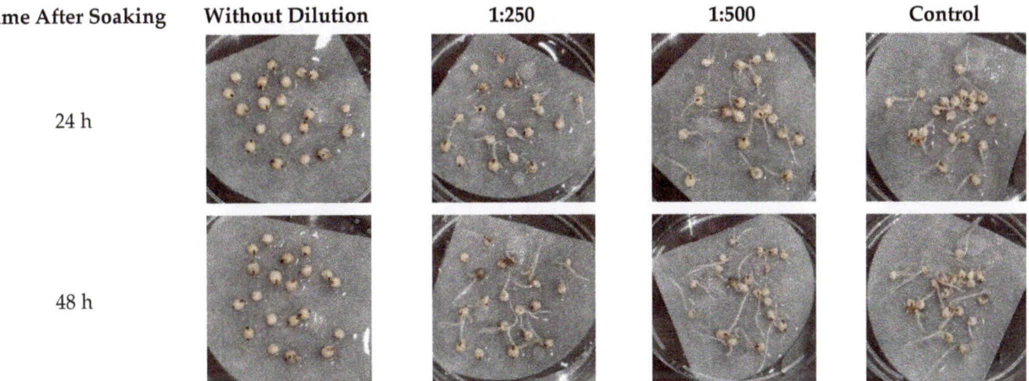

Figure 3. Photographs of sorghum seedlings at different dilutions of activated water and soaking times.

Thus, the optimal dilution ratio of the activated solution for sorghum seeds for long periods of soaking corresponds to 1:500. When diluted optimally, the activated water stimulates the development of sorghum and to some extent, increases the vitality of the plants. The high concentration of activated water and the long exposure time inhibit the growth of sorghum. With a long exposure time, the activated solution acts as an antiseptic.

The variant with the use of activated water without dilution may be promising for disinfection and protection of plants from diseases. In the selection and seed production of agricultural crops, or in areas related to microclonal reproduction of plants. The variant with the dilution of activated water is preferable in the conditions of industrial cultivation of crops. This option significantly reduces the need for activated water, eliminating the effect of inhibition.

2.4. Study of the Effectiveness of the Action of Activated Water on the Yield of Sorghum Crops (Field Trials)

The effect of the pre-sowing treatment of seeds with activated water was observed during the entire cycle of plant growth. At the first stage, seed germination was monitored. After sowing, performed on 19 May, from 20.05 to 26.05 precipitation with an intensity of 1–2 mm fell several times. In the control variant, the crops emerged on the 6th day (standard 7–8 days). The seeds, treated with activated water, sprouted more amicably a day earlier.

Taking into account the specificity of sorghum growth, the slow development of the aerial part in the first weeks with enhanced development of the root system, observations of the second stage were carried out 20 days after sowing (08.06), during the formation of the 5th leaf. Measurements of the height of the aboveground part recorded a statistically significant increase in the growth of sorghum in the activation variant by about 25%, which indirectly indicated a more intensive development of the root system.

The third stage of observations was carried out at the end of the growing season (16.08) before harvesting with a combine using the standard sheaf selection method. According to the results of processing sheaves, a positive effect of pre-sowing seed treatment on the growth of biomass and grain yield was also recorded (Figure 4). As can be seen from the graph, grain yield increased according to biological accounting of sheaves by 10.6%; bunker accounting after harvesting with a combine by 10.4%. The mass of the aboveground part of sorghum plants (air-dry weight) upon activation increased by 19.6%, from 5.04 t/ha to 6.03 t/ha. The most effective treatment of seeds influenced the yield of hay, higher by 58.3% compared to the control. We explain this circumstance by the fact that the plants developed a deeper and more powerful root system, which intensified the growth of the shoots after a late rain shower (after 2 months of drought; at the end of July, about 50 mm fell). Intensive precipitation and the specificity of sorghum to give a fit had a very good effect on the yield of hay, control option—1.37 t/ha, version with activated water—2.19 t/ha. These indicators were quite high compared to the average.

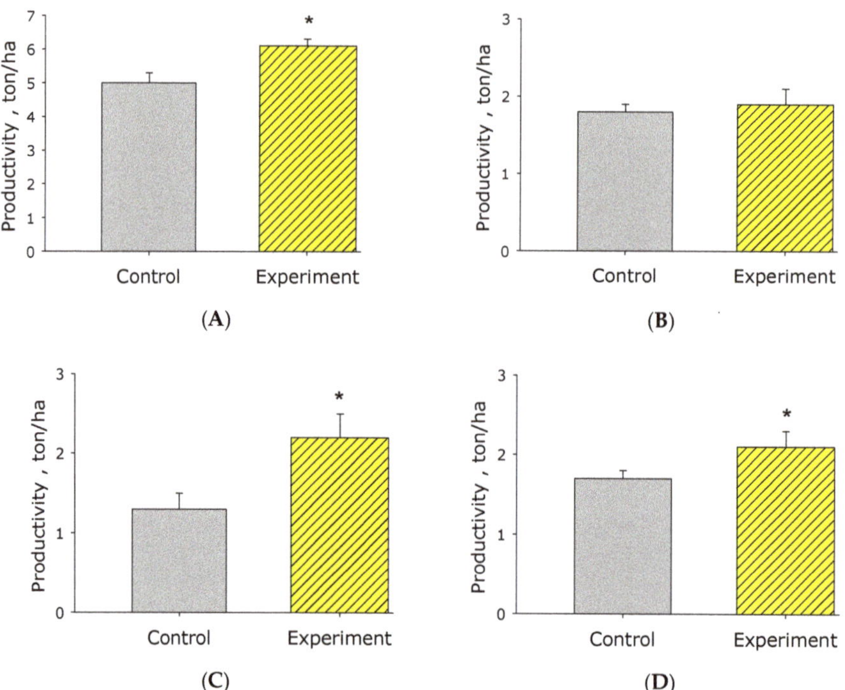

Figure 4. Histogram of the main results of the analysis of sheaves of biomass of grain sorghum "Eltonskoye" soaked in activated water (Experiment) and control water (Control). (**A**)—terrestrial biomass, ton/hectare; (**B**)—biomass of stubble 15 cm high, ton/hectare; (**C**)—hay biomass, ton/hectare; (**D**)—biomass of grain, ton/hectare. *—statistical differences relative to control ($p < 0.05$).

3. Discussion

We have created a setup for the activation of aqueous solutions by a glow discharge plasma (Figures 5 and 6). During the operation of the installation, a significant change in the physicochemical properties of aqueous solutions is observed (Table 1). With prolonged processing of liquids, the rate of change in the parameters of the solution is significantly slowed down. This is probably due to the following circumstances. As a result of the interaction of plasma with water vapor in water, atomic hydrogen and a hydroxyl radical are formed [21]. As a result of the recombination of these products in a liquid, molecular hydrogen ($H^\bullet + {}^\bullet H \rightarrow H_2$), hydrogen peroxide ($OH^\bullet + {}^\bullet HO \rightarrow H_2O_2$), and water ($OH^\bullet + {}^\bullet H \rightarrow H_2O$) are formed [22]. With the accumulation of hydrogen peroxide in the solution, the reaction of its decomposition begins to play a significant role, which leads to the establishment of a stationary concentration of H_2O_2. In some cases, the equilibrium can be mixed, the process of decomposition of the formed hydrogen peroxide will be observed. Obviously, this process depends both on the type of the selected salt and on the storage method of the prepared solution [23].

In the process of activation of an aqueous solution by low-temperature plasma, hydrogen peroxide and nitrate anion are generated at the plasma-solution interface [24,25]. These compounds have significant biological activity, impart antiseptic, and disinfecting properties to an aqueous solution [26], which is shown in the manuscript (Table 1). It should be noted that hydrogen peroxide plays an important role in a number of processes important for plants [27]. In particular, hydrogen peroxide modulates the process of photosynthesis, affects the synthesis of starch and chlorophyll. If we go to the level of the whole plant, then hydrogen peroxide has a significant effect on the growth and development of plants [28–31]. It is known that pre-sowing treatment of seeds and foliar treatment of green plants with solutions of hydrogen peroxide in certain concentrations increase the stability and safety of agricultural plants in severe conditions of moisture deficit, frost, and soil salinity [32–34]. For this reason, plasma-activated aqueous solutions are effective and environmentally friendly stimulators of plant growth and vitality (Tables 2–5, Figures 1–4). After treatment with activated water, seeds have a greater energy potential, structural and functional rearrangements of membrane formations and macromolecules occur in them, as a result of which various biological changes occur in plants (Tables 4 and 5). One of the main effects of biochemical exposure to an activated aqueous solution and other stressful influences is the generation of free radicals in plant seeds, which affect the enzymatic properties of seeds [35]. In addition, it was previously noted that this leads to an increase in the resistance of agricultural crops to fungal and bacterial diseases [36]. In addition, plants treated with activated water become more resistant not only to phytopathogens but also to drought [37]. It is the positive effect of activated water on the growth and resistance (survival) of plants in arid conditions that determined the feasibility of setting up a production experiment in extreme semi-desert conditions (Figure 4). We assume that the success of the experiment is related, among other things, to the peculiarities of the growth and development of the plant root system. It is known that hydrogen peroxide significantly affects the growth and development of the root system. This is due to the modification of the signal of the plant hormone auxin and changes in the expression of genes of the cell cycle [38]. The expediency of the experiment was also dictated by the well-known weak efficiency of fertilizers and growth stimulants in arid conditions due to a lack of moisture [39]. Previously, several laboratory studies were carried out on the effect of plasma-activated aqueous solutions on the drought resistance of plants. It was found that under conditions of water shortage when treating with plasma-activated water, the germination of *Vigna mungo* L. seeds increased by 10–15% [40]. Andreev et al. showed that plasma-activated water in laboratory conditions could improve the drought tolerance of seeds (10–20%) of radish [41], wheat [42], and other crops, as well as increase the rate of seed germination under drought conditions. A number of other laboratory examples are provided in the review article [43]. The activated solution obtained by the described method can be used in various sectors of the national economy, where activated water is traditionally used in agriculture for the treatment of seeds and adult plants, for foliar treatment of plants by spraying, as an antibacterial agent in animal husbandry, the food industry, etc.

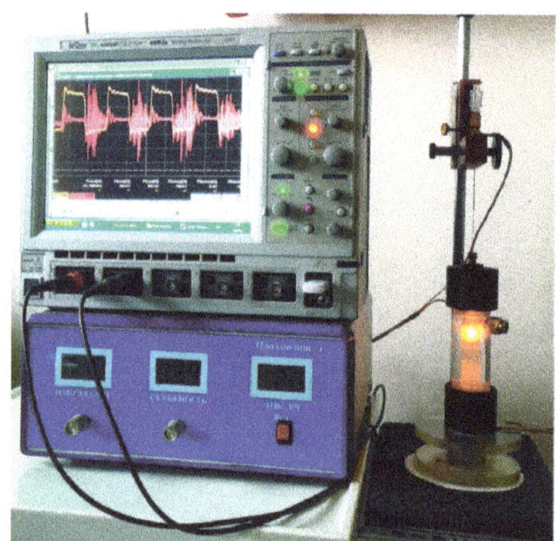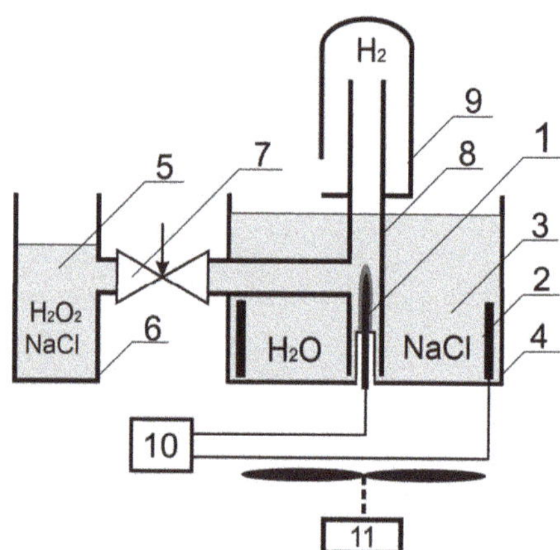

Figure 5. Photographs and schematic diagram of a reactor for obtaining an activated solution. Active electrode—1, neutral electrode—2, electrolyte solution—3, container for electrolyte—4, activated solution—5, container for activated Scheme—6. adjustable leak—7, quartz tube for overflow of activated solution—8, container for collecting hydrogen—9, high-frequency generator—10, magnetic stirrer—11.

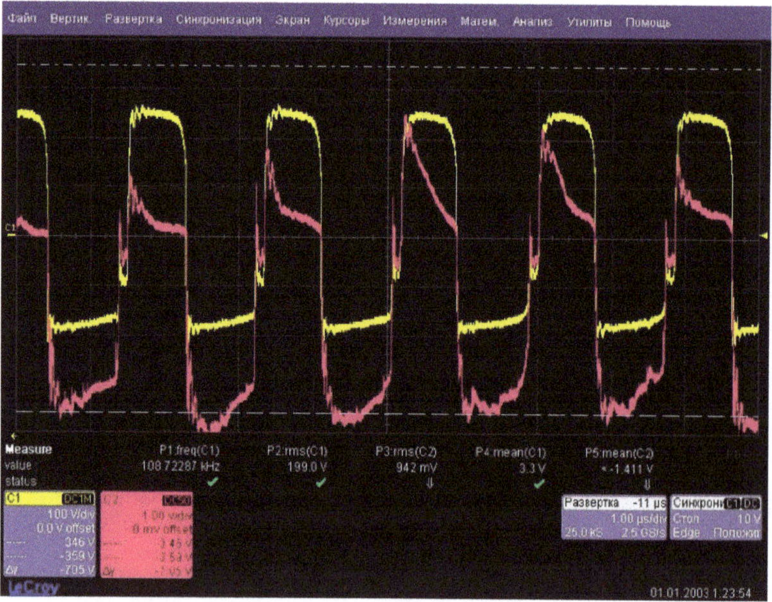

Figure 6. Oscillogram of current (red curve) and voltage (yellow curve) at the reactor input.

4. Materials and Methods

4.1. Activated Water Generation Method

To activate the water, a reactor was used in which 2 activation methods were implemented, plasma-chemical and electrochemical. The schematic diagram of the reactor is shown in Figure 5. The design of the reactor was based on an electrochemical cell, including a vessel with electrolyte, in the volume of which 2 electrodes were immersed—active and passive. The active electrode was made of platinum and had the shape of a cylinder with a diameter of 0.5 mm and a length of up to 20 mm. The passive electrode was made of pyrolytic graphite and had an area of 8 cm^2. The electrodes were connected to a specially designed high-frequency generator operating at 110 kHz. The working voltage on the electrodes was maintained in the range 250–350 V. In this case, the peak power consumed from the generator could reach 1500 W for a short time, which was enough to form a vapor-gas bubble on the active electrode and ignite a glow discharge in the vapor phase. The power of the device in a stationary mode was estimated depending on the composition of the aqueous solution at 100–300 W.

The oscillogram of the current flowing through the reactor is shown in Figure 2. It can be seen from the oscillogram that the current in the negative half-cycle (cathodic polarity) noticeably exceeded the current in the positive half-cycle (anodic polarity), while their shape was significantly different. This leads to the fact that the current flowing through the reactor had a constant component.

Thus, in the used reactor, 2 regions can be distinguished. The first region was formed by the gas phase, in which a high-frequency glow discharge burns in water vapor and plasma-chemical reactions take place. The electrodes for the plasma-chemical part of the reactor were the metal active electrode (cathode) and the plasma-electrolyte boundary (electrolyte anode). The second area was the electrolyte part, with a pulsating current with a constant component flowing through it, the electrodes of which were the plasma—electrolyte (electrolyte cathode) and passive (graphite) anode boundary. The presence of a constant current component leads to the appearance of electrochemical processes occurring in the electrolyte part of the reactor. As an electrolyte, a solution of sodium chloride with a concentration of 0.14 M was used. The volume of the experimental reactor was 200 cm^3. The electrolysis of the solution proceeded without using a diaphragm. During electrolysis, the solution was intensively stirred using a magnetic stirrer. Upon reaching the operating mode, the temperature of the solution in the reactor stabilized at the level of 45 ± 2 °C.

We estimate the market value of the installation (generator, reactor) at \$4200. The production costs of 1 L of activated water (electricity, distilled water, consumables) were \$3–\$5 (depending on the country, for example, in the Russian Federation, the price of one kilowatt of electricity is on average \$0.05). Dilution of one liter of activated water makes it possible to obtain 500 L of working solution for seed dressing using the moistening method.

4.2. Plants Samples

In the experiments, *Sorghum bicolor* (L.) *Moench* (grain sorghum of the Eltonskoe variety), *Triticum aestivum* (wheat of the Moskovskaya 40 variety), and *Fragaria* L. (strawberry of the Ruyana variety) were used as test objects. In a number of experiments, seeds were treated with water activated by smoldering plasma. In a number of experiments, activated water was diluted with distilled water. In laboratory experiments, 100 seeds were used per group. The seeds were stratified at 4 °C and relatively high humidity for 2 weeks. The seeds were germinated in Petri dishes with filter paper at a constant temperature of 20 °C.

In the study of fusarium, we used seeds of *Sorghum bicolor* (L.) *Moench*, wheat *Triticum aestivum*, and strawberry *Fragaria* L. healthy (−) and affected by fusarium (+). The proportion of affected seeds was 98, 92, and 76% for wheat, sorghum, and strawberry, respectively. Fusarium blight was identified by the presence of the pathogens *Fusarium* sp. by microscopy and real-time PCR analysis. Samples were provided by All-Russian Phytopathology Research Institute, Russia.

4.3. DNA Extraction

To verify the microscopic data, we performed diagnostics of fusarium by real-time PCR. Isolation of genomic DNA from samples was performed using cetyl trimethylammonium bromide (CTAB method). A detailed description of the method is given in [44].

4.4. Real-Time PCR

Primers specific for these pathogens were used to identify Fusarium avenaceum, Fusarium graminearum, and Fusarium oxysporum in the respective samples (Table 6). All primers were synthesized at Evrogen (Moscow, Russia). The reaction mixture was prepared by mixing 5 µL of the ready-mixed qPCRmix-HS SYBR (Evrogen, Moscow, Russia) with a pair of target primers (1 µL each), 1 µL of the template DNA solution (1.28×10^2 ng/mL), and Milli-Q water to a volume of 25 µL. The real-time PCR reaction was performed in an O-DTLITE 4S1 amplifier (DNA technology, Moscow, Russia). PCR for the identification of pathogens infestans was carried out according to the following protocol: denaturation 85 s at 94 °C, then 25 cycles 35 s at 95 °C, 30 s at 53 °C, and 30 s at 72 °C. Fluorescence intensity measurements were performed at the end of the 72 °C cycle. Ct values, standard curves, and the corresponding correlation coefficients (R^2) were automatically obtained using the Sequence Detection System v.1.2 software (Waltham, MA, USA) by interpolating Ct values against the decimal logarithms of the initial DNA concentrations [45]. As a negative control, 2 µL of Milli-Q water was added to the reaction mixture instead of the DNA template.

Table 6. Primers used to *Fusariun* sp. identification.

Species and Target	Primers (F and R)	Ref
F. graminearu Intergenic Spacer of rDNA (IGS region)	5′-GTTGATGGGTAAAAGTGTG-3′ 5′-CTCTCATATACCCTCCG-3′	[46]
F. avenaceum гентranslation elongation factor 1-alpha (TEF1)	5′-ATGGGTAAGGARGACAAGAC-3′ 5′-GGARGTACCAGTSATCATG-3′	[46]
F. oxysporum specific fragment between the transcription factors Han and Skippy	5′-CAGACTGGGGTGCTTAAAGTT-3′ 5′-AACGCTAGGGTCGTAACAAA-3′	[47]

4.5. Physicochemical Properties of Aqueous Solutions

Redox potential, pH, and electrical conductivity were measured on an S470 SevenExcellence high-precision measuring station (Mettler Toledo, Columbus, OH, USA). The recommended sensor electrodes InLab Expert Pro-ISM and InLab731-ISM (Mettler Toledo) were used. During measurements, aqueous solutions were mixed in a laminar mode using a magnetic stirrer (rotation frequency 3 Hz). All measurements were carried out at a solution temperature of 20 ± 1 °C. The experimental measurement details were described previously [48].

The concentration of molecular oxygen dissolved in water solutions was measured using an AKPM-1-02 polarograph (Bioanalytical systems and sensors, Moscow, Russia) [49]. The measurements took into account the atmospheric pressure, measured with a PRX-7001t (Casio, Tokyo, Japan), and the temperature of the samples, measured with a thermocompensating electrode. All measurements were carried out at a solution temperature of 20 ± 1 °C. The experimental measurement details were described previously [50].

The content of nitrate anions in the samples was determined using the Griss reagent according to the method described earlier [12]. To determine the concentrations of nitrate anions, the test solutions in a volume of 100 µL were applied to the wells of a 96-well flat-bottom polystyrene plate. Thereafter, a freshly prepared saturated solution of VCl_3 (8 g/L in 1 M HCl) in a volume of 100 µL. Immediately thereafter, 100 µL of Griss reagent (1 M HCl containing 10 g/L of sulfanilamide and 1 g/L of N-1-naphthylethylenediamine hydrochloride) was added. The reaction was carried out in the dark at 37 °C for 1 h. After that, the optical density of the medium was measured at a wavelength of 546 nm using a Multiscan FC plate reader (TermoScintific, Vaanta, Finland). Sodium nitrate solutions of known concentration were used for calibration.

For the quantitative determination of hydrogen peroxide in aqueous solutions, a highly sensitive method of enhanced chemiluminescence in the luminol-p-iodophenol-horseradish peroxidase system was used [51]. The luminescence intensity was determined using a Biotox-7A chemiluminometer (ANO ICE, Moscow, Russia). For calibration, hydrogen peroxide solutions of known concentration were used. The initial concentration of hydrogen peroxide used for calibration was determined spectrophotometrically at a wavelength of 240 nm with a molar absorption coefficient of 43.6 ($M^{-1} \times cm^{-1}$). The "counting solution" contained: 1 cM Tris-HCl buffer pH 8.5, 50 μM p-iodophenol, 50 μM luminol, 10 nM horseradish peroxidase. The sensitivity of the method makes it possible to determine hydrogen peroxide at a concentration of less than 1 nM [52].

4.6. Field Experiment

The experiment was carried out on a sorghum crop of the "Eltonskoye" variety in a saline clay semi-desert of the Northern Caspian region, in the Pallasovsky district of the Volgograd region. The advanced farm of the region—SPK Plemzavod Krasny Oktyabr—was used as a production base. The sum of the average annual precipitation in this region corresponds to 280–300 mm. The soils were represented by a 3-membered solonetzic complex, including up to 50–60% of solonetzes. The average long-term productivity of virgin vegetation, according to research, was 12–15 c/ha. The yield of hay from the solonetz complex when mowing was 3–4 c/ha, with normalized grazing by livestock—5.6 c/ha.

The treatment of inoculum with activated water and the setting up of the experiment was carried out as follows. A solution for pre-sowing seed treatment was prepared by diluting activated water in a ratio of 1:500 with water from an open reservoir. The seeds were moistened by watering and mixing the herd of seeds and then keeping them covered for 14 h (for 1000 kg of seeds—5 L of solution with a working concentration). The sowing of the treated seeds was carried out without allowing them to dry out. Sowing area of the option using activated water—100 hectares; control—100 hectares. The sowing rate of sorghum seeds was 10 kg/ha.

4.7. Statistics

Data were analyzed using SigmaPlot 11 software (Palo Alto, CA, USA) and were presented as means ± SEM. Data from at least 3 independent experiments were used for averaging.

5. Conclusions

The results obtained allow us to draw the following conclusions:

1. Activated water can be attributed to the category of plant growth regulators of grain and fruit crops, simultaneously contributing to an increase in plant resistance to negative biotic and abiotic environmental factors.
2. High concentrations of activated water can partially or completely block the growth and development of plants. The positive effect of the drug is usually manifested when the stock solution is diluted 250–500 times.
3. Pre-sowing treatment of seeds with activated water affects the growth and development of sorghum during the entire growing season.
4. The effectiveness of the drug is manifested even when grown on poor saline soils and in drought conditions, in which the effect of fertilizers and most growth stimulants is weak due to lack of moisture.
5. Activated water in optimal concentration is an inexpensive, promising, environmentally friendly preparation for production. In extreme growing conditions, the device developed by us is of particular value and promising for use.

Author Contributions: Conceptualization, Y.K.D.; methodology, K.N.K. and A.P.G.; formal analysis, R.M.S., S.V.B. and S.V.; investigation, S.V.B., A.B.E., V.I.L., V.A.S., L.M.A., V.Y.D., M.I.B., A.M.B., D.V.Y., M.E.A. and V.P.K.; resources, A.P.G.; writing—original draft preparation, Y.K.D.; writing—review and editing, S.V.G.; visualization, V.I.L.; supervision, Y.K.D.; project administration, S.V.G. All authors have read and agreed to the published version of the manuscript.

Funding: This work was supported by a grant from the Ministry of Science and Higher Education of the Russian Federation for large scientific projects in priority areas of scientific and technological development (Grant No. 075-15-2020-774).

Institutional Review Board Statement: Not applicable.

Informed Consent Statement: Not applicable.

Data Availability Statement: Not available.

Acknowledgments: The authors are grateful to the director and chief agronomist of the SPK-Plemzavod "Krasny Oktyabr" (Volgograd region) A.V. Fomin and S.P. Goremykin for organizing and helping in the production experiment. The authors are grateful to the Center for Collective Use of the GPI RAS for the equipment provided.

Conflicts of Interest: The authors declare no conflict of interest.

References

1. Judée, F.; Simon, S.; Bailly, C.; Dufour, T. Plasma-activation of tap water using DBD for agronomy applications: Identification and quantification of long lifetime chemical species and production/consumption mechanisms. *Water Res.* **2018**, *133*, 47–59. [CrossRef] [PubMed]
2. Thirumdas, R.; Kothakota, A.; Annapure, U.; Siliveru, K.; Blundell, R.; Gatt, R.; Valdramidis, V.P. Plasma activated water (PAW): Chemistry, physico-chemical properties, applications in food and agriculture. *Trends Food Sci. Technol.* **2018**, *77*, 21–31. [CrossRef]
3. Belov, S.V.; Lobachevsky, Y.P.; Danilejko, Y.K.; Egorov, A.B.; Simakin, A.V.; Maleki, A.; Temnov, A.A.; Dubinin, M.V.; Gudkov, S.V. The Role of Mitochondria in the Dual Effect of Low-Temperature Plasma on Human Bone Marrow Stem Cells: From Apoptosis to Activation of Cell Proliferation. *Appl. Sci.* **2020**, *10*, 8971. [CrossRef]
4. Belov, S.V.; Danyleiko, Y.K.; Glinushkin, A.P.; Kalinitchenko, V.P.; Egorov, A.V.; Sidorov, V.A.; Konchekov, E.M.; Gudkov, S.V.; Dorokhov, A.S.; Lobachevsky, Y.P.; et al. An Activated Potassium Phosphate Fertilizer Solution for Stimulating the Growth of Agricultural Plants. *Front. Phys.* **2021**, *8*, 618320. [CrossRef]
5. Sysolyatina, E.V.; Lavrikova, A.Y.; Loleyt, R.A.; Vasilieva, E.V.; Abdulkadieva, M.A.; Ermolaeva, S.A.; Sofronov, A.V. Bidirectional mass transfer-based generation of plasma-activated water mist with antibacterial properties. *Plasma Process Polym.* **2020**, *17*, e2000058. [CrossRef]
6. Kučerová, K.; Henselová, M.; Slováková, Ľ.; Bačovčinová, M.; Hensel, K. Effect of Plasma Activated Water, Hydrogen Peroxide, and Nitrates on Lettuce Growth and Its Physiological Parameters. *Appl. Sci.* **2021**, *11*, 1985. [CrossRef]
7. Maniruzzaman, M.; Sinclair, A.J.; Cahill, D.M.; Wang, X.; Dai, X.J. Nitrate and Hydrogen Peroxide Generated in Water by Electrical Discharges Stimulate Wheat Seedling Growth. *Plasma Chem. Plasma Process.* **2017**, *37*, 1393–1404. [CrossRef]
8. Aider, M.; Gnatko, E.; Benali, M.; Plutakhin, G.; Kastyuchik, A. Electro-activated aqueous solutions: Theory and application in the food industry and biotechnology. *Innov. Food Sci. Emerg. Technol.* **2012**, *15*, 38–49. [CrossRef]
9. Machala, Z.; Tarabova, B.; Hensel, K.; Spetlikova, E.; Sikurova, L.; Lukes, P. Formation of ROS and RNS in Water Electro-Sprayed through Transient Spark Discharge in Air and their Bactericidal Effects. *Plasma Process. Polym.* **2013**, *10*, 649–659. [CrossRef]
10. Locke, B.R. Environmental Applications of Electrical Discharge Plasma with Liquid Water: A Mini Review. *Inter. J. Plasma Env. Sci. Technol.* **2012**, *6*, 194–203. [CrossRef]
11. Kim, J.H.; Hong, Y.C.; Kim, H.S.; Uhm, H.S. Simple Microwave Plasma Source at Atmospheric Pressure. *J. Korean Phys. Soc.* **2003**, *42*, S876–S879.
12. Sergeichev, K.F.; Lukina, N.A.; Sarimov, R.M.; Smirnov, I.G.; Simakin, A.V.; Dorokhov, A.S.; Gudkov, S.V. Physicochemical Properties of Pure Water Treated by Pure Argon Plasma Jet Generated by Microwave Discharge in Opened Atmosphere. *Front. Phys.* **2021**, *8*, 614684. [CrossRef]
13. Dai, X.J.; Corr, C.S.; Ponraj, S.B.; Maniruzzaman, M.; Ambujakshan, A.T.; Chen, Z.; Kviz, L.; Lovett, R.; Rajmohan, G.D.; de Celis, D.R.; et al. Efficient and Selectable Production of Reactive Species Using a Nanosecond Pulsed Discharge in Gas Bubbles in Liquid. *Plasma Process. Polym.* **2016**, *13*, 306–310. [CrossRef]
14. Hamdan, A.; Cha, M.S. The effects of gaseous bubble composition and gap distance on the characteristics of nanosecond discharges in distilled water. *J. Phys. D Appl. Phys.* **2016**, *49*, 245203. [CrossRef]
15. Cejas, E.; Mancinelli, B.; Prevosto, L. Modelling of an Atmospheric–Pressure Air Glow Discharge Operating in High–Gas Temperature Regimes: The Role of the Associative Ionization Reactions Involving Excited Atoms. *Plasma* **2020**, *3*, 3. [CrossRef]
16. Sainct, F.P.; Durocher-Jean, A.; Gangwar, R.K.; Gonzalez, N.Y.M.; Coulombe, S.; Stafford, L. Spatially-Resolved Spectroscopic Diagnostics of a Miniature RF Atmospheric Pressure Plasma Jet in Argon Open to Ambient Air. *Plasma* **2020**, *3*, 5. [CrossRef]
17. Shutov, D.A.; Ivanov, A.N.; Rakovskaya, A.V.; Smirnova, K.V.; Manukyan, A.S.; Rybkin, V.V. Synthesis of oxygen-containing iron powders and water purification from iron ions by glow discharge of atmospheric pressure in contact with the solution. *J. Phys. D Appl. Phys.* **2020**, *53*, 445202. [CrossRef]
18. Ozen, E.; Singh, R. Atmospheric cold plasma treatment of fruit juices: A review. *Trends Food Sci. Technol.* **2020**, *103*, 144–151. [CrossRef]
19. Baburin, N.V.; Belov, S.V.; Danileĭko, Y.K.; Egorov, A.B.; Lebedeva, T.P.; Nefedov, S.M.; Osiko, V.V.; Salyuk, V.A.; Danileiko, Y.K. Heterogeneous recombination in steam plasma as a mechanism of affecting biological tissues. *Dokl. Phys.* **2009**, *54*, 259–261. [CrossRef]

20. Rusanov, V.D.; Fridman, A.A.; Sholin, G.V. The physics of a chemically active plasma with nonequilibrium vibrational excitation of molecules. *Sov. Phys. Uspekhi* **1981**, *24*, 447–474. [CrossRef]
21. Attri, P.; Ishikawa, K.; Okumura, T.; Koga, K.; Shiratani, M. Plasma Agriculture from Laboratory to Farm: A Review. *Processes* **2020**, *8*, 1002. [CrossRef]
22. Le Caër, S. Water Radiolysis: Influence of Oxide Surfaces on H2 Production under Ionizing Radiation. *Water* **2011**, *3*, 235–253. [CrossRef]
23. Bruskov, V.I.; Chernikov, A.V.; Gudkov, S.V.; Masalimov, Z.K. Thermal Activation of the Reducing Properties of Seawater Anions. *Biofizika* **2003**, *48*, 1022–1029.
24. Konchekov, E.M.; Glinushkin, A.P.; Kalinitchenko, V.P.; Artem'Ev, K.V.; Burmistrov, D.E.; Kozlov, V.A.; Kolik, L.V. Properties and Use of Water Activated by Plasma of Piezoelectric Direct Discharge. *Front. Phys.* **2021**, *8*, 616385. [CrossRef]
25. Chen, H.; Yuan, D.; Wu, A.; Lin, X.; Li, X. Review of low-temperature plasma nitrogen fixation technology. *Waste Dispos. Sustain. Energy* **2021**, *3*, 201–217. [CrossRef] [PubMed]
26. Wang, Q.; Salvi, D. Evaluation of plasma-activated water (PAW) as a novel disinfectant: Effectiveness on Escherichia coli and Listeria innocua, physicochemical properties, and storage stability. *LWT* **2021**, *149*, 111847. [CrossRef]
27. Quan, L.-J.; Zhang, B.; Shi, W.-W.; Li, H.-Y. Hydrogen Peroxide in Plants: A Versatile Molecule of the Reactive Oxygen Species Network. *J. Integr. Plant Biol.* **2008**, *50*, 2–18. [CrossRef] [PubMed]
28. Dunand, C.; Crèvecoeur, M.; Penel, C. Distribution of superoxide and hydrogen peroxide in Arabidopsis root and their influence on root development: Possible interaction with peroxidases. *New Phytol.* **2007**, *174*, 332–341. [CrossRef] [PubMed]
29. Barba-Espín, G.; Hernández, J.A.; Diaz-Vivancos, P. Role of H_2O_2 in pea seed germination. *Plant Signal. Behav.* **2012**, *7*, 193–195. [CrossRef]
30. Gorbanev, Y.; O'Connell, D.; Chechik, V. Non-Thermal Plasma in Contact with Water: The Origin of Species. *Chem. A Eur. J.* **2016**, *22*, 3496–3505. [CrossRef] [PubMed]
31. Ismail, S.Z.; Khandaker, M.M.; Mat, N.; Boyce, A.N. Effects of Hydrogen Peroxide on Growth, Development and Quality of Fruits: A Review. *J. Agron.* **2015**, *14*, 331–336. [CrossRef]
32. Fedina, I.S.; Nedeva, D.; Cicek, N. Pre-treatment with H_2O_2 induces salt tolerance in Barley seedlings. *Biol. Plant.* **2009**, *53*, 321–324. [CrossRef]
33. Wang, Y.; Li, J.; Wang, J.; Li, Z. Exogenous H_2O_2 improves the chilling tolerance of manilagrass and mascarenegrass by activating the antioxidative system. *Plant Growth Regul.* **2010**, *61*, 195–204. [CrossRef]
34. Li, J.T.; Zhao, P.P.; Qiu, Z.B.; Zhang, Y.H.; Wang, M.M.; Bi, Z.Z. Effects of Exogenous Hydrogen Peroxide on the Physiological Index of Wheat Seedlings under Salt Stress. *Acta Bot. Booreali-Occident. Sin.* **2012**, *32*, 1796–1801. [CrossRef]
35. Das, K.; Roychoudhury, A. Reactive oxygen species (ROS) and response of antioxidants as ROS-scavengers during environmental stress in plants. *Front. Environ. Sci.* **2014**, *2*, 53. [CrossRef]
36. Ka, D.H.; Priatama, R.A.; Park, J.Y.; Park, S.J.; Kim, S.B.; Lee, I.A.; Lee, Y.K. Plasma-Activated Water Modulates Root Hair Cell Density via Root Developmental Genes in *Arabidopsis thaliana* L. *Appl. Sci.* **2021**, *11*, 2240. [CrossRef]
37. Hossain, M.A.; Fujita, M. Hydrogen peroxide priming stimulates drought tolerance in mustard (*Brassica juncea* L.). *Plant Gene Trait.* **2013**, *4*, 109–123. [CrossRef]
38. Zhao, F.-Y.; Han, M.-M.; Zhang, S.-Y.; Wang, K.; Zhang, C.-R.; Liu, T.; Liu, W. Hydrogen Peroxide-Mediated Growth of the Root System Occurs via Auxin Signaling Modification and Variations in the Expression of Cell-Cycle Genes in Rice Seedlings Exposed to Cadmium Stress. *J. Integr. Plant Biol.* **2012**, *54*, 991–1006. [CrossRef]
39. Baral, B.R.; Pande, K.R.; Gaihre, Y.K.; Baral, K.R.; Sah, S.K.; Thapa, Y.B.; Singh, U. Increasing nitrogen use efficiency in rice through fertilizer application method under rainfed drought conditions in Nepal. *Nutr. Cycl. Agroecosyst.* **2020**, *118*, 103–114. [CrossRef]
40. Sajib, S.A.; Billah, M.; Mahmud, S.; Miah, M.; Hossain, F.; Omar, F.B.; Roy, N.C.; Hoque, K.M.F.; Talukder, M.R.; Kabir, A.H.; et al. Plasma activated water: The next generation eco-friendly stimulant for enhancing plant seed germination, vigor and increased enzyme activity, a study on black gram (Vigna mungo L.). *Plasma Chem. Plasma Process.* **2020**, *40*, 119–143. [CrossRef]
41. Andreev, S.N.; Apasheva, L.M.; Ashurov, M.K.; Lukina, N.A.; Sapaev, I.B.; Sapaev, I.; Sergeichev, K.F.; Shcherbakov, I.A. Production of Pure Hydrogen Peroxide Solutions in Water Activated by the Plasma of Electrodeless Microwave Discharge and Their Application to Control Plant Growth. *Phys. Wave Phenom.* **2019**, *27*, 145–148. [CrossRef]
42. Andreev, S.N.; Apasheva, L.M.; Ashurov, M.K.; Lukina, N.A.; Shcherbakov, I.A. Production of pure hydrogen peroxide solutions in water activated by plasma of an electrodeless microwave discharge and their application for controlling plant growth. *Dokl. Phys.* **2019**, *64*, 222–224. [CrossRef]
43. Guo, D.; Liu, H.; Zhou, L.; Xie, J.; He, C. Plasma-activated water production and its application in agriculture. *J. Sci. Food Agric.* **2021**, *101*, 4891–4899. [CrossRef] [PubMed]
44. Aboul-Maaty, N.A.-F.; Oraby, H.A.-S. Extraction of high-quality genomic DNA from different plant orders applying a modified CTAB-based method. *Bull. Natl. Res. Cent.* **2019**, *43*, 25. [CrossRef]
45. Sharapov, M.G.; Novoselov, V.I.; Penkov, N.V.; Fesenko, E.E.; Vedunova, M.V.; Bruskov, V.I.; Gudkov, S.V. Protective and adaptogenic role of peroxiredoxin 2 (Prx2) in neutralization of oxidative stress induced by ionizing radiation. *Free. Radic. Biol. Med.* **2019**, *134*, 76–86. [CrossRef] [PubMed]

46. Jurado, M.; Vázquez, C.; Patiño, B.; González-Jaén, M.T. PCR detection assays for the trichothecene-producing species Fusarium graminearum, Fusarium culmorum, Fusarium poae, Fusarium equiseti and Fusarium sporotrichioides. *Syst. Appl. Microbiol.* **2005**, *28*, 562–568. [CrossRef] [PubMed]
47. Pastrana, A.M.; Basallote-Ureba, M.J.; Aguado, A.; Capote, N. Potential Inoculum Sources and Incidence of Strawberry Soilborne Pathogens in Spain. *Plant Dis.* **2017**, *101*, 751–760. [CrossRef]
48. Shcherbakov, I.A.; Baimler, I.V.; Gudkov, S.V.; Lyakhov, G.A.; Mikhailova, G.N.; Pustovoy, V.I.; Sarimov, R.M.; Simakin, A.V.; Troitsky, A.V. Influence of a Constant Magnetic Field on Some Properties of Water Solutions. *Dokl. Phys.* **2020**, *65*, 273–275. [CrossRef]
49. Baymler, I.V.; Gudkov, S.V.; Sarimov, R.M.; Simakin, A.V.; Shcherbakov, I.A. Concentration Dependences of Molecular Oxygen and Hydrogen in Aqueous Solutions. *Dokl. Phys.* **2020**, *65*, 5–7. [CrossRef]
50. Baimler, I.V.; Lisitsyn, A.B.; Serov, D.A.; Astashev, M.E.; Gudkov, S.V. Analysis of Acoustic Signals During the Optical Breakdown of Aqueous Solutions of Fe Nanoparticles. *Front. Phys.* **2020**, *8*, 622551. [CrossRef]
51. Gudkov, S.V.; Lyakhov, G.A.; Pustovoy, V.; Shcherbakov, I.A. Influence of Mechanical Effects on the Hydrogen Peroxide Concentration in Aqueous Solutions. *Phys. Wave Phenom.* **2019**, *27*, 141–144. [CrossRef]
52. Baimler, I.V.; Lisitsyn, A.B.; Gudkov, S.V. Water Decomposition Occurring During Laser Breakdown of Aqueous Solutions Containing Individual Gold, Zirconium, Molybdenum, Iron or Nickel Nanoparticles. *Front. Phys.* **2020**, *8*, 620938. [CrossRef]

Article

The Effect of Plasma Activated Water on Maize (*Zea mays* L.) under Arsenic Stress

Zuzana Lukacova [1],*, Renata Svubova [1], Patricia Selvekova [2] and Karol Hensel [2]

[1] Department of Plant Physiology, Faculty of Natural Sciences, Comenius University, Ilkovičova 6, 842 15 Bratislava, Slovakia; renata.subova@uniba.sk

[2] Division of Environmental Physics, Faculty of Mathematics, Physics and Informatics, Comenius University, 842 48 Bratislava, Slovakia; ivanova.patricia@gmail.com (P.S.); karol.hensel@fmph.uniba.sk (K.H.)

* Correspondence: zuzana.lukacova@uniba.sk

Abstract: Plasma activated water (PAW) is a source of various chemical species useful for plant growth, development, and stress response. In the present study, PAW was generated by a transient spark discharge (TS) operated in ambient air and used on maize corns and seedlings in the 3 day paper rolls cultivation followed by 10 day hydroponics cultivation. For 3 day cultivation, two pre-treatments were established, "priming PAW" and "rolls PAW", with corns imbibed for 6 h in the PAW and then watered daily by fresh water and PAW, respectively. The roots and the shoot were then analyzed for guaiacol peroxidase (G-POX, POX) activity, root tissues for their lignification, and root cell walls for in situ POX activity. To evaluate the potential of PAW in the alleviation abiotic stress, ten randomly selected seedlings were hydroponically cultivated for the following 10 days in 0.5 Hoagland nutrient solutions with and without 150 μM As. The seedlings were then analyzed for POX and catalase (CAT) activities after As treatment, their leaves for photosynthetic pigments concentration, and leaves and roots for As concentration. The PAW improved the growth of the 3 day-old seedlings in terms of the root and the shoot length, while roots revealed accelerated endodermal development. After the following 10 day cultivation, roots from PAW pre-treatment were shorter and thinner but more branched than the control roots. The PAW also enhanced the POX activity immediately after the imbibition and in the 3 day old roots. After 10 day hydroponic cultivation, antioxidant response depended on the PAW pre-treatment. CAT activity was higher in As treatments compared to the corresponding PAW treatments, while POX activity was not obvious, and its elevated activity was found only in the priming PAW treatment. The PAW pre-treatment protected chlorophylls in the following treatments combined with As, while carotenoids increased in treatments despite PAW pre-treatment. Finally, the accumulation of As in the roots was not affected by PAW pre-treatment but increased in the leaves.

Keywords: antioxidants; arsenic; maize; plasma activated water

Citation: Lukacova, Z.; Svubova, R.; Selvekova, P.; Hensel, K. The Effect of Plasma Activated Water on Maize (*Zea mays* L.) under Arsenic Stress. *Plants* **2021**, *10*, 1899. https://doi.org/10.3390/plants10091899

Academic Editor: Magda Pál

Received: 18 August 2021
Accepted: 9 September 2021
Published: 14 September 2021

1. Introduction

Atmospheric plasma has shown promising potential in various agricultural applications, where it is applied to seeds or plants to stimulate germination, or to modulate growth, and fruit yield [1,2]. The plasma can be applied either directly or indirectly, i.e., its effect is mediated by a gas or a liquid exposed to plasma. The plasma produces various gaseous reactive species, which may dissolve into liquid/water to produce so-called plasma activated liquid or plasma activated water (PAW). The composition and activity of PAW can be tuned by various parameters, e.g., discharge type and its power, gas and water composition, and flow rate. An interesting feature of PAW, having potential also in commercial use, is that it contains various reactive oxygen and nitrogen species (RONS), such as nitrates (NO_3^-), nitrites (NO_2^-), and hydrogen peroxide (H_2O_2) [3], and is able to preserve its antibacterial activity for several days [4]. This feature predicts its use in agriculture where an increase in human population is reflected in a higher food demand. Nowadays, the use of chemicals to decrease bacterial contamination of seed surface, or the use of pesticides and herbicides to avoid pathogens and weeds, brings secondary contamination of soils and water. Additionally, the use of fertilizers, especially those with low quality and control, to increase the crop yield often results in the heavy metal soil contamination [5]. On the contrary, with the use of PAW or plasma activated ammonia solutions, such contamination can be avoided, as these solutions may serve as an effective source of nitrogen with nitrates (NO_3^-) and ammonium ions (NH_4^+) being the most important compounds for plant growth and development [6]. Further, hydrogen peroxide (H_2O_2) can serve to sterilize seeds and can also enhance seed priming. It may also act as a signaling molecule and activate proteins or genes responsible for plant growth and development. As a result of several reported positive effects of plasma use toward seed and plants, a new field has been established and is referred as "plasma agriculture" [1]. Atmospheric plasma and PAW in agriculture have been studied in recent years for their effects on seeds, mostly to improve their germination, growth, and subsequent yield [7–10]. They were also reported being able to change enzymatic activity in seeds [11–15], to alter secondary metabolites content [16,17], to induce structural modification of seed surface and associated changes in affinity towards water [18], and to reduce numbers of phytopathogenic microorganisms on the seed surface [19–21]. The effect of plasma and PAW have been also intensively studied for various plant species cultivated in different ways and analyzed by various methods. Their positive effects on macroscopic physical characteristics of seedlings and plants have been reported, including number and quality of leaves, length of above-ground parts and roots, and fresh and dry weight. The broad range of analytical methods have been used also to investigate the effects of plasma and PAW on physiological processes and metabolism, e.g., water uptake, photosynthetic pigments content, photosynthetic rate, enzymatic activity, and protein contents or DNA damage [22–25].

Plants are often challenged by various stresses. One of the most common is a stress from toxic elements contaminating soils. Among them, one of the most dangerous is metalloid arsenic (As), absorbed and translocated in the plant bodies into the edible parts where it threatens the highest trophic levels [26]. A common plant reaction to As stress is an overproduction of various reactive oxygen species (ROS) within the plant body [27] triggering antioxidant systems [28]. H_2O_2 plays a central role in stress signal transduction [27,29]; a delicate balance between its production and scavenging must be maintained in the plant cells. Too high level of ROS causes damage, especially to cell macromolecules, and this leads to cell death [30]. On the other side, the non-toxic H_2O_2 concentration, acting as signal molecules, activate multiple plant cell responses, especially via MAPK (mitogen-activated protein kinases) cascades, leading to ROS detoxification and surviving stress situation [29]. Systems joined with ROS in terms of their metabolism are either enzymatic or non-enzymatic. The most active enzymatic scavengers of H_2O_2 are catalase (CAT), decomposing it directly, and peroxidases (POX), reducing it by oxidizing various substrate, e.g., monolignols in the cell wall, which is an important step in lignin formation [28,31–33]. A promising method of inducing stress resistance in plants is the pre-treatment (priming) of seeds or plants by exposure to a chemical compound acting as a stressor [27,34–36]. Studies have revealed that priming phenomenon modulates the plant response positively to the followed-up stress. However, the molecular mechanism associated with priming is still to be elucidated, although there is evidence suggesting the role of agents such as H_2O_2 (and others) making plants more tolerant [27,37–39], especially by modulating ROS detoxification pathways [40,41].

The aim of the present study was to investigate the effect of PAW generated by a transient spark discharge (TS) operated in ambient air in contact with tap water on maize corns and young seedlings. Several treatments of PAW were established, and their potential in the priming of seedlings subsequently exposed to stress from arsenic (As) was evaluated. This was documented for the first time to our best knowledge and broadens the understanding of PAW interaction with the plant defensive systems. Maize seeds and seedlings were pre-treated in PAW for three days and subsequently analyzed for their POX and CAT activities, lignification of the root tissues, and in situ POX activity in roots. The pre-treatment was followed by 10 day hydroponic cultivation in 0.5 Hoagland nutrient solution with and without As stress. Subsequently, POX and CAT activity in seedlings, chlorophyll and carotenoid concentrations in leaves, and concentrations of As in leaves and roots were determined.

2. Results

The plasma activated water (PAW) produced in the present study had pH 7.5 and contained approximately 0.5 ± 0.1 mM of H_2O_2, 0.6 ± 0.1 mM of NO_2^-, and 1.7 ± 0.3 mM of NO_3^-. Exposure of water solutions, such as deionized water, or physiological solution to air plasmas usually leads to their acidification and a decrease in pH. However, the pH of tap water after plasma exposure remained fairly constant or changed very mildly due to its natural hydrocarbon buffer system. To investigate the effect of PAW containing these RONS, the growth parameters, antioxidant enzyme (G-POX, POX) activity, and the development of the young root of maize seedlings were assessed. First, activity of G-POX was measured in the corns imbibed in PAW. The enzymatic activity increased by more than four times in the maize corns after only 6 h imbibition in the PAW in comparison with corns in control (tap water) (Figure 1a). To investigate young seedlings exposed to PAW, corns were germinated and left growing in paper rolls for three days; the PAW treatment was watered every day with freshly produced PAW and the control treatment with tap water. A significant increase in G-POX activity for the PAW treatment was noticed in the young roots (Figure 1b), but the increase was not as big as in the corns; the difference between the control and PAW was an 80% increase (Figure 1b). On the other hand, no significant change in POX activity after 3 days of cultivation was noticed in the shoot.

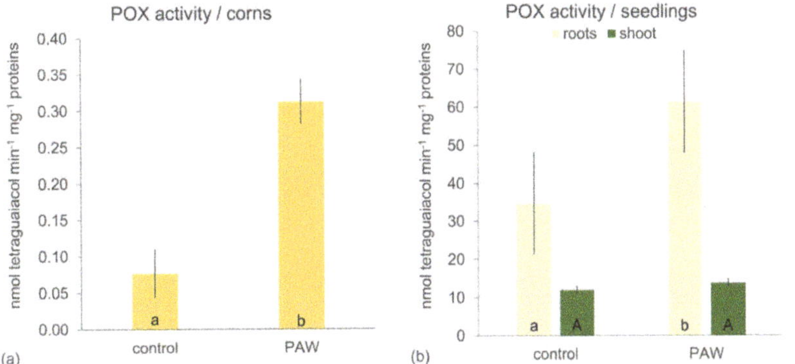

Figure 1. Activity of guaiacol peroxidase (G-POX, POX) in maize corns (hybrid Bielik) after 6 h imbibition (**a**) and in the roots and the shoot after 3 days of cultivation in the paper rolls; and (**b**) in the control (tap water) or treated with PAW. Values are means of four replicates $\pm$ SD. Different letters denote a significant difference between the treatments.

The growth of primary seminal root and the shoot after the treatment with PAW improved in comparison with control plants (Figure 2). The increase in the PAW treatment was 13% in both roots and the shoot.

Besides POX activity and growth, root tissue development was also accelerated after the PAW treatment (Figure 3). The developmental stages of the cell wall in terms of its lignification or suberization of selected tissues, such as exo- and endodermis and xylem vessels, were compared in the roots exposed to PAW and the control roots. At a distance of 10% from the root apex, Casparian bands in endodermis and exodermis were detected in the roots treated with the PAW; cell wall lignification was also observed in the early metaxylems (Figure 3a).

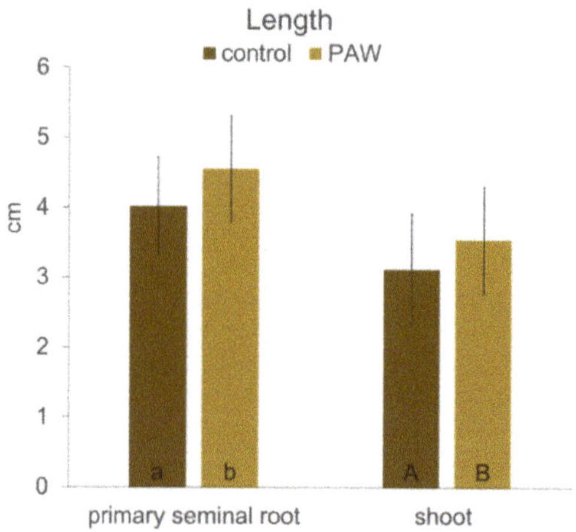

Figure 2. Growth of the primary seminal root and the shoot of maize (hybrid Bielik) after 3 days of cultivation in the paper rolls as control or treated with PAW. Values are means of four replicates ± SD. Different letters denote a significant difference between the treatments.

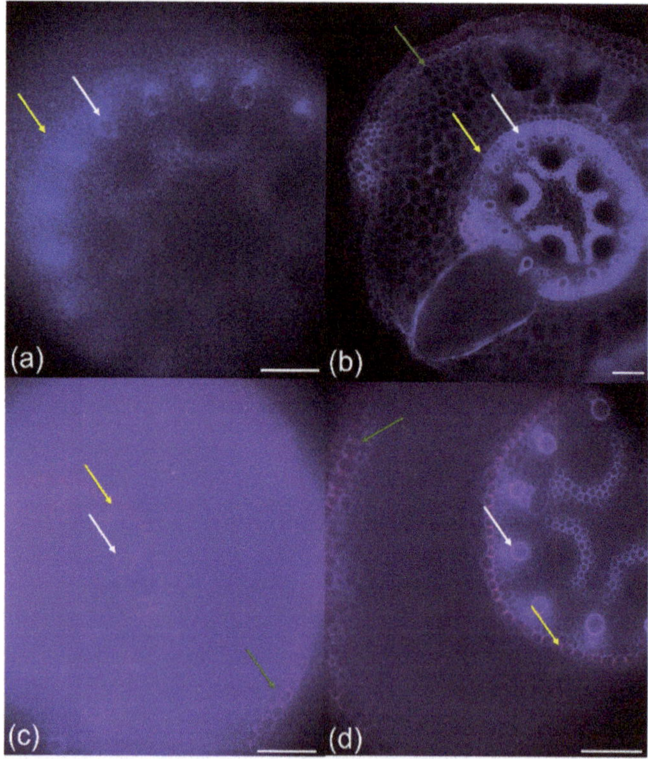

Figure 3. Lignification of the maize roots (hybrid Bielik) after 3 days of cultivation in the paper rolls as control (**a,c**) or treated with PAW (**b,d**). The hand cross sections were 10% from the root apex (**a,b**) or on the root base (**c,d**) and stained with phluoroglucinol-HCl. The arrows point at the exodermis (green), endodermis (yellow), and early metaxylem (white). Scale bars = 100 μm.

On the contrary, in the control roots, the development of the cell walls in terms of their lignification was obviously decelerated (Figure 3b). These differences disappeared at the root base, where similar developmental stages of the exo- and endodermis and xylem elements were observed (Figure 3c,d). The findings were confirmed by staining with 4 MN detecting POX activity in situ in the cell walls, which was also associated with lignification (Figure 4). A blue color was present in the early metaxylems and endodermis of the PAW treatment (Figure 4a) and only in early metaxylems in the control roots (Figure 4b) indicating the lignification process of the cell walls.

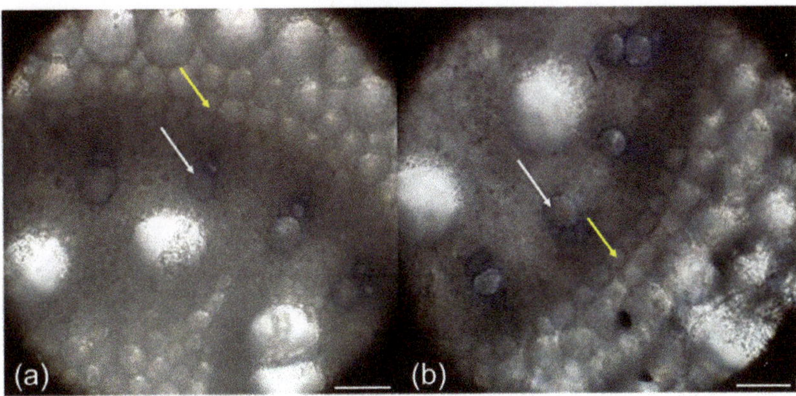

Figure 4. In situ POX activity in the maize roots (hybrid Bielik) after 3 days of cultivation in the paper rolls as control (**a**) or treated with PAW (**b**). The hand cross sections were 10% from the root apex and stained with 4 MN. The arrows point at the endodermis (yellow) and early metaxylem (white). Scale bars 100 μm.

To evaluate the potential of RONS in PAW in priming of the plants facing abiotic stress from arsenic (As^{V+}), the pre-treatment in the paper rolls was broadened by the treatment named priming PAW (as defined in chapter 4.8). In this case, the corns were only imbibed in the PAW and then cultivated in paper rolls with tap water for three days. Contrary to this, in the rolls PAW treatment, the corns were imbibed in PAW and the paper rolls were watered every day with the freshly prepared PAW. The control was imbibed and cultivated in tap water. After this pre-cultivation, 3 day-old seedlings were grown in the hydroponics for another 10 days without and with As and were subsequently analyzed. Interestingly, cultivation with 150 μM As in the As treatment did not influence the growth of maize roots and the shoot (FW per one plant) negatively in comparison with the control (Figure 5a). Contrary to this, roots in the priming PAW, priming PAW As, and rolls PAW had higher fresh biomass (FW per one plant), probably associated with the water management, because the roots of these three treatments did not achieve a higher percentage of the dry biomass accumulation (% DW) than the control (Figure 5b). Shoots accumulated significantly more fresh weight (FW per one plant) than the control only in the priming PAW As and the rolls PAW treatments, but the accumulation of the dry weight (% DW) decreased in all treatments in comparison with the control. The length of the primary seminal root was affected mostly negatively in comparison with the control (Figure 6a,b). Only roots of the rolls PAW treatment achieved the control root length. The overall worst shoot habitus was, however, noticed in the As treatment; the leaves were bigger in the priming PAW As and the rolls PAW As treatment than in the As treatment (Figure 6a).

Changes in the root morphology were also confirmed by software analysis (Table 1). The observed characteristics of roots in all PAW and As treatments were different in comparison with the control. Arsenic, in all As treatments, caused a highly significant decrease in the number of root tips and the average diameter of the root, but increased the branching frequency in comparison with the control. The least root tips were found for the priming PAW As treatment and the tiniest roots (their average diameter) for the As and the rolls PAW As treatments. A comparison of the number of root tips between the PAW treatments and its corresponding As treatments showed the decreasing tendency in the case of priming, but an increase in the case of rolls. On the contrary, branching frequency increased due to the As treatment in the priming As and decreased in the rolls PAW As treatment.

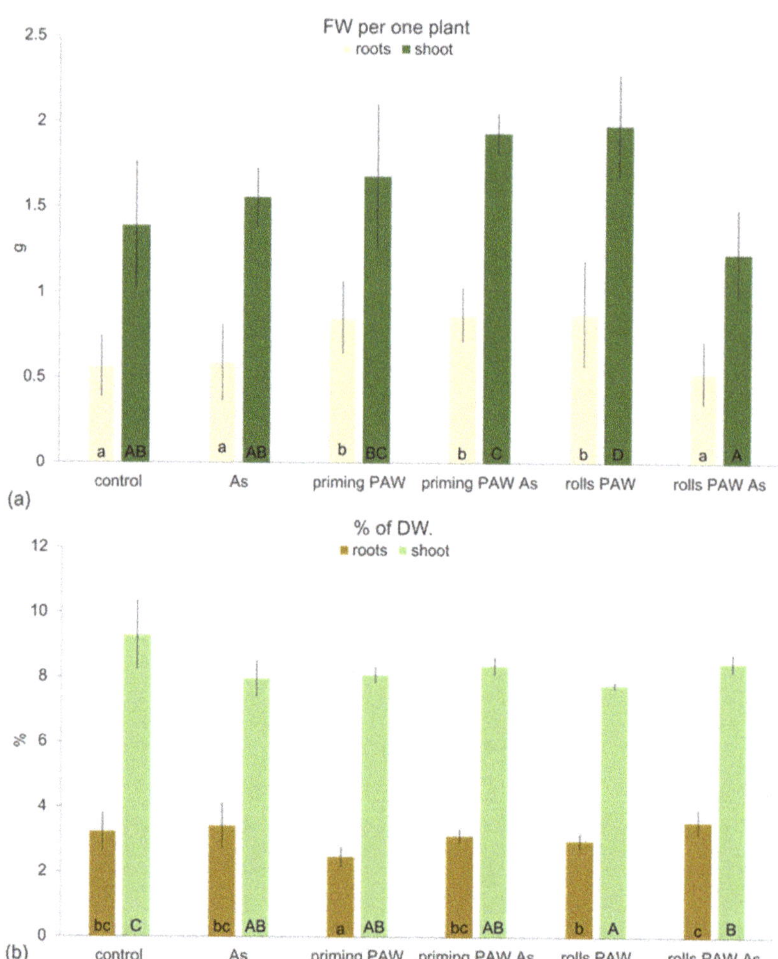

Figure 5. Growth parameters of the maize plants (hybrid Bielik) after 10 days of hydroponic cultivation as control or pre-treated with PAW (priming PAW and rolls PAW treatments) without and with As; the fresh weight per one plant (**a**) and the % of the dry weight (**b**). Values are means of four replicates ± SD. Different letters denote a significant difference between the treatments.

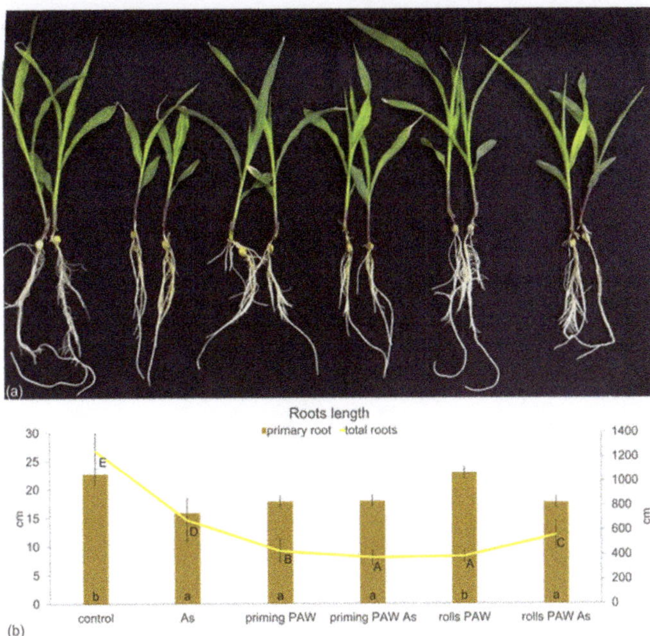

Figure 6. Habitus (**a**) of the maize plants (hybrid Bielik) and roots growth parameters (**b**) after 10 days of hydroponic cultivation as control or pre-treated with PAW (priming PAW and rolls PAW treatments) without and with As. Values (columns for primary root, line for total roots) are means of four replicates ± SD. Different letters denote a significant difference between the treatments.

Table 1. Maize root (hybrid Bielik) morphological characteristics after 10 days of hydroponic cultivation as control or pre-treated with PAW (priming PAW and rolls PAW treatments) without and with As. Values are means of four replicates ± SD. Different letters denote a significant difference between the treatments.

Table	Number of Root Tips	Branching Frequency per mm	Average Diameter (mm)
control	785 ± 12 e	0.65 ± 0.02 a	0.76 ± 0.04 d
As	447 ± 25 d	0.87 ± 0.11 d	0.49 ± 0.05 a
priming PAW	191 ± 13 b	0.71 ± 0.05 b	0.6 ± 0.01 c
priming PAW As	146 ± 18 a	0.87 ± 0.08 d	0.53 ± 0.08 b
rolls PAW	185 ± 9 b	0.86 ± 0.05 d	0.51 ± 0.07 b
rolls PAW As	264 ± 14 c	0.81 ± 0.06 c	0.46 ± 0.1 a

Using two-way ANOVA analysis, we compared the significance of the two selected factors on the measured POX and CAT activities; the first one was a plant organ (roots versus the second leaf) and the second was a treatment type. After hydroponic cultivation, roots were identified as organs with significantly higher POX activity in comparison with the second leaf (Figure 7), and the priming PAW was the treatment with the highest POX activity followed by the As treatment and the control. The rolls PAW, the priming PAW As and the rolls As treatments had the lowest POX activity. One-way ANOVA of G-POX activity was performed and compared separately in the roots and in the second leaf. In the roots, the only significant increase in the POX activity was observed in the priming PAW treatment in comparison with the control. On the contrary, in the priming PAW As, the rolls PAW, and the rolls PAW As treatments, a significant decrease was achieved (Figure 7). In the second leaf, a significant increase in the POX activity was detected only in the As treatment in comparison with the control.

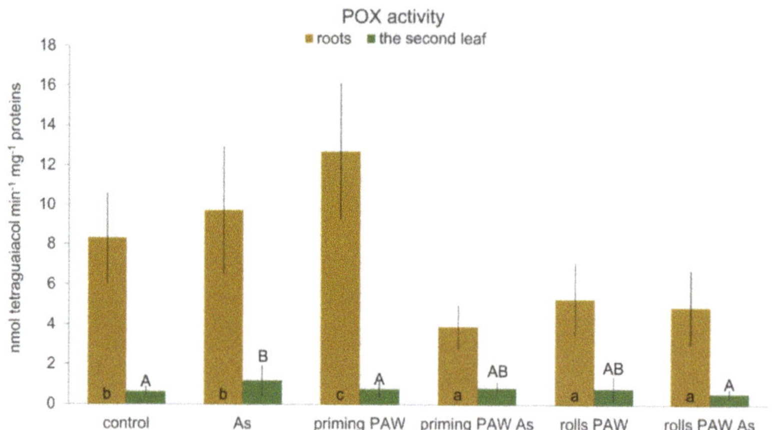

Figure 7. Activity of guaiacol peroxidase (G-POX, POX) of the maize roots and the second leaf (hybrid Bielik) after 10 days of hydroponic cultivation as control or pre-treated with PAW (priming PAW and rolls PAW treatments) without and with As. Values are means of four replicates ± SD. Different letters denote a significant difference between the treatments.

Contrary to POX activity, the two-way ANOVA with the same two selected factors did not show any difference between catalase (CAT) activities when comparing roots and the second leaf (Figure 8). When the factor of treatment type was assessed, the As, the rolls PAW As, and the priming PAW As treatments had significantly higher CAT activity than the priming treatment. The content of photosynthetic pigments Chl *a*, Chl *b*, and carotenoids was, in all treatments, affected negatively in comparison with the control. The only exception was, interestingly, the rolls PAW As treatment, where a significant increase in all tested pigments was noticed (Figure 9). The As treatment had the most noticeable decrease in all pigment concentrations. In the priming PAW, the priming PAW As, and the rolls PAW treatments, the concentration of Chl *a* and carotenoids was statistically the same and was decreased in comparison with the control.

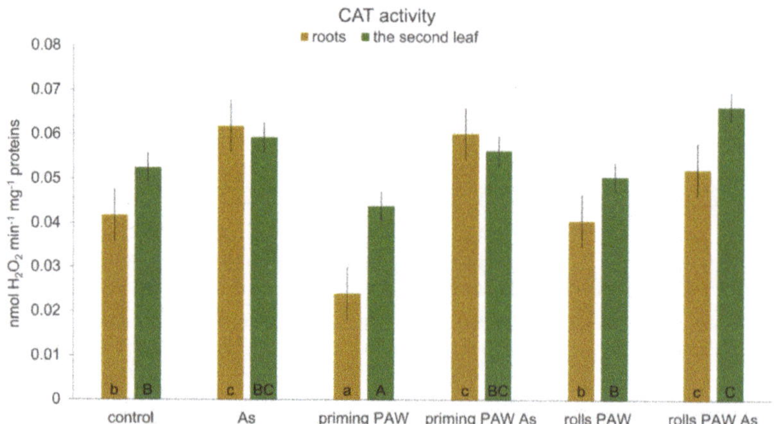

Figure 8. The catalase (CAT) activity of the maize roots and the second leaf (hybrid Bielik) after 10 days of hydroponic cultivation as control or pre-treated with PAW (priming PAW and rolls PAW treatments) without and with As. Values are means of four replicates ± SD. Different letters denote a significant difference between the treatments.

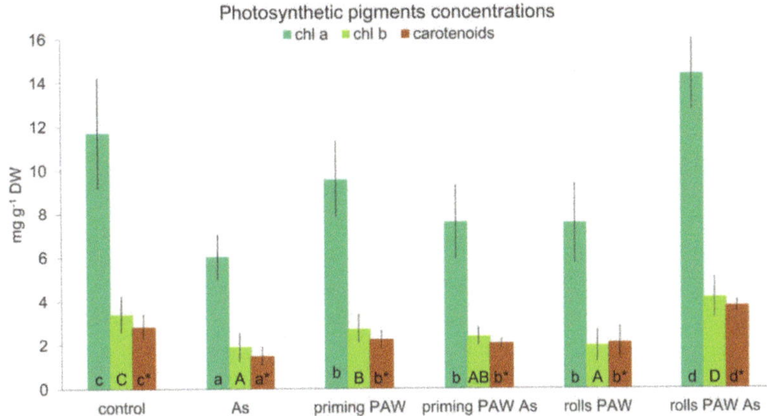

Figure 9. The photosynthetic pigments concentration per dry weight of the second leaf (hybrid Bielik) after 10 days of hydroponic cultivation as control or pre-treated with PAW (priming PAW and rolls PAW treatments) without and with As. Values are means of four replicates ± SD. Different letters denote a significant difference between the treatments.

Plants from the treatments without As (control, priming PAW, and rolls PAW) accumulated only a trace amount of As (data not shown), while all As treatments accumulated a significant amount of As, especially the roots, when compared to the first and the second leaves (Figure 10). In the roots, the highest concentration of As was found in the As treatment compared to the priming PAW As and the rolls PAW As treatments. Plants of the As and the priming PAW As treatments accumulated more As in the first leaf than in the second leaf. An opposite effect was found in the rolls PAW As treatment, where significantly more As was deposited into the second leaf. When calculating the ratios between As concentrations in the root and the selected leaves, we noticed 34.8, 21, and 31 between the root and the first leaf in the As, the priming PAW As, and the rolls PAW As treatments, respectively. This ratio was very different when considering the second leaf, where it was 80, 63, and only 6.6 in the As, the priming PAW As, and the rolls PAW As treatments, respectively. Another interesting fact arising from these results was the overall As accumulation in the roots and the selected leaves; the highest concentration was noticed in the As treatment.

After cluster analysis, groups with similar characteristics were grouped (Figure 11). To form the clusters, the procedure began with each observation in a separate group. Observations of the As and the priming PAW As treatments were the closest, forming a group along with the rolls PAW treatment. Interestingly, the rolls PAW As treatment and the control formed another group, although not so closely related according to the distance between them.

Another instructive result of this analysis is the separation of the priming PAW treatment, indicating a special position of this treatment in the observed experiments. The PCA analysis revealed a clustering of the plants from the rolls PAW, the As, and the priming PAW As treatments, while POX and CAT activities in the second leaf, CAT activity in the roots, and the % DW of the roots seem to play the most important roles as measured variables in these treatments. The rolls PAW As and the priming PAW treatments and the control were separated from each other, with different variables having the greatest impact. This analysis confirmed the correlation between some of the observed variables (Figure 12). A positive correlation was observed between the As concentration in the roots and in the first leaf ($R^2 = 0.89$), between the As concentration in the second leaf and the carotenoids ($R^2 = 0.71$), between the CAT activity in the second leaf and the roots ($R^2 = 0.90$), between the Chl *a* and Chl *b* ($R^2 = 0.84$), and between the Chl *a* or Ch *b* concentration and the carotenoids ($R^2 = 0.81$ and 0.74, respectively). The only two significantly negative correlations were confirmed between the FW of the roots and CAT activity in the roots ($R^2 = -0.88$) and between the % DW of the roots and POX activity in the roots ($R^2 = -0.67$).

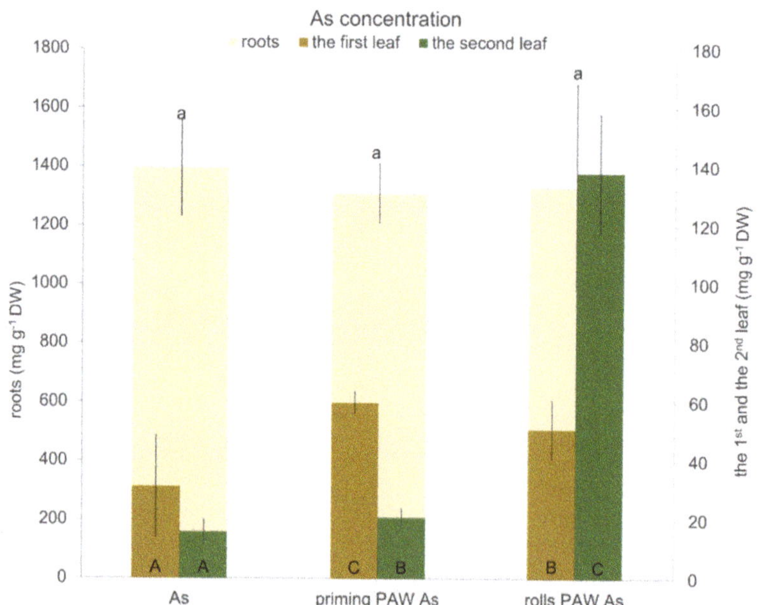

Figure 10. The As concentration per dry weight of the roots, the first leaf, and the second leaf (hybrid Bielik) after 10 days of hydroponic cultivation as As treatment or As treatments pre-treated with PAW (priming PAW As and rolls PAW As treatments).

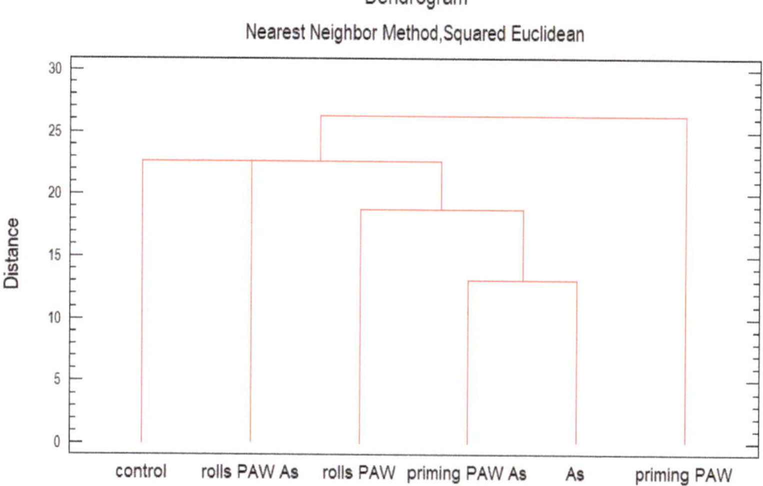

Figure 11. Cluster analysis. Clusters are groups of observations with similar characteristics according to the treatments.

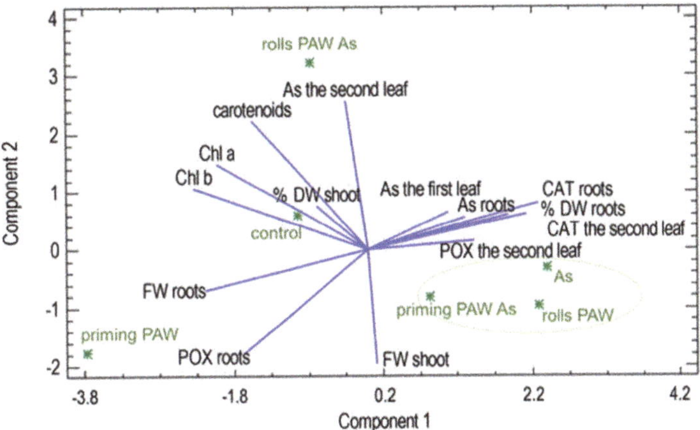

Figure 12. Biplot of the principal component analysis (PCA). The first two components extracted from the analysis together account for 61.2% of the variability in the original data.

3. Discussion

Plasma activated water is a source of various reactive oxygen and nitrogen species (ROS and RNS) that can improve plant growth in stress conditions and partially replace the use of fertilizers. It is the priming effect of H_2O_2 that provokes plants to react faster and stronger to a potential stress, and NO_3^- and NO_2^- as a source of critically important microelement essential for building proteins and other macromolecules. Some authors describe the effects of PAW as alternatives to chemical biostimulators in very early embryo development, e.g., during seed germination (e.g., [42]), although the effect of PAW depends on several factors, such as plant species, plasma activated water activity (its chemical composition), and other experimental conditions [24]. Maize used in the present study has, in general, a high percentage of germination, and the difference between the PAW treatments and the control with respect to germination was non-significant (data not shown). However, what is shown here is that POX activity was strongly enhanced after just 6 h of corn imbibition in PAW in comparison with those imbibed in tap water (Figure 1a). Corona-Carrillo et al. [43] described the paradoxical role of POX in the developing maize embryo and claim that these enzymes can either produce or decompose ROS via their peroxidative and hydroxylic cycles, maintaining ROS and nutrition at optimum levels. Enhanced POX, an important antioxidant enzyme, avoids potential embryo damage and counteracts the stress occurring during germination or seeds storage. This can increase the embryo vigor and its capacity to establish seedlings [43]. Improved growth of seedlings after imbibition in PAW was confirmed in the present study (Figure 2), and significantly enhanced POX activity in the PAW treatment remained in the roots (Figure 1b). This is very important for plant survival in the subsequent development, because roots are most often the first contact with potential soil stressors [44,45]. The roots of seedlings treated with PAW developed faster (Figure 3), which corresponds to tissue lignification, which was delayed in the control (Figure 3a,b). The lignification of exo-, endodermis, and xylem vessels is one of the most important processes in the roots, because it controls apoplasmic ion flow and protects the central stele from entering the toxic elements in the xylem followed by shoot translocation [32,46–50]. On the other side, xylem lignification is an assumption of water transpiration that is essential for plant survival. In the 3 day-old seedlings, we detected in situ POX activity in the root cell walls (Figure 4) with elevated reactions in the endodermis of the PAW treatment (Figure 4b), which is associated with lignification. Peroxidases are key players in this process, polymerizing lignin precursors, monolignols [31]. Improved growth after the PAW treatment was also reported by [24] on lettuce (*Lactuca sativa*); however, changes depended on the harvest date and on the used PAW composition. A positive effect of PAW on the growth of another important crop, a wheat (*Triticum aestivum*), was documented by [42], which also confirmed a positive effect of PAW on photosynthetic pigment concentration but detected lowered activities of antioxidant enzymes.

Another objective of the present study was to document the effect of the PAW pretreatment either at the level of 6 h corn imbibition (priming PAW treatment) or stimulation of seedlings during the 3 days of cultivation with PAW (rolls PAW treatment) followed by As stress during the other 10 days of hydroponic cultivation (Figures 5–12). Thus, the tested plants in the priming PAW and the rolls PAW treatments were exposed to different doses of PAW, containing H_2O_2 and NO_x^-. Hydrogen peroxide has a special position among ROS. It has a dual role; at low concentrations it acts as a signal molecule triggering the antioxidant systems, but at high concentrations, it has destroying effects on cells [51]. Roots of the priming PAW and the rolls PAW treatments and shoot of the rolls PAW treatment accumulated more water than the controls (Figure 5a), but the % DW was not improved in comparison with the control (Figure 5b). Similarly, an acceleration in water uptake was documented by [12] on pea seeds exposed by atmospheric plasma. The length of the primary root of PAW treated plants was not greater than that in the control. However, plant root morphology was altered dramatically (Table 1). Due to the PAW treatment, roots had significantly less total length, less lateral roots, the least in the rolls PAW treatment, but the branching frequency increased in both PAW treatments. Additionally, the average diameter decreased in comparison with the control root. All these morphological changes are important characteristics and influence the accessibility and mobility of ions with different properties [52]. Bafoil et al. [53] tested the effects of PAW on the model plant *Arabidopsis thaliana* and confirmed a significant increase in various plant growth parameters.

Plants, with their sessile way of life, must face various stresses in their environment by activating the defense mechanisms. These reactions, unfortunately, often contribute to a decreased yield, a problem that can be solved by various approaches. One of them could be a use of PAW, which was reported to improve the tolerance against stress from low temperature and hypoxia during barley germination [54]. In the present study, As stress reaction was evaluated. Basic growth characteristics were negatively changed by As itself, which is a common phenomenon of this toxic metalloid [28,55] (Figures 5 and 6, Table 1). Plants of the priming PAW As treatment accumulated more water than As-treated plants (Figure 5a), but no improvement of the FW was detected in the rolls PAW treatment. Similarly, the % DW was the same when comparing the As and other treatments (Figure 5b). In general, in several cases we found different plant response to the As treatment when the priming PAW As and the rolls PAW As treatments were compared to each other, to the As treatment or to the control. This phenomenon could be explained by various doses of H_2O_2 and NO_x^- given to corns or plants during pre-treatment, where the priming treatments received PAW only during imbibition, and the rolls treatments also received it during the following 3 days of cultivation. The divergent reactions were finally confirmed, also using multivariate statistical analysis (Figures 11 and 12), where treatments separated from each other in a specific way. Maintaining cell homeostasis during a stress reaction is a key in enabling plants to survive any sub-optimal conditions. To keep ROS content at the nontoxic level, the involvement of antioxidant enzymes is essential; however, stimulation of their activity is not always obvious [28,32,56,57]. A common phenomenon of significantly higher POX activity in the roots than in the leaves was also confirmed in the present study (Figure 7). Interestingly, when results were analyzed by two-way ANOVA, and factor treatment was evaluated, the rolls PAW As, the priming PAW As, and the rolls PAW had the lowest POX activities. The control and the As treatment were in the second group and the highest activity had plants of the priming PAW treatment. POX activity showed that roots reacted to the As treatment non-significantly; however, a significant increase was detected in the second leaf in comparison with the control (Figure 7). A significant increase in POX activity in the roots was detected only in the priming PAW treated roots, while all other roots had decreased POX activity. The production of antioxidants in the plants challenged by As is a common phenomenon [58]. However, a divergent reaction of POX activity and pattern in POX expression was documented, caused by different As doses in tobacco (*Nicotiana benthamiana*) plants [28]. It is clear that plants reacted with elevated POX activities on moderate As stress, but the reaction was different for low and high As doses. We suppose that, in the present study, a similar phenomenon occurred, i.e., plants experienced stress not only from As, but also from PAW containing H_2O_2. Contrary to POX, when we documented the activity of the collaborating antioxidant enzyme CAT [59], no difference was noticed when roots and the second leaf were compared (Figure 8). The obvious cooperation between the two enzymes' decreasing H_2O_2 levels was detected in the priming PAW treatment. It was the only case of decreased CAT activity in comparison with other treatments and, we suppose, the major role of POX reducing H_2O_2 in this treatment. Changes in CAT activity after the PAW treatment was also observed in other plant species [53,60]; however, Gierczik et al. [54] documented the time-dependent increase or decrease. It is obvious that the relationship between the CAT (de)activating and H_2O_2 content is time dependent, and another regulator, the ascorbate-glutathione cycle, could also be involved [61].

Plants also possess non-enzymatic antioxidants, such as pigments–carotenoids, preventing oxidative burst [62]. The rolls PAW As was the only treatment where we detected a significant increase in these pigments (Figure 9), and at the same time, the second leaf of this treatment had the highest chlorophyll contents; in this case, they were probably the best protected by carotenoids. These three pigments were closely correlated (Figure 12) in the present study, and the relationship between carotenoids and As acting was also confirmed by the detection of a significant negative correlation between As content in the second leaf and carotenoids (Figure 12). The alleviation of a salt-induced damage by PAW through its effect on carotenoids content has also been described [54]. Authors also proved a slight increase in the glutathione metabolism-related genes due to the PAW treatment, which indicates changes triggered by signal molecules produced in PAW at the DNA expression level.

The accumulation of As in the aboveground edible plant part is always dangerous due to the carcinogenic character of this metalloid [63]. The accumulation of As in the present study was not affected by the PAW treatments (Figure 10). Unfortunately, higher amounts were accumulated in both tested leaves of the priming PAW As and the rolls PAW As treatments. Changes in the As uptake, radial transport, and its translocation are a combination of the altered root morphology, anatomy, water transport management, and As transporters capacity. To explain this phenomenon completely, additional research in this field is necessary in the future. Here, we have only partially analyzed the formation of apoplasmic barriers in the roots due to PAW treatments, but we did not get enough data to draw further conclusions.

4. Materials and Methods

4.1. Plant Material

Corns of maize (*Zea mays* L.) (hybrid Bielik) used in the experiments were obtained from Sempol spol. s.r.o., Bratislava, Slovakia. Hybrid Bielik was selected after screening of several maize hybrids exposed to PAW. The corns were stored in fridge at 8 °C in the dark. All treatments were repeated at least three times, independently.

4.2. Production of Plasma Activated Water (PAW)

The plasma activated water was generated by a transient spark discharge operated in atmospheric air in a contact with tap water. The plasma reactor was of a point-to-plane geometry and consisted of a needle used a high voltage electrode placed above an inclined grounded electrode embedded in a polytetrafluoroethylene gutter. Tap water was driven down the gutter, repetitively circulated, and exposed to the transient spark discharge for a given time. The transient spark (TS) discharge is a repetitive streamer-to-spark transition discharge. It is a DC-driven, self-pulsing discharge typical of current pulses of high amplitude (order of several tens of amps), very short duration (10–100 ns), and frequency of orders of several kHz. The details on physical properties of TS [64,65] as well as on a system for plasma activated water generation [42,66,67] have been previously published. In the presented study, the transient spark discharge was operated at the applied voltage 11–13 kV, amplitude, and frequency of current pulses ~3 A and 1.5–3 kHz, respectively. The water flow was set to 15 mL min^{-1}. The amount of tap water exposed to plasma varied; however the plasma exposure/activation time was set to 1 mL min^{-1}, i.e., 20 mL of water was exposed to plasma for 20 min, etc. The experiments were performed at ambient air temperature ~22 $\pm$ 2 °C. The temperature of water was maintained by an ice bath to avoid unwanted heating of the produced PAW caused by the plasma exposure. The operating conditions were alike those we used in past experiments [42,67] and which turned out to be optimal for the stimulation of plants by PAW. Plasma activation of tap water did not affect its pH (7.5) due to natural hydrocarbon buffer system; however, it resulted in the formation of various RONS in water whose concentrations were evaluated.

4.3. Measurement of Hydrogen Peroxide (H_2O_2), Nitrate (NO_2^-) and Nitrite (NO_3^-)

The concentrations of RONS, namely hydrogen peroxide (H_2O_2), nitrate (NO_2^-), and nitrite (NO_3^-), in PAW were measured by UV–Vis absorption spectrometer UV-1900 (Shimadzu, Japan). Hydrogen peroxide (H_2O_2) is mostly produced in gas phase by a recombination of OH radicals and subsequently dissolves in water. Its analysis is based on its reaction with titanyl ions (Ti^{4+}) of titanium oxysulfate ($TiOSO_4$) [68]. The reaction results in the formation of a yellow-colored product of pertitanic acid (H_2TiO_4) with a maximum absorbance peak at 407 nm proportional to H_2O_2 concentration. The reaction is specific to H_2O_2 and does not interfere with other compounds. Prior to H_2O_2 analysis, the PAW was fixed with sodium azide (NaN_3) to eliminate its eventual decomposition by a mutual reaction with NO_2^-. NaN_3 reduces NO_2^- to molecular N_2 and does not interfere with the H_2TiO4. The nitrites (NO_2^-) and nitrates (NO_3^-) in PAW are mainly formed by the dissolution of gaseous HNO_2 and HNO_3 formed in a gas phase. They may also form by NO_2 dissolution, however, which is less efficient than the dissolution of the corresponding acids. The analysis of NO_2^- and NO_3^- was performed using commercially available kits based on reaction with sulfanilamide and N-1-naphthylethylenediamine, the so-called Griess reagents [69]. The Griess reagents react with NO_2^- to form a pink-colored azo-product with a maximum absorbance peak at 540 nm. To measure NO_3^- concentration, it must first be enzymatically reduced to NO_2^- and then analyzed by the same method as that of NO_2^-. The method is easy to perform and is approved as being precise for NO_2^- and NO_3^- measurements in PAW.

4.4. Maize Corns and Seedlings Treated with PAW

Dry maize corns were imbibed in tap water or plasma activated water (PAW) for 6 h at room temperature. Several corns (two to three) were randomly chosen for guaiacol peroxidase (G-POX, POX) activity measurement. The rest of imbibed grains (30 for each variant) were wrapped in wet sterile filter paper and cultivated. In this part, two treatments were established: control (corns imbibed in tap water, and then in paper rolls watered daily with tap water) and PAW treatment (corns imbibed in PAW, and then in paper rolls watered daily with freshly prepared PAW). Seedlings were cultivated for three days in the dark under controlled physical conditions in an incubator at the temperature 24 $\pm$ 2 °C and 60% relative humidity. At the end of the cultivation, young roots and shoot were measured and used for G-POX activity measurement, lignification of the root tissues, and in situ POX activity in roots.

4.5. G-POX Activity in Corns and Seedlings

After three days of cultivation, samples of the roots and the shoot were randomly chosen from at least four plants (~1.5 g), were ground with a mortar and pestle in liquid nitrogen, and suspended in 50 mM Na-phosphate protein extraction buffer with 1 mM

EDTA (ethylenediaminetetraacetic acid), pH 7.8. After 15 min centrifugation (12,000× *g*) at 4 °C, the supernatant was used for spectrophotometrical determination of total soluble protein concentration at 595 nm, according to [70]. Protein content was calculated as the total number of proteins per gram of fresh matter from the calibration curve with bovine serum albumin (BSA) as the protein standard. Guaiacol peroxidase (G-POX, POX, E.C.1.11.1.7) was measured according to standardized assays, with a minimum of three measurements and three technical replications per each sample. The activity of G-POX was established according to [71] and measured spectrophotometrically at 440 nm. The G-POX activity was expressed in nM of tetraguaiacol min^{-1} mg^{-1} multiplied by the molar extinction coefficient of tetraguaiacol 26.6, as follows:

$$specific\ G-POX\ activity = \frac{\Delta\ A\ \mathrm{min}^{-1} \times 1000}{26.6\ \frac{protein\ content\ in\ sample\ (\mu g)}{volume\ of\ extraction\ solution\ (mL)}}\ (nM)$$

4.6. Lignification of the Root Tissues

In the three-day-old seedlings, influence of the PAW treatment was observed on the lignin deposition on the free hand sections of the roots at a distance of 10% from the root apex and at the root base. This approach allows analysis and comparison of the same developmental stages of the root, irrespective of the root length differing between treatments. The sections were stained with fluoroglucinol–HCl for visualization of lignin deposition. Fluorescence of the lignin deposits was observed by an Axioskop 2 plus microscope (Carl Zeiss, Jena, Germany), equipped with excitation filter TBP 400 + 495 + 570 nm, chromatic beam splitter TFT 410 + 505 + 585, and emission filter TBP 460 + 530 + 610 nm, documented by an Olympus DP 72 camera system and analyzed with Lucia imaging software (Lim, Prague, Czech Republic). Four roots per treatment were analyzed in each experimental run.

4.7. In Situ POX Activity in Roots

To document the activity of cell wall peroxidases in situ, the hand cross root sections (~0.5 mm thick) 10% from the root apex were observed. The sections were incubated in 100 mM Na-acetate buffer (pH 5.2) with 5 mM 4-methoxy-1-naphthol (4-MN) in 96% ethanol for 15 min at 30 °C [72,73], observed in the bright field and documented as described in chapter 4.6. Four roots per treatment were analyzed in each experimental run.

4.8. Maize Seedlings Treated with Arsenic (As) after PAW Pre-Treatment

To document a potential of PAW in priming maize corns and seedlings subsequently exposed to stress from arsenic (As), the previous two treatments in the paper rolls (chapter 4.4) were broadened to three treatments: the control (corns imbibed in tap water, seedlings cultivated for three days in paper rolls with daily fresh tap water); priming PAW (corns imbibed in PAW, seedlings cultivated as control); and rolls PAW (corns imbibed in PAW, seedlings cultivated for three days in paper rolls with daily freshly prepared PAW—in chapter 2.4 called PAW treatment). Ten randomly selected 3 day-old seedlings from each treatment were transferred into 3 L containers and cultivated for other 10 days as hydroponic cultures in 0.5 Hoagland nutrient solution (pH 5.8). The concentration of arsenic (150 μM As) was established based on a previous concentration screening. It was added in the form of As^{5+} (Na$_2$HAsO$_4$.7H$_2$O) and, finally, six treatments in total were established: the control (0.5 Hoagland solution, seedlings from control treatment), As (150 μM As, seedlings from control treatment), priming PAW (seedlings from priming PAW treatment), priming PAW As (150 μM As, seedlings from priming PAW treatment), rolls PAW (seedlings from rolls PAW treatment), and rolls PAW As (150 μM As, seedlings from rolls PAW treatment). Hoagland nutrient solution without or with As was changed every 3 days. At the end of the cultivation, plants were evaluated in terms of their growth parameters—fresh weight (FW), % of the dry weight (DW) expressed as the ratio of FW and DW multiplied by 100, and length of the primary seminal root—and other biochemical characteristics were detected. Macrophotography images of plants in individual treatments were taken using a camera Nikon D90 with an AF-S Micro Nikkor 60 mm lens system. Roots of all treatments were also scanned in high quality, and the images were analyzed by RhizoVision Explorer to obtain other root characteristics, such as total root length, diameter, branching frequency, and number of root tips.

4.9. POX and CAT Activity in Maize Seedlings

The below- and above-ground plant parts of the 10 day-old seedlings were detached, and the roots were thoroughly washed three times in distilled water. Roots and the 2nd leaf of at least 4 randomly selected plants were used for assays of POX activity in the same way as described in chapter 4.5. Extract of proteins from roots and the 2nd leaf were used also for catalase (CAT) detection [74]. Its activity was calculated after spectrophotometrical measurement at 240 nm, based on the decomposition rate of H_2O_2 in time [73].

$$specific\ catalase\ activity = \frac{\Delta\ A\ min^{-1} \times 1000}{39.1\ \frac{protein\ content\ in\ sample\ (\mu g)}{volume\ of\ extraction\ solution\ (mL)}}$$

4.10. Evaluation of Chlorophylls and Carotenoids Concentration in Leaves

The 2nd leaf of every treatment was used for the determination of photosynthetic pigments concentration; leaves were randomly selected from at least four plants. Chlorophyll a, b (Chl *a* and Chl *b*), and carotenoids were extracted with the cooled mortar and pestle on the 10th day of cultivation (ca 500 mg of FW) of each treatment with cooled 80% acetone (10–15 mL) with 200 mg of $MgCO_3$ mixed with a little sea sand to prevent phaeophytin formation. The pigment concentrations were determined spectrophotometrically (Jenway 6400, London, UK) as follows: Chl *a* at 663.2 nm, Chl *b* at 646.8 nm, and carotenoids at 470 nm. The concentrations were calculated after [74] and expressed as mg of pigment per 1 g of plant material fresh weight.

$$\text{Concentration of Chl } a = ((12.25\ A663.2 - 2.79\ A646.8) \times y)/60\ (\text{mg g}^{-1}\ \text{FW})$$

$$\text{Concentration of Chl } b = ((21.50\ A646.8 - 5.10\ A663.2)) \times y)/60\ (\text{mg g}^{-1}\ \text{FW})$$

$$\text{Concentration of carotenoids} = ((1000 \times y^* \ A470 - 1.82\ \text{Chl } a\ \text{mg L}^{-1} - 85.02\ \text{Chl } b\ \text{mg L}^{-1})/198))/60\ (\text{mg L}^{-1})$$

where

$$y = (\text{the volume of acetone used for extraction} \times 0.06)/\text{FW of material (g)},$$

$$\text{Chl } a\ \text{mg L}^{-1} = (12.25\ A663.2 - 2.79\ A646.8) \times y$$

$$\text{Chl } b\ \text{mg L}^{-1} = (21.50\ A646.8 - 5.10\ A663.2) \times y$$

4.11. Determination of As Concentration in Leaves and Roots

Concentrations of As in the 1st oldest leaf, in the 2nd leaf, and in the roots were determined in each As treatment by atomic absorption spectrometry (AAS). Dry plant samples taken randomly from the plants were dried at 70 °C until constant weight. At least 200 mg of DW was used for As determination. Samples were dissolved in concentrated HNO_3. After heating at 160 °C for 3 h, concentrated HF was added. Thereafter the samples were dried, and a mixture of concentrated HNO_3 and H_3BO_3 was added. The control and the PAW treatments were also checked for As content—plants accumulated only trash amounts of this metalloid.

4.12. The Statistical Analysis

All results were evaluated using PC software Excel with XLSTAT (Microsoft Office 365, Redmond, Washington, DC, USA) and statistic software Statgraphics Centurion XVI. Analysis of variance (ANOVA, One-Way, and Multifactorial) with LSD test, standardized clustering method—nearest neighbor (distance metric: squared Euclidean), and principal component analysis (PCA) with the level of significance $p < 0.05$ (significant) were performed. All experiments were conducted independently three times and results were expressed as means of four replicates $\pm$ standard deviations (SD) in the figures.

5. Conclusions

The effect of PAW generated by a transient spark discharge (TS) operated in ambient air on maize corns and young seedlings was investigated and confronted with the effect of As stress in several different treatments. PAW treatment provoked the enhancement of POX activity immediately after the corn imbibition, which points to the influence of highly active molecules (RONS) within this solution. PAW itself improved the growth parameters only in the young seedlings; however, in continuous hydroponic cultivation, plants also achieved positive changes in the dry biomass accumulation. The primary root morphology also changed due to PAW treatments, which significantly influence the plant nutrients uptake. Together with the enhanced photosynthetic pigments, it can be concluded that treatments with PAW contribute to better survival of the plant. We also found a different pattern of plant response to the subsequent As treatment. Plants from different PAW treatments reacted to the As stress by elevating their antioxidant capacities; depending on PAW pre-treatment, antioxidant enzymes, POX and CAT, or non-enzymatic molecules, carotenoids were elevated, reflecting the active defense system activated due to PAW pre-treatment. The study confirmed that the use of PAW in the pre-treatment (priming) of maize corns or young plants may improve their tolerance against As stress, which can be utilized when plants are grown in toxic elements contaminating soils.

Author Contributions: Methodology, Z.L., R.S. and K.H.; formal analysis Z.L.; investigations, Z.L., R.S., P.S. and K.H.; data curation, Z.L.; formal analysis, Z.L., R.S. and K.H.; writing—original draft preparation Z.L. and K.H.; writing—editing, R.S.; funding acquisition, K.H. All authors have read and agreed to the published version of the manuscript.

Funding: This work was supported by Slovak Research and Development Agency APVV 17-0382 and APVV-17-0164 grants and Scientific Grant Agency VEGA grant 1/0419/18.

Institutional Review Board Statement: Not applicable.

Informed Consent Statement: Not applicable.

Data Availability Statement: Not applicable.

Conflicts of Interest: The authors declare no conflict of interest.

References

1. Puač, N.; Gherardi, M.; Shiratani, M. Plasma agriculture: A rapidly emerging field. *Plasma Process. Polym.* **2018**, *15*, 1700174. [CrossRef]
2. Misra, N.N.; Schlüter, O.; Cullen, P.J. *Cold Plasma in Food and Agriculture*; Academic Press: London, UK, 2016; p. 380.
3. Oehmigen, K.; Hähnel, M.; Brandenburg, R.; Wilke, C.; Weltmann, K.-D.; von Woedtke, T. The role of acidification for antimicrobial activity of atmospheric pressure plasma in liquids. *Plasma Process. Polym.* **2010**, *7*, 250. [CrossRef]
4. Traylor, M.J.; Pavlovich, M.J.; Karim, S.; Hait, P.; Sakiyama, Y.; Clark, D.S.; Graves, D.B. Long-term antibacterial efficacy of air plasma-activated water. *J. Phys. D Appl. Phys.* **2011**, *44*, 472001. [CrossRef]
5. Mortvedt, J.J. Heavy metal contaminants in inorganic and organic fertilizers. In *Fertilizers and Environment: Proceedings of the International Symposium "Fertilizers and Environment", Held in Salamanca, Spain, 26–29 September 1994*; Rodriguez-Barrueco, C., Ed.; Developments in Plant and Soil Sciences; Springer Netherlands: Dordrecht, The Netherlands, 1996; pp. 5–11. ISBN 978-94-009-1586-2.
6. Sharma, R.K.; Patel, H.; Mushtaq, U.; Kyriakou, V.; Zafeiropoulos, G.; Peeters, F.; Welzel, S.; van de Sanden, M.C.M.; Tsampas, M.N. Plasma Activated Electrochemical Ammonia Synthesis from Nitrogen and Water. *ACS Energy Lett.* **2021**, *6*, 313–319. [CrossRef]
7. Puač, N.; Petrović, Z.L.; Radetić, M.; Djordjević, A. Low Pressure RF Capacitively Coupled Plasma Reactor for Modification of Seeds, Polymers and Textile Fabrics. *Mater. Sci. Forum* **2005**, *494*, 291–296. [CrossRef]
8. Dhayal, M.; Lee, S.-Y.; Park, S.-U. Using low-pressure plasma for *Carthamus tinctorium* L. seed surface modification. *Vacuum* **2006**, *5*, 499–506. [CrossRef]

9.   Šerá, B.; Šerý, M.; Štraňák, V.; Špatenka, P.; Tichý, M. Does Cold Plasma Affect Breaking Dormancy and Seed Germination? A Study on Seeds of Lamb's Quarters (*Chenopodium album* agg.). *Plasma Sci. Technol.* **2009**, *11*, 750–754. [CrossRef]

10. Šerá, B.; Stranák, V.; Serý, M.; Tichý, M.; Spatenka, P. Germination of Chenopodium Album in Response to Microwave Plasma Treatment. *Plasma Sci. Technol.* **2008**, *10*, 506–511. [CrossRef]

11. Henselová, M.; Slováková, L.; Martinka, M.; Zahoranová, A. Growth, anatomy and enzyme activity changes in maize roots induced by treatment of seeds with low-temperature plasma. *Biologia* **2012**, *67*, 490–497. [CrossRef]

12. Stolárik, T.; Henselová, M.; Martinka, M.; Novák, O.; Zahoranová, A.; Černák, M. Effect of low-temperature plasma on the structure of seeds, growth and metabolism of endogenous phytohormones in pea (*Pisum sativum* L.). *Plasma Chem. Plasma Process.* **2015**, *35*, 659–676. [CrossRef]

13. Sadhu, S.; Thirumdas, R.; Deshmukh, R.R.; Annapure, U.S. Influence of cold plasma on the enzymatic activity in germinating mung beans (*Vigna radiate*). *LWT* **2017**, *78*, 97–104. [CrossRef]

14. Švubová, R.; Kyzek, S.; Medvecká, V.; Slováková, L.; Gálová, E.; Zahoranová, A. Novel insight at the Effect of Cold Atmospheric Pressure Plasma on the Activity of Enzymes Essential for the Germination of Pea (*Pisum sativum* L. cv. Prophet) Seeds. *Plasma Chem Plasma Process.* **2020**, *40*, 1221–1240. [CrossRef]

15. Švubová, R.; Slováková, L.; Holubová, L.; Rovňanová, D.; Gálová, E.; Tomeková, J. Evaluation of the Impact of Cold Atmospheric Pressure Plasma on Soybean Seed Germination. *Plants* **2021**, *10*, 177. [CrossRef] [PubMed]

16. Mildaziene, V.; Pauzaite, G.; Naucienė, Z.; Malakauskiene, A.; Zukiene, R.; Januskaitiene, I.; Jakstas, V.; Ivanauskas, L.; Filatova, I.; Lyushkevich, V. Pre-sowing seed treatment with cold plasma and electromagnetic field increases secondary metabolite content in purple coneflower (*Echinacea purpurea*) leaves. *Plasma Process. Polym.* **2018**, *15*, 1700059. [CrossRef]

17. Park, Y.; Oh, K.S.; Oh, J.; Seok, D.C.; Kim, S.B.; Yoo, S.J.; Lee, M.-J. The biological effects of surface dielectric barrier discharge on seed germination and plant growth with barley. *Plasma Process. Polym.* **2016**, *15*, 1600056. [CrossRef]

18. Štepánová, V.; Slavíček, P.; Kelar, J.; Prášil, J.; Smékal, M.; Stupavská, M.; Jurmanová, J.; Černák, M. Atmospheric pressure plasma treatment of agricultural seeds of cucumber (*Cucumis sativus* L.) and pepper (*Capsicum annuum* L.) with effect on reduction of diseases and germination improvement. *Plasma Process. Polym.* **2017**, *15*, 1700076. [CrossRef]

19. Randeniya, L.K.; de Groot, G.J.J.B. Non-Thermal Plasma Treatment of Agricultural Seeds for Stimulation of Germination, Removal of Surface Contamination and Other Benefits: A Review. *Plasma Process. Polym.* **2015**, *12*, 608–623. [CrossRef]

20. Zahoranová, A.; Henselová, M.; Hudecová, D.; Kaliňáková, B.; Kováčik, D.; Medvecká, V.; Černák, M. Effect of Cold Atmospheric Pressure Plasma on the Wheat Seedlings Vigor and on the Inactivation of Microorganisms on the Seeds Surface. *Plasma Chem. Plasma Process.* **2016**, *36*, 397–414. [CrossRef]

21. Szőke, C.; Nagy, Z.; Gierczik, K.; Székely, A.; Spitkól, T.; Zsuboril, Z.T.; Galiba, G.; Marton, C.L.; Kutasi, K. Effect of the afterglows of low pressure Ar/N$_2$-O$_2$ surface-wave microwave discharges on barley and maize seeds. *Plasma Process. Polym.* **2018**, *15*, 1700138. [CrossRef]

22. Li, L.; Jiang, J.; Li, J.; She, M.; He, X.; Shao, H.; Dong, Y. Effects of cold plasma treatment on seed germination and seedling growth of soybean. *Sci. Rep.* **2014**, *4*, 5859.

23. Ranieri, P.; Sponsel, N.; Kizer, J.; Rojas-Pierce, M.; Hernández, R.; Gatiboni, L.; Grunden, A.; Stapelmann, K. Plasma agriculture: Review from the perspective of the plant and its ecosystem. *Plasma Process. Polym.* **2021**, *18*, 2000162. [CrossRef]

24. Stoleru, V.; Burlica, R.; Mihalache, G.; Dirlau, D.; Padureanu, S.; Teliban, G.-C.; Astanei, D.; Cojocaru, A.; Beniuga, O.; Patras, A. Plant growth promotion effect of plasma activated water on *Lactuca sativa* L. cultivated in two different volumes of substrate. *Sci. Rep.* **2020**, *10*, 20920. [CrossRef]

25. Peťková, M.; Švubová, R.; Kyzek, S.; Medvecká, V.; Slováková, L.; Ševčovičová, A.; Zahoranová, A.; Gálová, E. The effects of cold atmospheric pressure plasma on germination parameters, enzymes activities and DNA damage of barley grains. *Int. J. Mol. Sci.* **2021**, *22*, 2833. [CrossRef]

26. Khalid, S.; Shahid, M.; Niazi, N.K.; Rafiq, M.; Bakhat, H.F.; Imran, M.; Abbas, T.; Bibi, I.; Dumat, C. Arsenic Behaviour in Soil-Plant System: Biogeochemical Reactions and Chemical Speciation Influences. In *Enhancing Cleanup of Environmental Pollutants: Volume 2: Non-Biological Approaches*; Anjum, N.A., Gill, S.S., Tuteja, N., Eds.; Springer: Cham, Switzerland, 2017; pp. 97–140. ISBN 978-3-319-55423-5.

27. Hossain, M.A.; Bhattacharjee, S.; Armin, S.-M.; Qian, P.; Xin, W.; Li, H.-Y.; Burritt, D.J.; Fujita, M.; Lam-Son, P.; Tran, L.-S.P. Hydrogen peroxide priming modulates abiotic oxidative stress tolerance: Insights from ROS detoxification and scavenging. *Front. Plant Sci.* **2015**, *6*, 420. [CrossRef]

28. Lukacova, Z.; Bokor, B.; Vavrova, S.; Soltys, K.; Vaculik, M. Divergence of reactions to arsenic (As) toxicity in tobacco (*Nicotiana benthamiana*) plants: A lesson from peroxidase involvement. *J. Hazard. Mater.* **2021**, *417*, 126049. [CrossRef]

29. Mielecki, J.; Gawroński, P.; Karpiński, S. Retrograde Signaling: Understanding the Communication between Organelles. *Int. J. Mol. Sci.* **2020**, *21*, 6173. [CrossRef]

30. Anjum, N.A.; Sofo, A.; Scopa, A.; Roychoudhury, A.; Gill, S.S.; Iqbal, M.; Lukatkin, A.S.; Pereira, E.; Duarte, A.C.; Ahmad, I. Lipids and proteins–major targets of oxidative modifications in abiotic stressed plants. *Environ. Sci. Pollut. Res. Int.* **2015**, *22*, 4099–4121. [CrossRef] [PubMed]

31. Passardi, F.; Cosio, C.; Penel, C.; Dunand, C. Peroxidases have more functions than a Swiss army knife. *Plant Cell Rep.* **2005**, *24*, 255–265. [CrossRef]

32. Lukačová, Z.; Švubová, R.; Kohanová, J.; Lux, A. Silicon mitigates the Cd toxicity in maize in relation to cadmium translocation, cell distribution, antioxidant enzymes stimulation and enhanced endodermal apoplasmic barrier development. *Plant Growth Regul.* **2013**, *70*, 89–103. [CrossRef]

33. Kapoor, D.; Sharma, R.; Handa, N.; Kaur, H.; Rattan, A.; Yadav, P.; Gautam, V.; Kaur, R.; Bhardwaj, R. Redox homeostasis in plants under abiotic stress: Role of electron carriers, energy metabolism mediators and proteinaceous thiols. *Front. Environ. Sci.* **2015**, *3*, 13. [CrossRef]

34. Beckers, G.J.M.; Conrath, U. Priming for stress resistance: From the lab to the field. *Curr. Opin. Plant Biol.* **2007**, *10*, 425–431. [CrossRef] [PubMed]

35. Wang, Y.; Zhang, J.; Li, J.-L.; Ma, X.-R. Exogenous hydrogen peroxide enhanced the thermotolerance of *Festuca arundinacea* and *Lolium perenne* by increasing the antioxidative capacity. *Acta Physiol. Plant.* **2014**, *36*, 2915–2924. [CrossRef]

36. Borges, A.A.; Jiménez-Arias, D.; Expósito-Rodríguez, M.; Sandalio, L.M.; Pérez, J.A. Priming crops against biotic and abiotic stresses: MSB as a tool for studying mechanisms. *Front. Plant Sci.* **2014**, *5*, 642. [CrossRef] [PubMed]

37. Chao, Y.-Y.; Hsu, S.Y.T.; Kao, S.C.H. Involvement of glutathione in heat shock– and hydrogen peroxide–induced cadmium tolerance of rice (*Oryza sativa* L.) seedlings. *Plant Soil* **2009**, *318*, 37–45. [CrossRef]

38. Liu, Z.-J.; Guo, Y.-K.; Bai, J.-G. Exogenous Hydrogen Peroxide Changes Antioxidant Enzyme Activity and Protects Ultrastructure in Leaves of Two Cucumber Ecotypes Under Osmotic Stress. *J. Plant Growth Regul.* **2010**, *29*, 171–183. [CrossRef]

39. Yan, W.; JianLong, L.; JiaZhen, W.; ZhengKui, L. Exogenous H$_2$O$_2$ improves the chilling tolerance of manilagrass and mascarenegrass by activating the antioxidative system. *Plant Growth Regul.* **2010**, *61*, 195–204.

40. Xu, F.J.; Jin, C.W.; Liu, W.J.; Zhang, Y.S.; Lin, X.Y. Pretreatment with H2O2 Alleviates Aluminum-induced Oxidative Stress in Wheat Seedlings. *J. Integr. Plant Biol.* **2010**, *53*, 44–53. [CrossRef]

41. Hossain, M.A.; Fujita, M. Hydrogen peroxide priming stimulates drought tolerance in mustard (*Brassica juncea* L.). *Plant Gene Trait.* **2013**, *4*, 109–123.

42. Kučerová, K.; Henselová, M.; Slováková, L.; Hensel, K. Effects of plasma activated water on wheat: Germination, growth parameters, photosynthetic pigments, soluble protein content, and antioxidant enzymes activity. *Plasma Process. Polym.* **2019**, *16*, 1800131. [CrossRef]

43. Corona-Carrillo, J.I.; Flores-Ponce, M.; Chávez-Nájera, G.; Díaz-Pontones, D.M. Peroxidase activity in scutella of maize in association with anatomical changes during germination and grain storage. *Springerplus* **2014**, *3*, 399. [CrossRef]

44. Lux, A.; Lukačová, Z.; Vaculík, M.; Švubová, R.; Kohanová, J.; Soukup, M.; Martinka, M.; Bokor, B. Silicification of Root Tissues. *Plants* **2020**, *9*, 111. [CrossRef]

45. Vaculík, M.; Lukačová, Z.; Bokor, B.; Martinka, M.; Tripathi, D.K.; Lux, A. Alleviation mechanisms of metal(loid) stress in plants by silicon: A review. *J. Exp. Bot.* **2020**, *71*, 6744–6757. [CrossRef]

46. Schreiber, L.; Hartmann, K.; Skrabs, M.; Zeier, J. Apoplastic barriers in roots: Chemical composition of endodermal and hypodermal cell walls. *J. Exp. Bot.* **1999**, *50*, 1267–1280. [CrossRef]

47. Martinka, M.; Lux, A. Response of roots of three populations of *Silene dioica* to cadmium treatment. *Biologia* **2004**, *59*, 185–189.

48. Vaculík, M.; Landberg, T.; Greger, M.; Luxová, M.; Stoláriková, M.; Lux, A. Silicon modifies root anatomy, and uptake and subcellular distribution of cadmium in young maize plants. *Ann. Bot.* **2012**, *110*, 433–443. [CrossRef] [PubMed]

49. Stoláriková, M.; Vaculík, M.; Lux, A.; Di Baccio, D.; Minnocci, A.; Andreucci, A.; Sebastiani, L. Anatomical differences of poplar (*Populus* × *euramericana* Clone I-214) roots exposed to zinc excess. *Biologia* **2012**, *67*, 483–489. [CrossRef]

50. Huang, L.; Li, W.C.; Tam, N.F.Y.; Ye, Z. Effects of root morphology and anatomy on cadmium uptake and translocation in rice (*Oryza sativa* L.). *J. Environ. Sci.* **2019**, *75*, 296–306. [CrossRef]

51. Ashraf, M. Biotechnological approach of improving plant salt tolerance using antioxidants as markers. *Biotechnol. Adv.* **2009**, *27*, 84–93. [CrossRef] [PubMed]

52. Fitter, A.H.; Atkinson, D.; Read, D.J.; Usher, M.B. Ecological interactions in soil: Plants, microbes and animals. In *Ecological Interactions in Soil*; Fitter, A.H., Atkinson, D., Read, D.J., Usher, M.B., Eds.; Blackwell Scientific Publications: Oxford, UK, 1985; pp. 87–106.

53. Bafoil, M.; Jemmat, A.; Martinez, Y.; Merbahi, N.; Eichwald, O.; Dunand, C.; Yousfi, M. Effects of low temperature plasmas and plasma activated waters on *Arabidopsis thaliana* germination and growth. *PLoS ONE* **2018**, *13*, e0195512. [CrossRef]

54. Gierczik, K.; Vukušić, T.; Kovács, L.; Székely, A.; Kocsy, G.; Szalai, G.; Milošević, S.; Kutasi, K.; Galiba, G. Plasma-activated water to improve the stress tolerance of barley. *Plasma Process. Polym.* **2020**, *17*, e1900123. [CrossRef]

55. Abbas, G.; Murtaza, B.; Bibi, I.; Shahid, M.; Niazi, N.K.; Khan, M.I.; Amjad, M.; Hussain, M. Natasha Arsenic Uptake, Toxicity, Detoxification, and Speciation in Plants: Physiological, Biochemical, and Molecular Aspects. *Int. J. Environ. Res. Public. Health* **2018**, *15*, 59. [CrossRef]

56. Bokor, B.; Vaculík, M.; Slováková, L.; Masarovič, D.; Lux, A. Silicon does not always mitigate zinc toxicity in maize. *Acta Physiol. Plant.* **2014**, *36*, 733–743. [CrossRef]

57. Shahid, M.; Pourrut, B.; Dumat, C.; Nadeem, M.; Aslam, M.; Pinelli, E. Heavy-Metal-Induced Reactive Oxygen Species: Phytotoxicity and Physicochemical Changes in Plants. In *Reviews of Environmental Contamination and Toxicology*; Whitacre, D.M., Ed.; Reviews of Environmental Contamination and Toxicology; Springer International Publishing: Cham, Switzerland, 2014; Volume 232, pp. 1–44. ISBN 978-3-319-06746-9.

58. Sharma, I. Arsenic induced oxidative stress in plants. *Biologia* **2012**, *67*, 447–453. [CrossRef]

59. Imlay, J.A. Cellular defenses against superoxide and hydrogen peroxide. *Annu. Rev. Biochem.* **2008**, *77*, 755–776. [CrossRef]
60. Sivachandiran, L.; Khacef, A. Enhanced seed germination and plant growth by atmospheric pressure cold air plasma: Combined effect of seed and water treatment. *RSC Adv.* **2017**, *7*, 1822–1832. [CrossRef]
61. Noctor, G.; Reichheld, J.P.; Foyer, C.H. ROS-related redox regulation and signaling in plants. *Semin. Cell Dev. Biol.* **2018**, *80*, 3–12. [CrossRef] [PubMed]
62. Jiang, H.M.; Yang, J.C.; Zhang, J.F. Effects of external phosphorus on the cell ultrastructure and the chlorophyll content of maize under cadmium and zinc stress. *Environ. Pollut.* **2007**, *147*, 750–756. [CrossRef]
63. Bhat, J.A.; Ahmad, P.; Corpas, F.J. Main nitric oxide (NO) hallmarks to relieve arsenic stress in higher plants. *J. Hazard. Mater.* **2021**, *406*, 124289. [CrossRef]
64. Janda, M.; Machala, Z.; Niklová, A.; Martišovitš, V. The streamer-to-spark transition in a transient spark: A dc-driven nanosecond-pulsed discharge in atmospheric air. *Plasma Sources Sci. Technol.* **2012**, *21*, 04500665. [CrossRef]
65. Kučerová, K.; Machala, Z.; Hensel, K. Transient Spark Discharge Generated in Various N2/O2 Gas Mixtures: Reactive Species in the Gas and Water and Their Antibacterial Effects. *Plasma Chem. Plasma Process.* **2020**, *40*, 749–773. [CrossRef]
66. Kučerová, K.; Henselová, M.; Slováková, L.; Bačovčinová, M.; Hensel, K. Effect of Plasma Activated Water, Hydrogen Peroxide, and Nitrates on Lettuce Growth and Its Physiological Parameters. *Appl. Sci.* **2021**, *11*, 1985. [CrossRef]
67. Eisenberg, G.M. Colorimetric determination of hydrogen peroxide. *Ind. Eng. Chem.* **1943**, *15*, 327–328. [CrossRef]
68. Bradford, M.M. A rapid and sensitive method for the quantitation of microgram quantities of protein utilizing the principle of protein-dye binding. *Anal. Biochem.* **1976**, *72*, 248–254. [CrossRef]
69. Frič, F.; Fuchs, W.H. Veränderungen der Aktivität einiger enzyme im Weizenblatt in Abhängigkeit von *Puccinia graministritici*. *J. Phytopathol.* **1970**, *67*, 161–174. [CrossRef]
70. Ferrer, M.A.; Calderón, A.A.; Muñoz, R.; Barceló, A.R. 4-Methoxy-α-naphthol as a specific substrate for kinetic, zymographic and cytochemical studies on plant peroxidase activities. *Phytochem. Anal.* **1990**, *1*, 63–69. [CrossRef]
71. Ďurčeková, K.; Huttová, J.; Mistrík, I.; Olle, M.; Tamás, L. Cadmium induces premature xylogenesis in barley roots. *Plant Soil* **2007**, *290*, 61–68. [CrossRef]
72. Hodges, D.M.; Andrews, C.J.; Johnson, D.A.; Hamilton, R.I. Antioxidant enzyme and compound responses to chilling stress and their combining abilities in differentially sensitive maize hybrids. *Crop Sci.* **1997**, *37*, 857–863. [CrossRef]
73. Claiborne, A. Catalase Activity. In *Handbook of Methods for Oxygen Radical Research*; Greenwald, R.A., Ed.; CRC Press: Boca Raton, FL, USA, 2018; pp. 283–284.
74. Lichtenthaler, H.K. Chlorophylls and carotenoids: Pigments of photosynthetic biomembranes. In *Methods in Enzymology*; Plant Cell Membranes; Academic Press: Cambridge, MA, USA, 1987; Volume 148, pp. 350–382.

Article

Physiological Responses of Young Pea and Barley Seedlings to Plasma-Activated Water

Dominik Kostoláni [1], Gervais B. Ndiffo Yemeli [2], Renáta Švubová [1], Stanislav Kyzek [3] and Zdenko Machala [2,*]

[1] Department of Plant Physiology, Faculty of Natural Sciences, Comenius University in Bratislava, Mlynská Dolina, Ilkovičova 6, 842 15 Bratislava, Slovakia; kostolani6@uniba.sk (D.K.); renata.svubova@uniba.sk (R.Š.)

[2] Division of Environmental Physics, Faculty of Mathematics, Physics and Informatics, Comenius University in Bratislava, Mlynská Dolina, 842 48 Bratislava, Slovakia; gbn.yemeli@fmph.uniba.sk

[3] Department of Genetics, Faculty of Natural Sciences, Comenius University in Bratislava, Mlynská Dolina, Ilkovičova 6, 842 15 Bratislava, Slovakia; kyzek2@uniba.sk

* Correspondence: machala@fmph.uniba.sk

Abstract: This study demonstrates the indirect effects of non-thermal ambient air plasmas (NTP) on seed germination and plant growth. It investigates the effect of plasma-activated water (PAW) on 3-day-old seedlings of two important farm plants—barley and pea. Applying different types of PAW on pea seedlings exhibited stimulation of amylase activity and had no inhibition of seed germination, total protein concentration or protease activity. Moreover, PAW caused no or only moderate oxidative stress that was in most cases effectively alleviated by antioxidant enzymes and proved by in situ visualization of H_2O_2 and $^{\cdot}O_2^{-}$. In pea seedlings, we observed a faster turn-over from anaerobic to aerobic metabolism proved by inhibition of alcohol dehydrogenase (ADH) activity. Additionally, reactive oxygen/nitrogen species contained in PAW did not affect the DNA integrity. On the other hand, the high level of DNA damage in barley together with the reduced root and shoot length and amylase activity was attributed to the oxidative stress caused by PAW, which was exhibited by the enhanced activity of guaiacol peroxidase or ADH. Our results show the glow discharge PAW at 1 min activation time as the most promising for pea. However, determining the beneficial type of PAW for barley requires further investigation.

Keywords: antioxidant enzymes; barley; non-thermal plasmas; dehydrogenases; DNA damage; lytic enzymes; pea; plasma-activated water; RONS

Citation: Kostoláni, D.; Ndiffo Yemeli, G.B.; Švubová, R.; Kyzek, S.; Machala, Z. Physiological Responses of Young Pea and Barley Seedlings to Plasma-Activated Water. *Plants* **2021**, *10*, 1750. https://doi.org/10.3390/plants10081750

Academic Editor: Roberta Masin

Received: 14 July 2021
Accepted: 19 August 2021
Published: 23 August 2021

Publisher's Note: MDPI stays neutral with regard to jurisdictional claims in published maps and institutional affiliations.

1. Introduction

In light of the rapidly increasing world population, a demand for sustainable food production becomes more and more critical [1]. Despite many efforts, the sufficient agricultural production generating high-quality crops depends on the use of commercial fertilizers. From all micro- and macronutrients contained in fertilizers, nitrogen represents the main compound promoting plant growth and production [2,3]. Considering a negative impact of chemicals on human and animal health and the environment, the recent research focuses on alternative ways to promote plant growth which are less harmful, more sustainable and efficient [3,4]. Besides organic fertilizers and biofertilizers, which also pose some risks and/or have limitations [5,6], physical methods such as static magnetic field or pulsed electric field indicate promising results [7,8]. The recent use of cold atmospheric gas plasmas is also considered promising but demands further investigations to understand the mechanisms of interactions with plants.

Plasma represents the fourth state of matter and occurs naturally in space or Earth's atmosphere or can be produced artificially under laboratory conditions [9,10]. According to its temperature and thermal equilibrium of charged and neutral particles, plasma can be categorised as thermal or non-thermal (cold, nonequilibrium) plasma [11]. Unlike in thermal plasma, the excitation, dissociation or ionization of molecules in non-thermal plasma (NTP) occurs with higher effectiveness and causes significantly lower temperature of heavy particles when compared to the electron temperature. Relatively low temperature of NTP allows its subsequent application for the treatment of thermo-sensitive and biological materials [11–14].

The application of NTP in agriculture includes both a direct and indirect way of plasma treatment. A major part of published data showed the effects of the direct way that represents an explicit interaction between the plasma and the plant material. This method has been approved in seed sterilization, germination enhancement and plant growth promotion [15–23]. Far less studies including this article have been focused on the indirect plasma treatment, where the interaction between plasma and plant material is mediated by plasma-activated water (PAW) [24–28]. The PAW is usually generated by performing the electrical discharges directly in water or on its surface [29–31]. Regarding the type of plasma discharge, technique of interaction with water and water solution buffering capacity, various types of PAW may strongly differentiate in chemical composition, especially in concentration of long-lifetime reactive oxygen and nitrogen species (RONS) delivered by the interaction with the plasma, such as hydrogen peroxide (H_2O_2), nitrites (NO_2^-) and nitrates (NO_3^-) [32].

Hydrogen peroxide can easily enter the plant cells via free diffusion or via water channels aquaporins. As the most stable reactive oxygen species, it plays a crucial role in the cascades of intracellular signalling. However, if abundant, it can be metabolised to water and oxygen by the catalase and peroxidase activities [21,33]. The NO_2^- and particularly NO_3^- contained in PAW represent an important source of nitrogen for plants, and their transport across the plasma membrane is provided via diffusion or via specific transporters [34,35]. Thanks to its composition, PAW has the potential to become an environment-friendly and sustainable alternative to classical fertilizers used in agriculture.

Maniruzzaman [36] documented the enhanced effect of PAW RONS on plant growth in comparison with chemical fertilizers. The study referred better results to long-life RONS in PAW, which can be in the solution with chemical fertilizers partially replaced by unstable intermediate species. Gierczik et al. [37] showed that pre-treatment of barley grains increased germination under abiotic stress through the improved signalling and activation of different defence mechanisms by H_2O_2 and nitric oxide (NO; formed from NO_2^- and NO_3^-). Nevertheless, a wide range of plasma sources and plant samples tested in scattered studies and resulting in different efficiencies of treatment requires further investigations.

The objective of this article is to study the influence of PAW generated by two sources of cold atmospheric air plasma, namely transient spark (TS) with water electrospray and glow discharge (GD) with water cathode on pea (*Pisum sativum* L. cv. Eso) seeds and barley (*Hordeum vulgare* L. cv. Kangoo) grains. Before applying PAW to the plants, we measured the concentration of long-life species (H_2O_2, NO_2^-, NO_3^-) in various types of PAW by UV–Vis absorption spectroscopy. Then, we analysed specific plant growth parameters (germination, root, shoot and seedling length) and physiological parameters, such as total soluble proteins concentration (TSP) and DNA damage. Since the RONS are the main monitored components in PAW, we also focused on antioxidant enzyme activities (superoxide dismutase [SOD], guaiacol peroxidase [G-POX] and catalase [CAT]) and in situ visualization of H_2O_2 and $^·O_2^-$. Finally, the activity of lytic enzymes (proteases, amylases) and dehydrogenases (alcohol and succinate dehydrogenases) were determined. A lack of studies investigating the activity of these enzymes after PAW treatment in relation to other growth and physiological parameters in early stages of plant development represents the novelty of this part of the study. Hopefully, our results will contribute to the better understanding of relations between certain plasma treatments and plant species, resulting in changes in the plant metabolism and plant growth.

2. Results and Discussion

2.1. Physiochemical Properties of PAW

Figure 1a shows the pH of the tap water control and PAW and 2 mM HNO_3. A slight decrease of pH from 8 to 7.8 was found between the control and TS PAW as well as from 7.5 and 7.4 for GD1 (PAW of glow discharge at activation time 1 min) and GD2 (PAW of glow discharge at activation time 2 min), respectively. The small difference of pH before and after the plasma treatment can be explained by the natural hydrocarbon buffer system, unlike when plasma treating deionized or distilled water where we observed a strong acidification [38–40]. Our previous articles also reported these slight pH changes using tap water [21,26,28]. It turns out that overall pH variations are negligible, making this parameter a non-disruptive factor of the germination process.

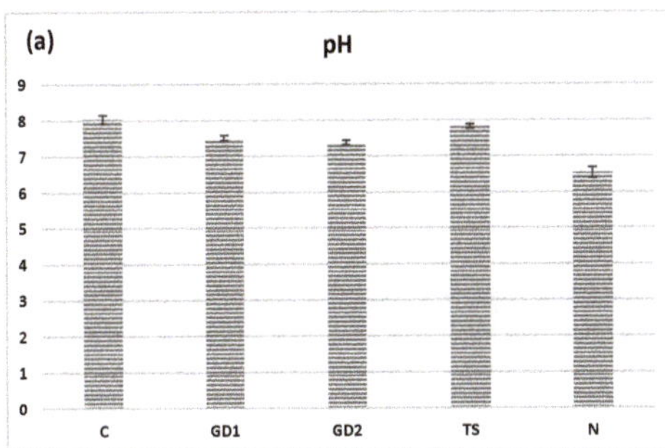

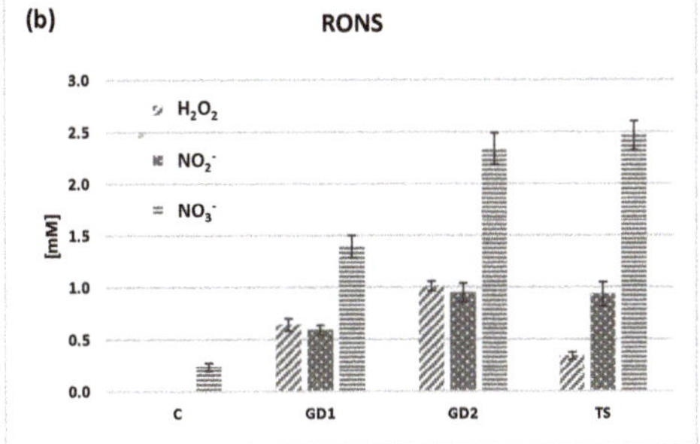

Figure 1. pH (**a**) and concentrations of hydrogen peroxide (H_2O_2), nitrite (NO_2^-) and nitrate (NO_3^-) (**b**) in tap water control (C), PAW generated by glow discharges for 1 and 2 min (GD1 and GD2, respectively), PAW generated by transient spark (TS) and 2 mM of nitric acid (N). Values are expressed as a mean ± SD, minimum ten repetitions.

Figure 1b shows the concentrations of H_2O_2, NO_2^- and NO_3^- generated in the three types of PAW generated by both plasma sources. The following concentrations were obtained for H_2O_2 ~0.33 mM, 0.64 mM and 1.00 mM for TS, GD1 and GD2, respectively. The concentrations of NO_2^- for TS, GD1 and GD2 were approximately 0.93 mM, 0.59 mM and 0.95 mM, respectively. The concentrations of NO_3^- for TS, GD1 and GD2 were approximately 2.46 mM, 1.40 mM and 2.34 mM, respectively. Both plasma sources are rich providers of RONS in water, as previously shown [28,38,39,41]. Thanks to the almost constant pH, these RONS do not decay after plasma treatment over a period of several hours, which facilitates their application on plants.

2.2. Germination Dynamics

The three types of plasma-activated water (PAW) have a slight impact on the percentage of germination (%) or germination dynamics (%) of barley grains and pea seeds (Figure 2). This is probably because the seeds/grains used in our study retained almost 100% germination rate even under control conditions. In the case of barley grains, we observed a slight decrease in the percentage of germination (%) in the variant B-TS on the first day of cultivation, but on the third day, the number of germinated grains was comparable to other variants (Figure 2a). The highest germination, compared to the control (89%), was recorded in the case of the B-GD1 variant (96%). Pea seeds started the germination processes most effectively in the case of the P-GD1 variant (first day of cultivation). Already on the second day of cultivation, germination levelled for all variants and reached approximately 98% (Figure 2b). In published studies, fairly contrast results considering PAW's effect on the germination can be found. These results suggest, on one hand, a stimulative effect, as shown for example by Kučerová et al. [26] on wheat plants, Zhang et al. [42] on lentils with an increased germination rate of about 50%, or Zhou et al. [25] on mung beans with the observed shortening of the germination time up to 36 h when compared to untreated plants. In these studies, authors referred the positive effect of PAW to the H_2O_2 ability to stimulate the respiration, react with germination inhibitors and cause erosion of seed coat, facilitating an improved imbibition, as confirmed also by other studies [43,44]. Moreover, reactive nitrogen species in PAW may be involved in intracellular signalling and interactions with endogenous phytohormones either by regulating abscisic acid concentration or by stimulating of gibberellic acid synthesis [45]. On the other hand, Lindsay et al. [46] found no significant changes in the germination rate of radishes, tomatoes and marigolds irrigated with PAW. However, during the growth phase, there were significant differences between treated and untreated plants. These results compared to ours indicate that PAW application could significantly affect germination rate in seeds with naturally low germination rate or accelerate germination, but it also strongly depends on the plant species, which makes the final effect difficult to predict without an experimental backup.

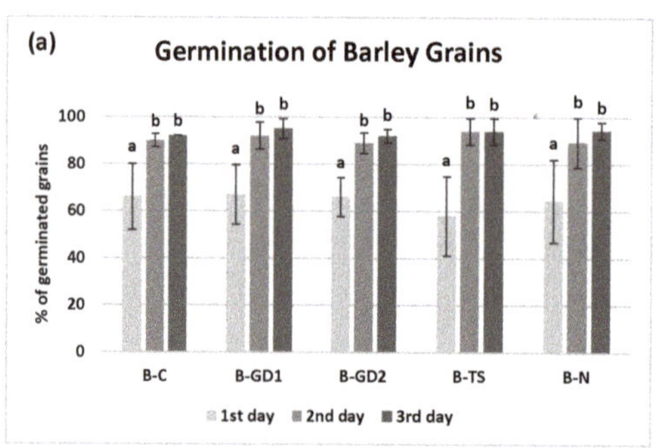
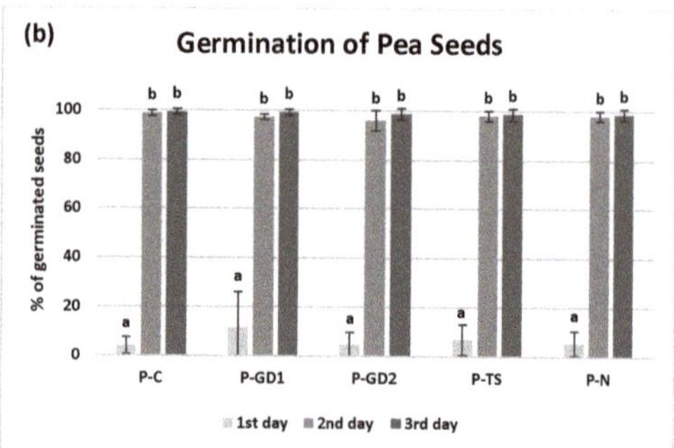

Figure 2. The percentage of barley grain (**a**) and pea seed (**b**) germination in 3-day horizon, watered with tap water (control: B/P-C), plasma-activated water (PAW) of glow discharge with water cathode and treatment time 1 and 2 min (B/P-GD1 and B/P-GD2), PAW of transient spark discharge with electrospray (B/P-TS) and 2-mM nitric acid (B/P-N). Values are expressed as a mean ± SD from three repeated experimental rounds (one run represents 50 grains/seeds per variant, n = 150). Different letters represent statistically significant difference at $p < 0.05$ according to LSD test.

2.3. Growth Parameters

When measuring production parameters (length of barley roots and shoots, length of pea seedlings) of 3-day-old seedlings, we found that the application of PAW and 2-mM nitric acid solution negatively affected the length of the roots of barley seedlings (Figure 3a): more than 3-fold reduction occurred for B-GD1, B-GD2 and B-TS variants and approximately one third reduction for B-N variant). It is important to mention that the number of adventitious roots was not negatively affected and shoot lengths (Figure 3b) in all treated variants were comparable to the untreated control. Shashikanthalu et al. [47] also reported changes in root length in 14-day-old *Cuminum cyminum* roots where the seeds were subjected to direct NTP treatment. In this study, seeds exposed to non-thermal plasma for 2 min and 4 min exhibited the root growth reduction of 12.6% and 8.78%, respectively. Otherwise, the seeds exposed to NTP for 3 min exhibited an increase in root length of about 41.79%. By the shoot length, they documented a positive effect under exposure times of 2 min and 3 min (20.82% and 34.5%, respectively). Longer exposure time (4 min) induced a decrease in the shoot length. On the other hand, Feizollahi et al. [48] using direct plasma treatment on barley grains documented root growth reduction in 7-day-old barley seedlings for an exposure time of 10 min, while exposure times of 1 min and 6 min caused no changes compared to control. Additionally, none of the mentioned exposure times caused a decrease in the shoot length. These findings indicate that certain exposure times (in direct plasma treatment) and different RONS contents in PAW, specific to different plant species, may be responsible for dramatic changes in growth parameters varying between roots and shoots.

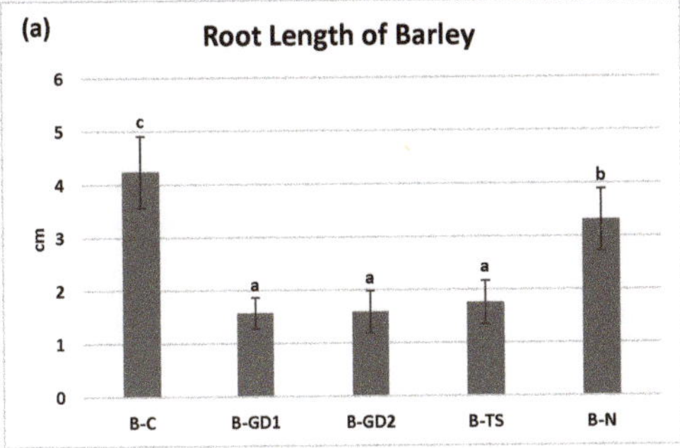
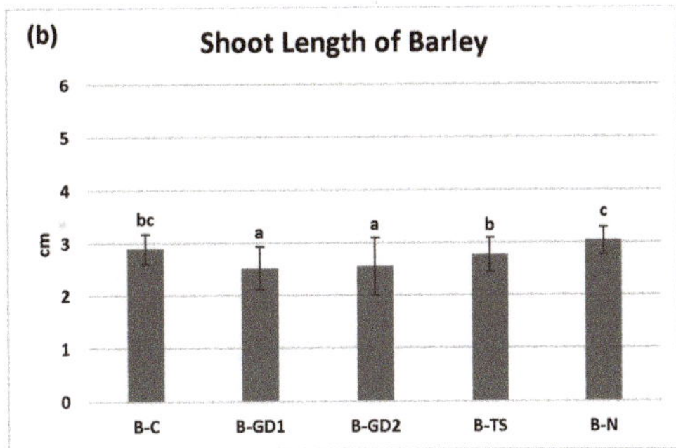
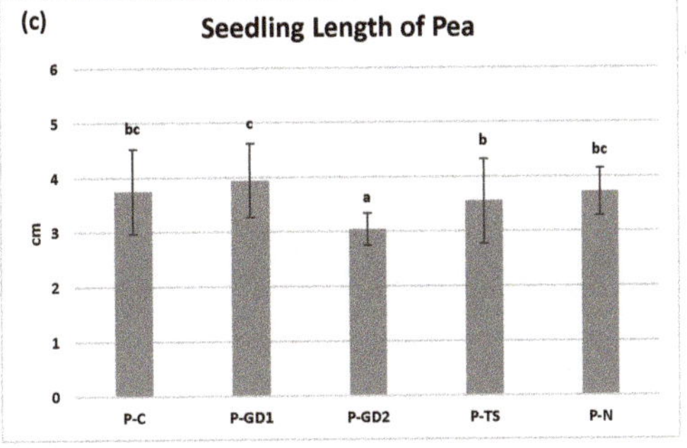

Figure 3. The growth parameters of 3-day-old seedlings of barley (**a,b**) and pea (**c**), watered with tap water (control: B/P-C), plasma-activated water (PAW) of glow discharge with water cathode and treatment time 1 and 2 min (B/P-GD1 and B/P-GD2), PAW of transient spark discharge with electrospray (B/P-TS) and 2-mM nitric acid (B/P-N). Values are expressed as a mean ± SD from five repeated experimental rounds (one run represents 10 seedlings per variant, n = 50). Different letters represent statistically significant difference at $p < 0.05$ according to LSD test.

In our experiments, the total length of pea seedlings in variants P-GD1 and P-TS did not change when compared to the negative and positive control (Figure 3c). On the other hand, the application of GD of 2 min caused a significant reduction in the length of pea seedlings (Figure 3c, P-GD2 variant). With respect to other results, we can conclude that pea is able to resist to the external supplementation of reactive species, without critical changes in growth parameters, more effectively than barley. Our results correspond with a study by Švubová et al. [49], who documented no or negative impact of direct NTP treatment (applicated on seeds) on 3-day-old pea seedling length under different exposure times (10–40 s) and gas atmospheres (ambient air, nitrogen and oxygen). On the contrary, a stimulative effect on the seedling length was shown by Stolárik et al. [18] when seeds were exposed to the direct plasma treatment for longer times (60 s and 180 s), possibly even regarding the effect of components other than RONS in the direct plasma treatment.

2.4. Total Soluble Proteins Concentration

Despite the moderate effect on growth parameters, the application of PAW and 2 mM nitric acid had no statistically significant impact on the total soluble protein (TSP) concentrations neither in barley (Figure 4a) nor in pea (Figure 4b). Almost two-fold higher values for TSP in pea (compared to barley) can be linked to different storage substances in grains and seeds of barley and pea, respectively. An increase in TSP in 4-week-old plants irrigated with PAW was documented by our preceding study by Yemeli et al. [28] for the same types of PAW in variants GD2 and N for barley and variants GD1 and GD2 for maize. Kučerová et al. [26] showed no changes in TSP in above-ground parts of wheat plants and only slightly increased concentration in roots after application of PAW generated by transient spark discharge from tap and deionized water. The effect of direct plasma exposure was also documented. Švubová et al. [49] documented an amelioration of TSP concentration under NTP treatment of pea in ambient air and oxygen with exposure times of 10 s and 40 s, respectively. On the contrary, plasmas generated in the same gases for 20 s caused a significant decrease of TSP. However, almost all combinations of gases and exposure times showed no changes in TSP concentration, as shown in the study by Švubová et al. [20]. With respect to these findings, we can state that even the same type of PAW can result in different metabolic responses when considering different plant age and application method. Additionally, it becomes even more complicated with various types of direct NTP applications, which brings additional physical effects into play (e.g., electric field, UV radiation, short-lived reactive species).

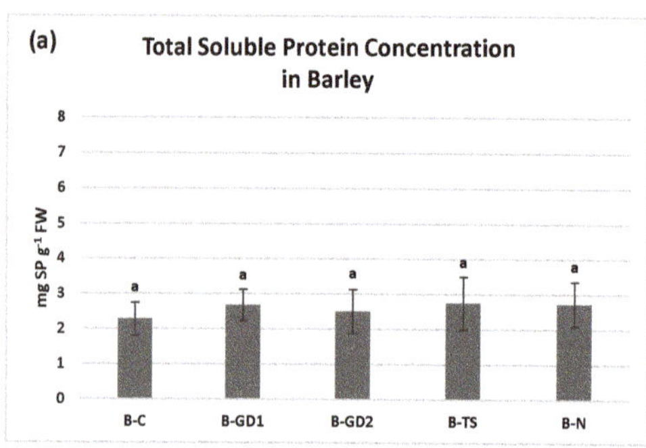

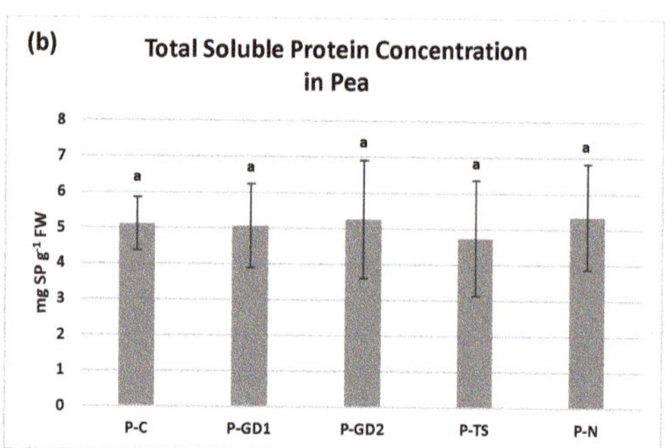

Figure 4. The total soluble protein concentration in the seedlings of barley (**a**) and pea (**b**) after 3 days of growth, watered with tap water (control: B/P-C), plasma-activated water (PAW) of glow discharge with water cathode and treatment time 1 and 2 min (B/P-GD1 and B/P-GD2), PAW of transient spark discharge with electrospray (B/P-TS) and 2-mM nitric acid (B/P-N). Values are shown as mean ± SD from three experimental rounds (one run represents five seedlings per variant, 1.5 g mixed samples were analysed per one experimental run and each variant for total soluble protein concentration). Different letters represent statistically significant difference at $p < 0.05$ according to LSD test.

2.5. Activity of Lytic Enzymes

The activity of protease, the enzyme that cleaves storage proteins, was slightly affected (Figure 5a,b). In the barley B-TS variant, its activity decreased significantly. In the B-GD1 variant, we observed a slight increase in protease activity, which, however, was not statistically significant compared to the untreated control (Figure 5a). In pea, we recorded no statistically significant differences when compared to the control (Figure 5b). Unlike protease, the activity of amylase, a starch-cleaving enzyme, was strongly and negatively affected in case of barley (Figure 5c). In the B-TS variant, the amylase activity decreased by about 90% compared to the untreated seedlings. Significant reduction was also recorded in the other three variants. In pea, P-N was the only negatively affected variant (Figure 5d). On the other hand, when compared to the control, the variant P-GD2 reached more than twofold increase. Several studies linking amylase activity to proper seed development directly affecting plant growth and yield, as summarised, i.e., in the review from Damaris et al. [50].

Despite the fact that many authors attribute the increased germination to the higher activity of lytic enzymes, the pool of published data examining the effect of NTP on these enzymes is considerably restricted. Moreover, the available data are focused mainly on the direct plasma treatment. For example, Peťková et al. [51] showed a positive effect of NTP generated in three different atmospheres with short exposure time (10–30 s) on barley grains, while longer exposure time (60 s and 180 s) caused an opposite effect on protease activity. These results were also confirmed by Švubová et al. [20,49] on pea seeds. They documented a positive effect under exposure times varying from 10 s to 40 s and no or negative effect under exposure times varying from 60 s to 300 s for all tested working gases. Furthermore, higher exposure times were responsible for a significant decrease in amylase activity, with the lowest activity in the plasma-treated samples at 300 s. Sadhu et al. [52] obtained a stimulative effect of direct plasma treatment on the amylase activity of mung bean under varying exposure times and power levels. Enhanced activity of amylase in brown rice grains was also documented by Chen et al. [53] after using direct plasma treatment. Regarding this knowledge based on direct NTP effects, we can assume that dramatically negative changes in amylase activity in barley indicate using PAW with RNS, especially NO_2^- and NO_3^- concentrations higher than the beneficial levels for this type of plant and application. A possible mechanism of action may be the signalling activity of NO derived from NO_2^- and NO_3^-. In this signal cascade, abundancy of NO inhibits the activity of specific protein kinase (SnRK1), which normally stimulates the activity of amylase in wheat [54,55]. Another mechanism of action may be the inhibition effect of ethanol on gibberellin-induced activity of amylase, as proved by Perata et al. [56], which correlates with our findings on ADH activity (as shown in Section 2.7).

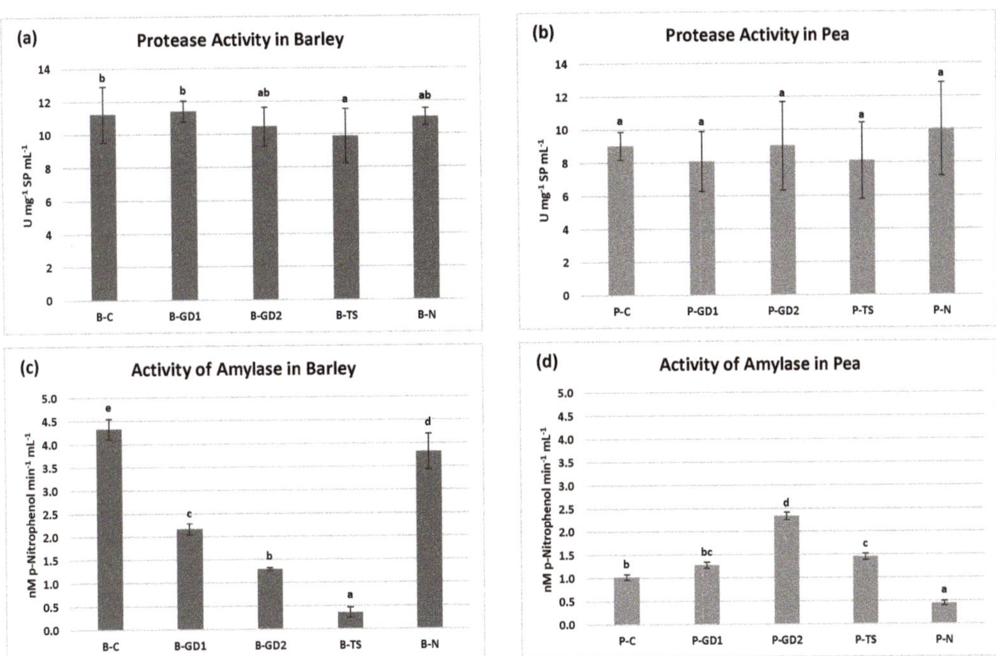

Figure 5. Protease and amylase activities in the barley (**a,c**) and pea (**b,d**) seedlings after 3 days of growth, watered with tap water (control: B/P-C), plasma-activated water (PAW) of glow discharge with water cathode and treatment time 1 and 2 min (B/P-GD1 and B/P-GD2), PAW of transient spark discharge with electrospray (B/P-TS) and 2-mM nitric acid (B/P-N). Values are shown as mean ± SD from three experimental rounds (one run represents five seedlings per variant, 1.5 g mixed samples were analysed per one experimental run, and each variant for protease activity and five 0.1 g mixed samples were analysed per one experimental run and each variant for amylase activity). Different letters represent statistically significant difference at $p < 0.05$ according to LSD test.

2.6. Activity of Antioxidant Enzymes and In Situ Visualisation of Reactive Oxygen Species

Although PAW is a reactive environment (containing hydrogen peroxide, nitrates and nitrites), it did not cause significant oxidative stress in the case of pea seedlings. In the case of barley, as suggested by other analyses, the situation was moderately different. The activity of SOD, the enzyme that detoxifies the superoxide radicals, slightly decreased in the case of PAW variants of barley compared to the untreated control. In the case of pea seedlings, it increased slightly in the P-GD2 and P-N variants. When monitoring the accumulation of $\cdot O_2^-$, we did not notice any significant differences between the individual variants (Figures 6a,b and 7). Our results clearly show that due to the application of PAW, barley and pea seedlings exogenously absorb increased concentrations of H_2O_2 (Figure 1b). We can state that barley seedlings reacted negatively to this exogenous application of H_2O_2, where we observed a significant shortening of the root length (Figure 3b).

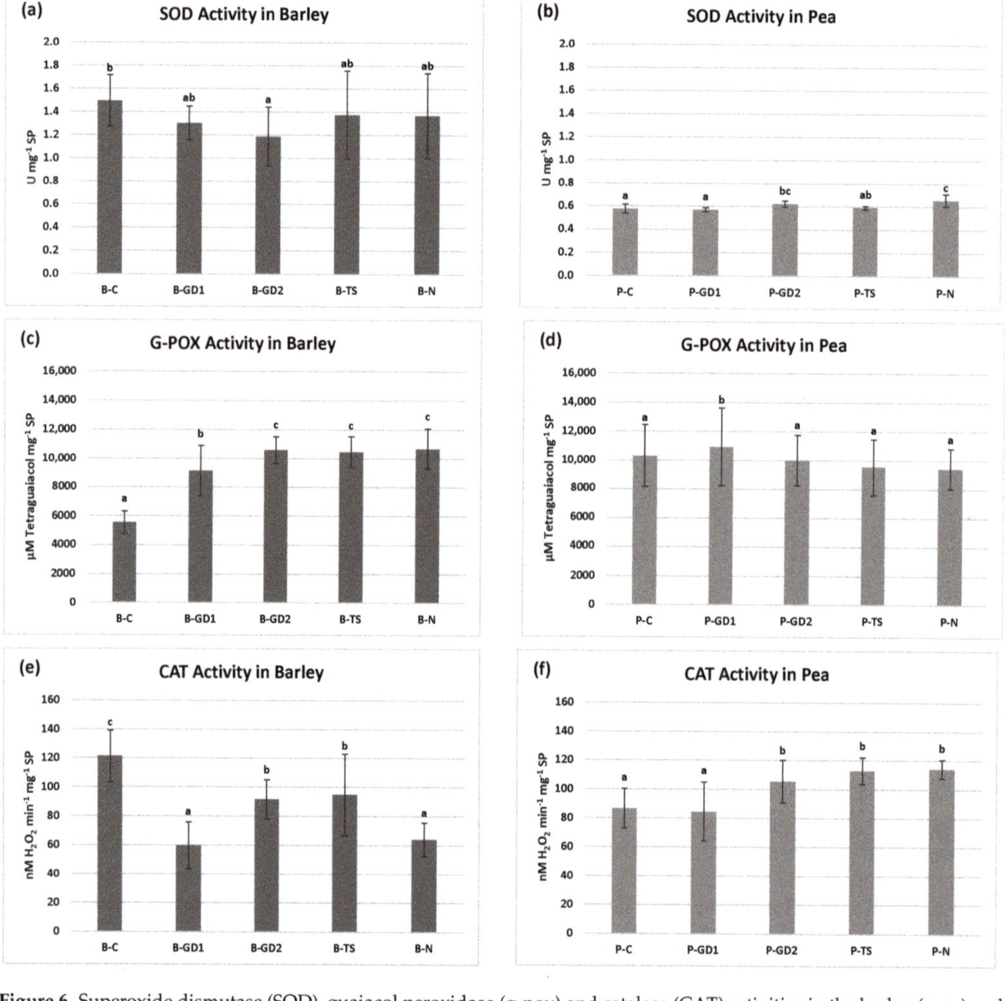

Figure 6. Superoxide dismutase (SOD), guaiacol peroxidase (g-pox) and catalase (CAT) activities in the barley (**a,c,e**) and pea (**b,d,f**) seedlings after 3 days of growth, watered with tap water (control: B/P-C), plasma-activated water (PAW) of glow discharge with water cathode and treatment time 1 and 2 min (B/P-GD1 and B/P-GD2), PAW of transient spark discharge with electrospray (B/P-TS) and 2-mM nitric acid (B/P-N). Values are shown as mean ± SD from three experimental rounds, in each round (one run represents five seedlings per variant, 1.5 g mixed samples were analysed per one experimental run and each variant for SOD, G-POX and CAT activity). Different letters represent statistically significant difference at $p < 0.05$ according to LSD test.

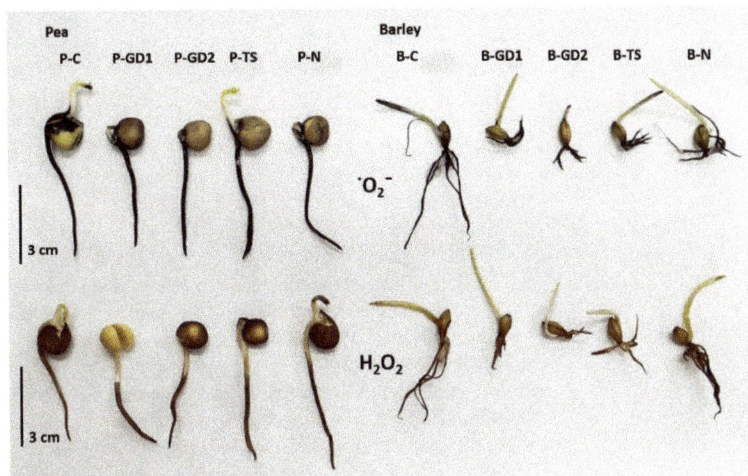

Figure 7. The visualisation of H_2O_2 and $^{\cdot}O_2^{-}$ in tissues of the barley and pea seedlings after 3 days of growth, watered with tap water (control: B/P-C), plasma-activated water (PAW) of glow discharge with water cathode and treatment time 1 and 2 min (B/P-GD1 and B/P-GD2), PAW of transient spark discharge with electrospray (B/P-TS) and 2-mM nitric acid (B/P-N). The experiment was repeated four times (one run = five seedlings from each variant were incubated in NBT solution and five seedlings from each variant were incubated in DAB solution, n = 20). The image represents a representative selection of seedlings after staining.

The G-POX activity was significantly increased and CAT activity correlated, and we observed the accumulation of H_2O_2 in the whole barley root system and the accumulation of $^{\cdot}O_2^{-}$ in roots as well as in seed coats (Figures 6c,e and 7). The growth and development of pea seedlings were not affected by exogenously applied H_2O_2 (Figure 3c). The activity of the enzymes that decompose H_2O_2 into water and oxygen was generally comparable to the control. The predominant accumulation of H_2O_2 in the root tips and the elongation zone of the roots of pea seedlings is probably associated with intensive cell division and elongation, thus having a physiological basis (Figures 6d,f and 7).

The activity of the antioxidant enzymes, as well as in situ visualisation of oxygen radicals, was a subject of interest also in other studies. In our previous study, we analysed the activity of SOD, G-POX and CAT in 4-week-old barley and maize plants watered with PAW [28]. The results from the barley's above-ground parts correlate with our current barley data in the activity of SOD and partially with G-POX and CAT activities. However, these findings are in agreement with studies that consider CAT as a dominant enzyme for scavenging H_2O_2 in the above-ground parts, while G-POX is considered dominant in the root system of plants [26,57]. Thus, we attribute the significantly higher activity of G-POX and different CAT tendency (in comparison with our previous study [28]) mostly to disparate plant parts that were analysed. In 6-day-old wheat seedlings treated with PAW, Kučerová et al. [26] documented the decrease in activity of all three enzymes in the above-ground parts as well as in the roots. In their other study, Kučerová et al. [21] investigated PAW application on SOD activity in lettuce with the same result. Decreased CAT activity in PAW-treated *Paulownia tomentosa* seedlings was also shown by Puač et al. [58].

On the contrary to the mentioned studies, publications dealing with direct plasma treatment showed mainly a stimulative effect of NTP on antioxidant enzyme activity that could possibly be referred to the different physical factors at play (electric field, UV radiation, short-lived species). For example, Švubová et al. [20,49] showed generally the increased SOD activity in 3-day-old pea seedlings treated with plasma generated in ambient air, oxygen and nitrogen for exposure times varying from 10 s to 300 s. On the other hand, the G-POX activity significantly increased only in variants with nitrogen atmosphere and exposures for 20 s and 40 s, respectively. They associated the $^{\bullet}O_2{}^-$ occurring in root tips of plants treated for 10 s and 20 s with an extensive root growth and development. Higher exposure times than 20 s caused the accumulation of $^{\bullet}O_2{}^-$ additionally in seed coats and cotyledons that the authors attributed to the increased oxidative stress. A similar conclusion has been also obtained in the study of Švubová et al. [59] on 3-day-old soybean seedlings, where exposure times exceeding 60 s induced a severe oxidative stress proven by inefficient decomposing of oxygen radicals.

2.7. Activity of Dehydrogenases

The anaerobic environment during the grain/seed imbibition requires an alternative way to metabolise pyruvate. In this weakly efficient pathway comprising conversion of pyruvate to ethanol, alcohol dehydrogenase (ADH) plays a crucial role. When germination starts, aerobic environment progressively takes control over anaerobic; thus, succinate dehydrogenase (SDH), an important enzyme of the Krebs cycle, becomes dominant in the metabolism of pyruvate [60,61]. Inhibition of SDH leads to retarded germination, blocked hypocotyl elongation and light-dependent seedling establishment as proven by Restovic et al. [62] after using SDH inhibitor (thenoyltrifluoroacetone). Our results indicate that using PAW in barley (Figure 8a) postponed the transition from anaerobic to aerobic conditions, as indicated by high ADH activity (increased ethanol concentration) in all PAW variants. This correlates with decreased amylase activity (Figure 5c) and thus inefficient energy metabolism. The SDH activity in barley, as well as in pea, stayed unchanged (Figure 8c,d). On the other hand, we observed a decrease in ADH activity in pea in all PAW variants and P-N, which refers to an accelerated transition to aerobic metabolism (Figure 8b). This correlates, similar to barley, with the amylase activity that reached its peak in the pea P-GD2 variant (Figure 5d). The studies of Švubová et al. [20,59] showed similar results corresponding with our hypothesis of a postponed turn to aerobic metabolism in the case of barley. In these publications dealing with pea and soybean seedlings, the authors showed a negative impact of high exposure times to direct NTP treatment (with respect to other analyses), resulting in an increase of ADH activity and thus in the suffocation of seeds. Moreover, they documented a positive effect for shorter exposure times that showed an increase in SDH activity simultaneously with a decrease in ADH activity.

2.8. DNA Damage

DNA damage in barley and pea seedlings was analysed using an alkaline comet assay (Figure 9). In pea seedlings, a slight increase in DNA damage was observed in all PAW-treated samples. However, only an increase in the P-TS variant was statistically significant. The highest DNA damage detected in PAW-treated samples was on the level of 17.4% (P-TS variant); however, it could be repaired during plant growth. Damages detected by comet assay are primary and may not have a negative impact on the plant life and plant growth parameters, such as germination (Figure 2) or seedling length (Figure 3). The effect of direct plasma treatment on DNA damage in pea seedlings was analysed in several studies. Švubová et al. [20,49] compared the effect of plasma generated in oxygen, nitrogen or ambient air on pea DNA using the same comet assay. In all plasma-treated samples, an increase in DNA damage was observed in all exposure times (15–300 s). However, none of these damages were caused by double-strand breaks, as was proven by constant field gel electrophoresis. No negative effects of plasma treatment on pea seedlings were observed in Kyzek et al. [63]. In this study, only a slight increase in DNA damage was observed on plasma-treated samples for 1–5 min. Tomeková et al. [64] tried to find the effect of different mixtures of oxygen and nitrogen on DNA damage in pea seedlings. Their results suggest that nitrogen addition to the working gas increased DNA damage in pea seedlings treated with plasma. However, the lowest DNA damage comparable to control samples was observed in seedlings treated with plasma generated in ambient air. Here in the current study, DNA damage in all PAW-treated samples of barley seedlings and B-N variant was statistically significant compared to control samples (B–C).

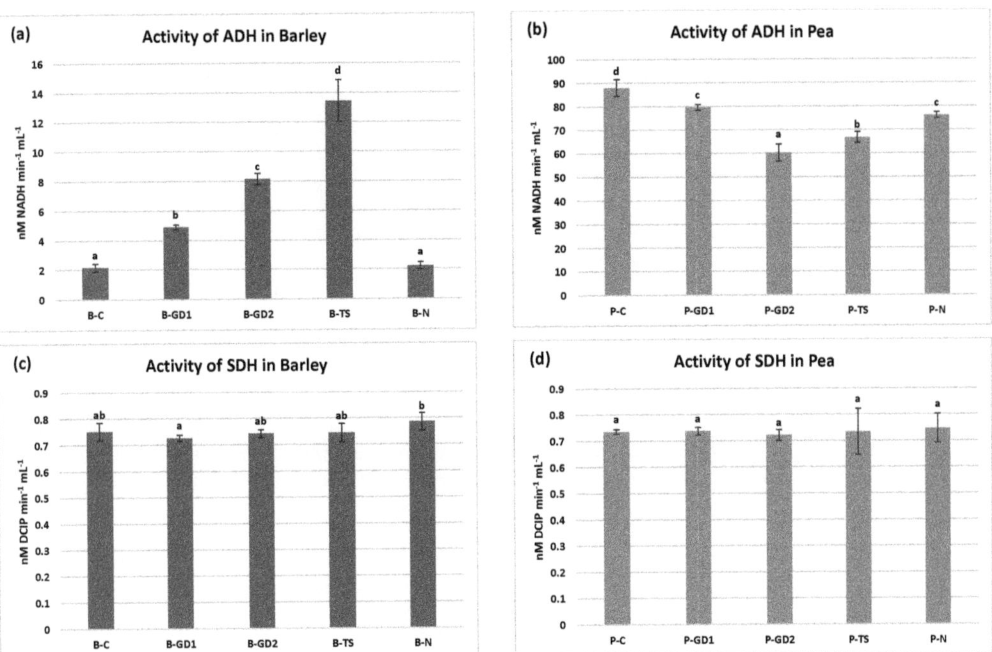

Figure 8. The activity of alcohol (ADH) and succinate (SDH) dehydrogenases in the barley (**a,c**) and pea (**b,d**) seedlings after 3 days of growth, watered with tap water (control: B/P-C), plasma-activated water (PAW) of glow discharge with water cathode and treatment time 1 and 2 min (B/P-GD1 and B/P-GD2), PAW of transient spark discharge with electrospray (B/P-TS) and 2-mM nitric acid (B/P-N). Values are shown as mean ± SD from three experimental rounds (one run represents five seedlings per variant; five 0.1 g mixed samples were analysed per one experimental run and each variant for SDH activity; five 0.05 g mixed samples were analysed per one experimental run and each variant for ADH activity). Different letters represent statistically significant difference at $p < 0.05$ according to LSD test.

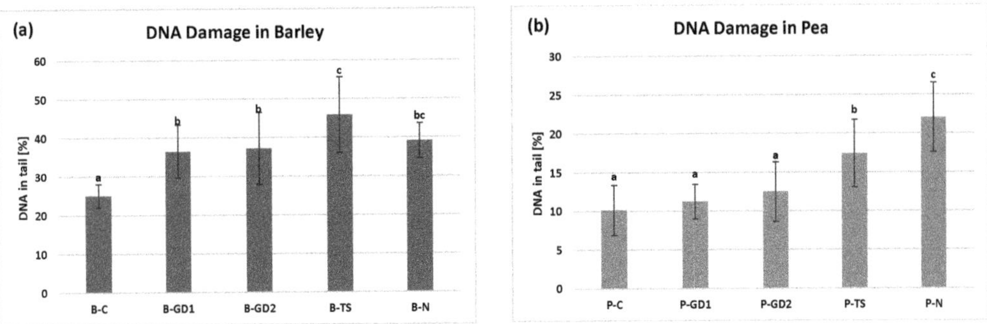

Figure 9. The DNA damage in the barley (**a**) and pea (**b**) seedlings after 3 days of growth, watered with tap water (control: B/P-C), plasma-activated water (PAW) of glow discharge with water cathode and treatment time 1 and 2 min (B/P-GD1 and B/P-GD2), PAW of transient spark discharge with electrospray (B/P-TS) and 2-mM nitric acid (B/P-N). Values are shown as mean ± SD from three experimental rounds. Different letters represent statistically significant difference at $p < 0.05$ according to LSD test.

Higher concentrations of hydrogen peroxide, nitrates and nitrites (Figure 1) could be responsible for the observed damages due to the induction of oxidative stress. All of these three RONS could generate hydroxyl radicals that could easily interact with DNA and cause its damage [65–67]. These damages could be repaired, however, only in the initial days of cultivation and could result in lower germination (Figure 2) or root length (Figure 3). Yemeli et al. [28] also analysed DNA damage in barley seedlings after 1-, 2- or 3-week treatment with water treated by the same sources as were used in this study and observed no harmful effects of PAW on barley DNA. However, in this study, younger plants (3-day-old seedlings) were used for experiments, and this could be the reason why the DNA damage was higher (also in control samples). DNA damage in 7-day-old barley seedlings after direct plasma treatment was analysed in the study of Peťková et al. [51]. A statistically significant increase in DNA damage was observed in all samples treated with plasma generated in nitrogen, oxygen or ambient air for 10–60 s. These damages were caused mainly by single-strand breaks and purine oxidation.

3. Materials and Methods

3.1. Plant Material

Dried barley (*Hordeum vulgare* L.) cv. Kangoo grains and pea (*Pisum sativum* L.) cv. Eso seeds, used as target material in the experiments, were obtained from the Slovenské farmárske družstvo, Slovakia. The pea seeds and barley grains were stored in a fridge at 8 °C in the dark. The experiments were performed in 2020 and 2021.

3.2. Experimental Setup, Plasma-Activated Water Production

Figures 10 and 11 show, respectively, the representation of the experimental setup for plasma water treatments used in this work and the typical waveforms of the applied discharge voltage and current. It also shows the photos of the plasma discharges taken during the activation of water. We used a transient spark (TS) discharge with water electrospray (Figure 10a) described in more detail in our previous articles [28,38,68] and glow discharge (GD) with water cathode (Figure 10b), described in detail in [28,69]. The TS and GD setups contain a high voltage (HV) DC power supply with the following parameters: U_{max} = 20 kV, I_{max} = 30 mA, P_{max} = 600 W. A positive HV is applied directly through the ballast resistors (R = 8.8 MΩ and R = 0.5 MΩ for TS and GD, respectively) on the HV electrode (4 needles for TS and 1 needle for GD, as shown in the setup). The HV probe Tektronix P6015A is used for both setups to measure the voltage. The discharge current is measured for TS plasma by a Rogowski current monitor Pearson electronics 2877 (1 V/1 A) and for GD by the ammeter. The time evolution of electrical parameters of the discharges (voltage and current) is recorded and processed by the digitising oscilloscope Tektronix TDS 2024 (parameters 200 MHz; 2.5 Gs/s; 4 channels).

In this work, tap water was used to produce PAW due to its availability. The discharges are generated in ambient air at atmospheric pressure. For TS, four HV hollow needle electrodes inject the tap water into the active zone of the TS discharge at a constant flow rate 0.5 mL/min per needle by the syringe pump. The treatment time for TS was 25 min to produce 50 mL of PAW used per one seed group in this work. The electrosprayed PAW is collected under the metallic mesh in a Petri dish (see setup). The TS discharge device is held in a Faraday cage due to the electromagnetic radiation.

The generation of PAW by GD was done at a constant activation time of 1 and 2 min. 12 mL of tap water are introduced in a Petri dish containing a metallic wire, and the discharge is initiated between the HV needle electrode and the surface of the water. The water activated by both plasma sources is collected and then subjected to chemical analysis before being used for seed experiments.

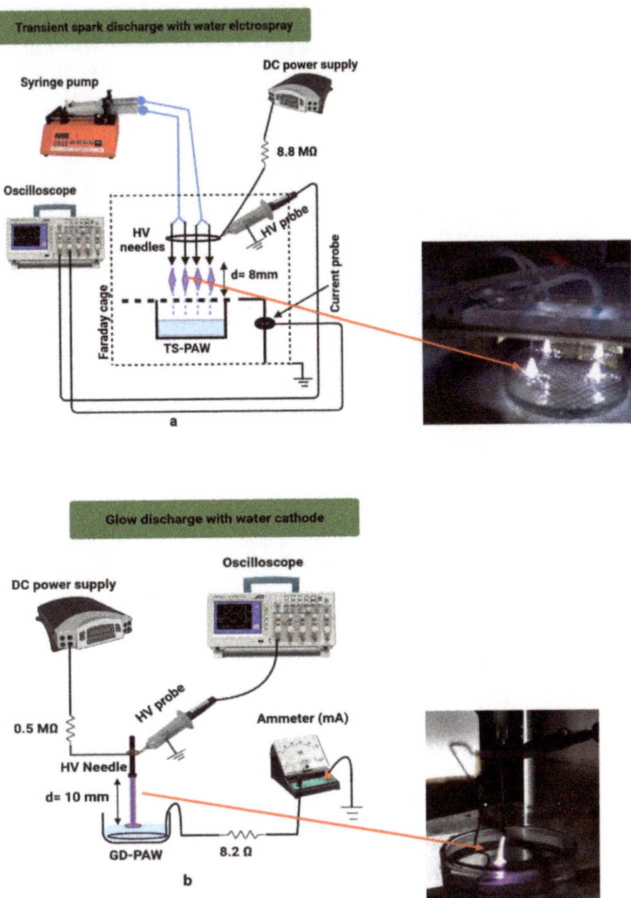

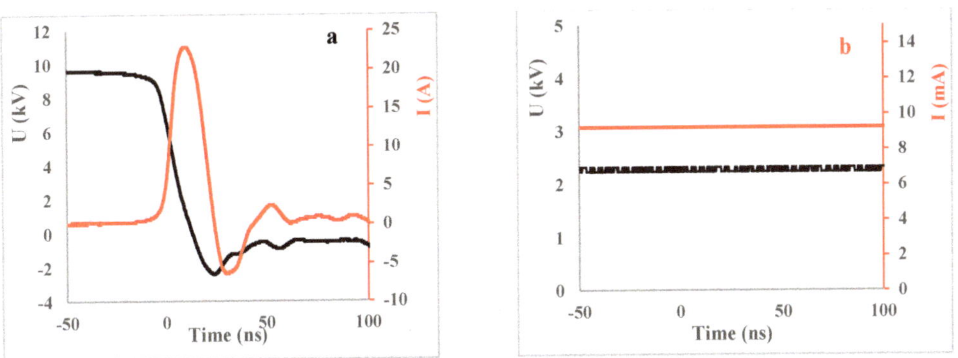

Figure 10. (**a**,**b**) Schematic diagrams of the experimental setups of TS and GD used in this work to generate PAW with photographs of the TS and GD discharges on the right side.

Figure 11. Typical waveforms of the applied discharge voltage and current of (**a**) TS with water electrospray and (**b**) GD with water cathode [28].

3.3. Measurement of Nitrites/Nitrates and Hydrogen Peroxide in PAW

The PAW generated were analysed by measuring the pH and the long-life reactive species (hydrogen peroxide, nitrites/nitrates). The RONS detection was performed by UV-Vis absorption spectroscopy colorimetric methods (Shimadzu UV-1800 Spectrophotometer) [70]. The measurement of hydrogen peroxide concentration in PAW was performed by the titanium oxysulfate assay. This colorimetric method is based on the reaction of H_2O_2 with the titanyl (IV) ions in strong acidic conditions. The maximum absorption of the yellow-coloured reaction product is at 407 nm. The concentration of NO_2^- and NO_3^- in the PAW was determined by colorimetric assay kit of Griess reagents. This colorimetric method is based on the reaction of nitrites with the Griess reagents, which, after reaction, form a pink-coloured azo product. Nitrates are converted to nitrites by nitrate reductase (with the help of coenzyme) and afterward are analysed the same way as nitrites. Both measurements are done at the maximum absorption at 540 nm. The detailed procedure of the RONS measurement can be found in our previous works [39,40,70]. The following figure shows the representation of the PAW generation and analysis.

3.4. Imbibition, Germination and Growth Conditions

Dry pea seeds (50 seeds for each variant) were soaked in 50 mL of tap water (P-C: control) or in PAW (P-TS, P-GD1 and P-GD2 variants) as well as in 2-mM nitric acid (P-N variant) for 1 h at room temperature. The 2 mM HNO_3 tap water solution was selected to mimic the average nitrate concentration measured in the three types of PAW and represents a positive control providing nitrogen. Imbibed seeds were wrapped in wet sterile filter paper.

Dry barley grains (50 grains for each variant) were sown on Petri dishes (ø 21.5 cm) and watered with 50 mL of tap water (B–C: control) or with PAW (P-TS, P-GD1 and P-GD2 variants) as well as 2-mM nitric acid (B-N variant). Rolls and Petri dishes were cultivated in dark conditions in an incubator at the temperature $24 \pm 2\ ^\circ C$ for 3 days and fresh tap water or PAW were added each day.

During cultivation, the number of germinated pea seeds/barley grains were counted, and after three days, the material for biochemical analyses was collected (total soluble proteins content, assay on lytic enzymes assessment, assays on antioxidant enzymes and visualisation of ROS, assays on dehydrogenase activities evaluation, comet assay). The length of seedlings (for pea) and length of shoots and roots (for barley) were measured. Percentage of germination was calculated according to Abdul-Baki [71].

3.5. Total Soluble Proteins Content Measurement

Samples (~1.5 g) were ground in liquid nitrogen with mortar and pestle and suspended in 50 mM Na-Phosphate protein extraction buffer with 1 mM EDTA, pH 7.8. After 15 min centrifugation ($12,000\times g$), the supernatant was used for determination of protein concentration according to Bradford [72]. Total soluble proteins content was calculated as the amount of total proteins per gram of fresh matter from the calibration curve. Bovine Serum Albumin (BSA) was used as a protein standard.

3.6. Assay on Lytic Enzymes Assessment

Changes in activity of protease in 3-day-old barley and pea seedlings were determined by incubating 150 µL of an extracted protein sample with 150 µL of 2% (w/v) BSA in 200 mM glycine-HCl (pH 3.0) at 37 °C for 1 h. The reaction was stopped by the addition of 450 µL of 5% (w/v) trichloroacetic acid. Samples were incubated on ice for 10 min and centrifuged at $20,000\times g$ for 10 min at 4 °C. Absorbance of the supernatant at 280 nm was measured by a spectrophotometer by Jenway 6705 UV/Vis (Bibby Scientific Ltd., Essex, UK) [73].

For determination of α-amylase activities, a commercially-available colorimetric assay kit purchased from Sigma-Aldrich Co. was used. One unit of α-amylase activity is the amount of amylase that cleaves ethylidene-pNP-G7 to generate 1.0 mM of p-nitrophenol per minute at 25 °C.

3.7. Assays on Antioxidant Enzymes, Superoxide Dismutase (SOD), Guaiacol Peroxidase (POX) and Catalase (CAT) Activities Assessment and Visualization of ROS (H_2O_2 and $\cdot O_2^-$)

The activities of enzymes that detoxify $\cdot O_2^-$ (SOD, E.C.1.15.1.1) and H_2O_2 (POX, E.C.1.11.1.7) were tested. The activity of superoxide dismutase was established according to Beauchamp and Fridovich [74], the guaiacol peroxidase according to Frič and Fuchs [75] and catalase according to Hodges et al. [76]. One unit of SOD activity is the amount of proteins required to inhibit 50% of initial reduction of Nitrotetrazolium Blue Chloride (NBT) under the light. Guaiacol peroxidase activity is expressed in µM of tetraguaiacol $min^{-1} \cdot mg^{-1}$ by molar extinction coefficient of tetraguaiacol 26.6. The specific activity of CAT (E.C. 1.11.1.6) was calculated according to Claiborne [77]. Chemicals were purchased from Sigma-Aldrich Co. The presence of H_2O_2 and $\cdot O_2^-$ was detected in 3-day-old barley and pea seedlings according to Kumar et al. [78]. Hydrogen peroxide was visualised as reddish-brown stain formed by the reaction of 3,3'-Diaminobenzidine (DAB) with the endogenous H_2O_2. The content of $\cdot O_2^-$ was detected as a dark blue stain of formazan compound, formed as a result of NBT reacting with the endogenous $\cdot O_2^-$.

3.8. Assays on Dehydrogenase Activities Evaluation

For determination of alcohol (ADH) and succinate (SDH) dehydrogenases in 3-day-old seedlings, commercially-available colorimetric assay kits purchased from Sigma-Aldrich Co. were used. Activities of enzymes were determined according to manufacturer's instructions. One unit of ADH represents the amount of enzyme that will generate 1.0 mM of NADH per minute at pH 8.0 at 37 °C. One unit of SDH is the amount of enzyme that generates 1.0 µM of 2,6-dichlorophenolindophenol (DCIP) per minute at pH 7.2 at 25 °C.

3.9. Comet Assay

Alkaline comet assay is a method used for measuring of DNA damage (single-strand breaks, double-strand breaks, cross-links, apyrimidine and apurine sites) in eukaryotic cells [79,80]. Our experiments were performed according to [20,81]. Briefly, two roots (in the case of pea seedlings) or leaves (in the case of barley seedlings) for each sample were cut by a razor blade in the dark and on ice ensuring DNA release in the 175 µL of 0.4 M Tris-HCl buffer solution (pH 7.5) (Sigma-Aldrich Co., St. Louis, MO, USA). After that, 100 µL of the DNA and buffer suspension was mixed with 100 µL of 1% low melting point agarose (Roth, Karlsruhe, Germany). The final solution was placed on a slide covered by 1% normal melting point agarose (Roth) and then covered by a coverslip. The coverslips were removed after 10 min and the slides were placed in the electrophoretic chamber filled with cold electrophoretic buffer solution containing 1 mM EDTA (Sigma-Aldrich Co.) and 300 mM NaOH (Centralchem, Bratislava, Slovakia) for 8 min. After that, electrophoresis was launched on 1.25 Vcm^{-1} for 15 min at 4 °C. Slides were then neutralised three times with 0.4 M Tris-HCl buffer solution (pH 7.5) and stained with fluorescent dye ethidium bromide (0.05 mM, 80 µL for each slide, Serva, Heidelberg, Germany) for 5 min. DNA damage was observed using fluorescent microscope OLYMPUS BX 51 with green excitation filter UMWIG3 under 400× magnification and evaluated by Comet visual software. Plasma-untreated seedlings were used as negative controls (P-C, B-C).

3.10. The Statistical Analysis

The data were analysed using Microsoft Excel (Microsoft Office 365) and Statgraphics Centurion 19 (Statgraphics Technologies, Inc., The Plains, VA, USA). Treatment effects were investigated by means of ANOVA single-step multiple comparisons of means by means of LSD tests, and comparisons between the mean values were considered significant at $p < 0.05$. All experimental data in this work are from at least three independent experiments.

4. Conclusions

As shown by many other studies, non-thermal plasma can in certain doses and application methods stimulate plant growth and physiological parameters, besides other positive effects that have been already demonstrated in various life science fields. This study investigated the indirect plasma effect of three types of plasma-activated water (PAW) generated by two different plasma discharges (transient spark with water electrospray and glow discharge with water cathode operated for 1 and 2 min) on 3-day-old seedlings of two important farm plants, barley (*Hordeum vulgare* L. cv. Kangoo) and pea (*Pisum sativum* L. cv. Eso), as a potential alternative to commercial fertilizers supplying plants with nitrogen. Table 1 schematically summarises the obtained results.

Table 1. An overview of PAW effects on barley (B-) and pea (P-) growth parameters, activity of enzymes and DNA damage. Horizontal arrow represents values without significant change when compared to control plants, upward arrow represents values with significant increase when compared to control plants and downward arrow represent values with significant decrease when compared to control plants.

	B-C	B-GD1	B-GD2	B-TS	B-N	P-C	P-GD1	P-GD2	P-TS	P-N
Germination	→	→	→	→	→	→	→	→	→	→
Root Length	→	↓	↓	↓	↓					
Shoot Length	→	↓	↓	→	→					
Seedling Length						→	→	↓	→	→
TSP	→	→	→	→	→	→	→	→	→	→
Protease	→	→	→	↓	→	→	→	→	→	→
Amylase	→	↓	↓	↓	↓	→	→	↑	↑	↓
SOD	→	→	↓	→	→	→	→	↑	→	↑
G-POX	→	↑	↑	↑	↑	→	↑	→	→	→
CAT	→	↓	↓	↓	↓	→	→	↑	↑	↑
ADH	→	↑	↑	↑	→	→	↓	↓	↓	↓
SDH	→	→	→	→	→	→	→	→	→	→
DNA Damage	→	↑	↑	↑	↑	→	→	→	↑	↑

PAW applied on pea seeds had a stimulative effect in multiple plant growth and physiological parameters with respect to the negative control (tap water) and positive control (chemically added nitrate in N-variants). Using different types of PAW on pea seedlings exhibits a positive effect on amylase activity and has no inhibition effect on seed germination, seedling length, total protein concentration or protease activity. Moreover, PAW caused no or only moderate oxidative stress, which was in most cases effectively alleviated by natural plant antioxidant enzymes (SOD, G-POX, CAT) and resulted in a very low DNA damage in PAW-treated samples. This effect was also proven by in situ visualisation of H_2O_2 and $^{\cdot}O_2{}^{-}$. In pea seedlings, we observed a faster turn-over from anaerobic metabolism (related to imbibition) to aerobic metabolism, proven by inhibition of alcohol dehydrogenase (ADH) activity. Screening among all variants, the most perspective PAW seems to be the one prepared by the glow discharge at 1 min exposure (variant P-GD1). This PAW contains the lowest concentration of $NO_2{}^{-}$ and $NO_3{}^{-}$ and an intermediate concentration of H_2O_2. With respect to the findings raised from our study and other available ones, we assume that this ratio of RONS could be responsible for effective intracellular signalling, speeding up the transition from anaerobic to aerobic metabolism (proven by inhibition of ADH activity).

Interpreting the results for barley imbibed in three types of PAW was more complex. PAW had no effect on grain germination, total soluble protein and SDH activity when compared to negative as well as positive control. However, after using PAW, we observed the high level of DNA damage together with reduced root and shoot length and decreased amylase activity. These negative effects were attributed to the oxidative stress caused by PAW, which was also exhibited by the enhanced activity of G-POX or ADH related to grain suffocation. Based on other related studies, we can conclude that barley either reacts differently with a delayed positive effect of PAW treatment (that we were not able to record in the early stages of its growth), or this application method combined with the timing does not represent the best way to improve its growth and physiological parameters.

In summary, the use of plasma pre-sowing technologies such as seed/grain imbibition in plasma-activated water seems to be important in the faster recovery of the metabolic activity in grains/seeds (activation of lytic and antioxidant enzymes), if the suitable RONS composition in the PAW is used. In the light of other available studies, it could lead to the faster growth and development of young seedlings and the increase of yield without using chemical fertilizers. Therefore, plasma-activated water exhibits indubitable potential in sustainable and environmentally-friendly agriculture. From the results of this study, we can conclude that concentrations of RONS in GD1 PAW, suitable for faster transition to aerobic metabolism in pea, may not be suitable for another plant species, such as barley, and further investigation needs to be done to answer arising questions on the mechanisms of plant responses to the NTP or PAW treatments.

Author Contributions: Conceptualisation, D.K., R.Š. and Z.M.; methodology, D.K., R.Š., G.B.N.Y. and S.K.; validation, D.K., R.Š. and S.K.; resources, D.K., G.B.N.Y. and S.K.; data curation, D.K., S.K. and G.B.N.Y.; writing—original draft preparation, D.K.; writing—review and editing, G.B.N.Y., S.K. and Z.M.; supervision, R.Š. and Z.M.; project administration and funding acquisition, Z.M. All authors have read and agreed to the published version of the manuscript.

Funding: This work was supported by the Slovak Research and Development Agency APVV-17-0382.

Conflicts of Interest: The authors declare no conflict of interest.

References

1. Green, H.; Broun, P.; Cakmak, I.; Condon, L.; Fedoroff, N.; Gonzalez-Valero, J.; Graham, I.; Lewis, J.; Moloney, M.; Oniang'o, R.K.; et al. Planting seeds for the future of food. *J. Sci. Food Agric.* **2016**, *96*, 1409–1414. [CrossRef]
2. Masclaux-Daubresse, C.; Daniel-Vedele, F.; Dechorgnat, J.; Chardon, F.; Gaufichon, L.; Suzuki, A. Nitrogen uptake, assimilation and remobilization in plants: Challenges for sustainable and productive agriculture. *Ann. Bot.* **2010**, *105*, 1141–1157. [CrossRef]
3. Sharma, A. A Review on the Effect of Organic and Chemical Fertilizers on Plants. *Int. J. Res. Appl. Sci. Eng. Technol.* **2017**, *5*, 677–680. [CrossRef]
4. Aktar, M.W.; Sengupta, D.; Chowdhury, A. Impact of pesticides use in agriculture: Their benefits and hazards. *Interdiscip. Toxicol.* **2009**, *2*, 1–12. [CrossRef]
5. Adegoke, A.A.; Awolusi, O.O.; Stenström, T.A. Organic Fertilizers: Public Health Intricacies. In *Organic Fertilizers—From Basic Concepts to Applied Outcomes*; Larramendy, M.L., Soloneski, S., Eds.; InTech: London, UK, 2016; pp. 343–374.
6. Shridhar, B.S. Review: Nitrogen Fixing Microorganisms. *Int. J. Microbiol. Res.* **2012**, *3*, 46–52. [CrossRef]
7. Reina, F.G.; Pascual, L.A.; Fundora, I.A. Influence of a stationary magnetic field on water relations in lettuce seeds. Part II: Experimental results. *Bioelectromagnetics* **2001**, *22*, 596–602. [CrossRef]
8. Sonoda, T.; Takamura, N.; Wang, D.; Namihira, T.; Akiyama, H. Growth Control of Leaf Lettuce Using Pulsed Electric Field. *IEEE Trans. Plasma Sci.* **2014**, *42*, 3202–3208. [CrossRef]
9. Crookes, W. Experiments on the dark space in vacuum tubes. *Proc. R. Soc. Lond. Ser. A* **1907**, *79*, 98–117. [CrossRef]
10. Chen, F.F. Plasma Applications. In *Introduction to Plasma Physics and Controlled Fusion*; Chen, F.F., Ed.; Springer: Cham, Switzerland, 2016; pp. 333–353.
11. Starič, P.; Vogel-Mikuš, K.; Mozetič, M.; Junkar, I. Effects of Nonthermal Plasma on Morphology, Genetics and Physiology of Seeds: A Review. *Plants* **2020**, *9*, 1736. [CrossRef]
12. Tendero, C.; Tixier, C.; Tristant, P.; Desmaison, J.; Leprince, P. Atmospheric pressure plasmas: A review. *Spectrochim. Acta Part B At. Spectrosc.* **2006**, *61*, 2–30. [CrossRef]
13. Randeniya, L.K.; De Groot, G.J.J.B. Non-Thermal Plasma Treatment of Agricultural Seeds for Stimulation of Germination, Removal of Surface Contamination and Other Benefits: A Review. *Plasma Process. Polym.* **2015**, *12*, 608–623. [CrossRef]
14. Moreau, M.; Orange, N.; Feuilloley, M.G.J. Non-thermal plasma technologies: New tools for bio-decontamination. *Biotechnol. Adv.* **2008**, *26*, 610–617. [CrossRef]
15. Zhou, Z.; Huang, Y.; Yang, S.; Chen, W. Introduction of a new atmospheric pressure plasma device and application on tomato seeds. *Agric. Sci.* **2011**, *2*, 23–27. [CrossRef]
16. Henselová, M.; Slováková, L.; Martinka, M.; Zahoranová, A. Growth, anatomy and enzyme activity changes in maize roots induced by treatment of seeds with low-temperature plasma. *Biologia* **2012**, *67*, 490–497. [CrossRef]
17. Zahoranová, A.; Henselová, M.; Hudecová, D.; Kaliňáková, B.; Kováčik, D.; Medvecká, V.; Černák, M. Effect of Cold Atmospheric Pressure Plasma on the Wheat Seedlings Vigor and on the Inactivation of Microorganisms on the Seeds Surface. *Plasma Chem. Plasma Process.* **2016**, *36*, 397–414. [CrossRef]
18. Stolárik, T.; Henselová, M.; Martinka, M.; Novák, O.; Zahoranová, A.; Černák, M. Effect of Low-Temperature Plasma on the Structure of Seeds, Growth and Metabolism of Endogenous Phytohormones in Pea (*Pisum sativum* L.). *Plasma Chem. Plasma Process.* **2015**, *35*, 659–676. [CrossRef]

19. Štěpánová, V.; Slavíček, P.; Kelar, J.; Prášil, J.; Smékal, M.; Stupavská, M.; Jurmanová, J.; Černák, M. Atmospheric pressure plasma treatment of agricultural seeds of cucumber (*Cucumis sativus* L.) and pepper (*Capsicum annuum* L.) with effect on reduction of diseases and germination improvement. *Plasma Process. Polym.* **2018**, *15*, 1700076. [CrossRef]

20. Švubová, R.; Kyzek, S.; Medvecká, V.; Slováková, Ľ.; Gálová, E.; Zahoranová, A. Novel insight at the Effect of Cold Atmospheric Pressure Plasma on the Activity of Enzymes Essential for the Germination of Pea (*Pisum sativum* L. cv. Prophet) Seeds. *Plasma Chem. Plasma Process.* **2020**, *40*, 1221–1240. [CrossRef]

21. Kučerová, K.; Henselová, M.; Slováková, L.; Bačovčinová, M.; Hensel, K. Effect of Plasma Activated Water, Hydrogen Peroxide, and Nitrates on Lettuce Growth and Its Physiological Parameters. *Appl. Sci.* **2021**, *11*, 1985. [CrossRef]

22. Terebun, P.; Kwiatkowski, M.; Starek, A.; Reuter, S.; YS, M.; Pawlat, J. Impact of Short Time Atmospheric Plasma Treatment on Onion Seeds. *Plasma Chem. Plasma Process.* **2021**, *41*, 559–571. [CrossRef]

23. Pawlat, J.; Starek, A.; Sujak, A.; Kwiatkowski, M.; Terebun, P.; Budzeń, M. Effects of atmospheric pressure plasma generated in GlidArc reactor on *Lavatera thuringiaca* L. seeds' germination. *Plasma Process. Polym.* **2018**, *15*, 1700064. [CrossRef]

24. Maniruzzaman, M.; Sinclair, A.J.; Cahill, D.M.; Wang, X.; Dai, X.J. Nitrate and Hydrogen Peroxide Generated in Water by Electrical Discharges Stimulate Wheat Seedling Growth. *Plasma Chem. Plasma Process.* **2017**, *37*, 1393–1404. [CrossRef]

25. Zhou, R.; Li, J.; Zhou, R.; Zhang, X.; Yang, S. Atmospheric-pressure plasma treated water for seed germination and seedling growth of mung bean and its sterilization effect on mung bean sprouts. *Innov. Food Sci. Emerg. Technol.* **2019**, *53*, 36–44. [CrossRef]

26. Kučerová, K.; Henselová, M.; Slováková, Ľ.; Hensel, K. Effects of plasma activated water on wheat: Germination, growth parameters, photosynthetic pigments, soluble protein content, and antioxidant enzymes activity. *Plasma Process. Polym.* **2019**, *16*, 1800131. [CrossRef]

27. Zhao, Y.; Patange, A.; Sun, D.; Tiwari, B. Plasma-activated water: Physicochemical properties, microbial inactivation mechanisms, factors influencing antimicrobial effectiveness, and applications in the food industry. *Compr. Rev. Food Sci. Food Saf.* **2020**, *19*, 3951–3979. [CrossRef]

28. Ndiffo Yemeli, G.B.; Švubová, R.; Kostolani, D.; Kyzek, S.; Machala, Z. The effect of water activated by nonthermal air plasma on the growth of farm plants: Case of maize and barley. *Plasma Process. Polym.* **2021**, *18*, 2000205. [CrossRef]

29. Foster, J.E.; Kovach, Y.E.; Lai, J.; Garcia, M.C. Self-organization in 1 atm {DC} glows with liquid anodes: Current understanding and potential applications. *Plasma Sources Sci. Technol.* **2020**, *29*, 34004. [CrossRef]

30. Adamovich, I.; Baalrud, S.D.; Bogaerts, A.; Bruggeman, P.J.; Cappelli, M.; Colombo, V.; Czarnetzki, U.; Ebert, U.; Eden, J.G.; Favia, P.; et al. The 2017 Plasma Roadmap: Low temperature plasma science and technology. *J. Phys. D Appl. Phys.* **2017**, *50*, 323001. [CrossRef]

31. Vanraes, P.; Bogaerts, A. Plasma physics of liquids—A focused review. *Appl. Phys. Rev.* **2018**, *5*, 31103. [CrossRef]

32. Zhou, R.; Zhou, R.; Wang, P.; Xian, Y.; Mai-Prochnow, A.; Lu, X.; Cullen, P.J.; Ostrikov, K.; Bazaka, K. Plasma-activated water: Generation, origin of reactive species and biological applications. *J. Phys. D Appl. Phys.* **2020**, *53*, 303001. [CrossRef]

33. Bienert, G.P.; Schjoerring, J.K.; Jahn, T.P. Membrane transport of hydrogen peroxide. *Biochim. Biophys. Acta Biomembr.* **2006**, *1758*, 994–1003. [CrossRef] [PubMed]

34. Goyal, S.S.; Huffaker, R.C. The uptake of NO3-, NO2-, and NH4+ by intact wheat (*Triticum aestivum*) seedlings. I. Induction and kinetics of transport systems. *Plant Physiol.* **1986**, *82*, 1051–1056. [CrossRef] [PubMed]

35. Kotur, Z.; Siddiqi, Y.M.; Glass, A.D.M. Characterization of nitrite uptake in *Arabidopsis thaliana*: Evidence for a nitrite-specific transporter. *New Phytol.* **2013**, *200*, 201–210. [CrossRef]

36. Maniruzzaman, M. Investigation of plasma-treated water for plant growth. Ph.D. Thesis, Deakin University, Melbourne, Australia, 2018.

37. Gierczik, K.; Vukušić, T.; Kovács, L.; Székely, A.; Szalai, G.; Milošević, S.; Kocsy, G.; Kutasi, K.; Galiba, G. Plasma-activated water to improve the stress tolerance of barley. *Plasma Process. Polym.* **2020**, *17*, 1900123. [CrossRef]

38. Machala, Z.; Tarabova, B.; Hensel, K.; Spetlikova, E.; Sikurova, L.; Lukes, P. Formation of ROS and RNS in Water Electro-Sprayed through Transient Spark Discharge in Air and their Bactericidal Effects. *Plasma Process. Polym.* **2013**, *10*, 649–659. [CrossRef]

39. Kučerová, K.; Machala, Z.; Hensel, K. Transient Spark Discharge Generated in Various N2/O2 Gas Mixtures: Reactive Species in the Gas and Water and Their Antibacterial Effects. *Plasma Chem. Plasma Process.* **2020**, *40*, 749–773. [CrossRef]

40. Machala, Z.; Tarabová, B.; Sersenová, D.; Janda, M.; Hensel, K. Chemical and antibacterial effects of plasma activated water: Correlation with gaseous and aqueous reactive oxygen and nitrogen species, plasma sources and air flow conditions. *J. Phys. D Appl. Phys.* **2019**, *52*, 34002. [CrossRef]

41. Jamróz, P.; Gręda, K.; Pohl, P.; Żyrnicki, W. Atmospheric Pressure Glow Discharges Generated in Contact with Flowing Liquid Cathode: Production of Active Species and Application in Wastewater Purification Processes. *Plasma Chem. Plasma Process.* **2014**, *34*, 25–37. [CrossRef]

42. Zhang, S.; Rousseau, A.; Dufour, T. Promoting lentil germination and stem growth by plasma activated tap water, demineralized water and liquid fertilizer. *RSC Adv.* **2017**, *7*, 31244–31251. [CrossRef]

43. Ogawa, K.; Iwabuchi, M. A mechanism for promoting the germination of *Zinnia elegans* seeds by hydrogen peroxide. *Plant Cell Physiol.* **2001**, *42*, 286–291. [CrossRef]

44. Barba-Espin, G.; Diaz-Vivancos, P.; Clemente-Moreno, M.J.; Albacete, A.; Faize, L.; Faize, M.; Pérez-Alfocea, F.; Hernández, J.A. Interaction between hydrogen peroxide and plant hormones during germination and the early growth of pea seedlings. *Plant Cell Environ.* **2010**, *33*, 981–994. [CrossRef]

45. Alboresi, A.; Gestin, C.; Leydecker, M.-T.; Bedu, M.; Meyer, C.; Truong, H.-N. Nitrate, a signal relieving seed dormancy in Arabidopsis. *Plant Cell Environ.* **2005**, *28*, 500–512. [CrossRef]

46. Lindsay, A.; Byrns, B.; King, W.; Andhvarapou, A.; Fields, J.; Knappe, D.; Fonteno, W.; Shannon, S. Fertilization of Radishes, Tomatoes, and Marigolds Using a Large-Volume Atmospheric Glow Discharge. *Plasma Chem. Plasma Process.* **2014**, *34*, 1271–1290. [CrossRef]

47. Shashikanthalu, P.S.; Ramireddy, L.; Radhakrishnan, M. Stimulation of the germination and seedling growth of *Cuminum cyminum* L. seeds by cold plasma. *J. Appl. Res. Med. Aromat. Plants* **2020**, *18*, 100259. [CrossRef]

48. Feizollahi, E.; Iqdiam, B.; Vasanthan, T.; Thilakarathna, M.S.; Roopesh, M.S. Effects of Atmospheric-Pressure Cold Plasma Treatment on Deoxynivalenol Degradation, Quality Parameters, and Germination of Barley Grains. *Appl. Sci.* **2020**, *10*, 3530. [CrossRef]

49. Švubová, R.; Válková, N.; Bathoova, M.; Kyzek, S.; Gálová, E.; Medvecká, V.; Slováková, L. Enhanced In situ Activity of Peroxidases and Lignification of Root Tissues after Exposure to Non-Thermal Plasma Increases the Resistance of Pea Seedlings. *Plasma Chem. Plasma Process.* **2021**, *41*, 903–922. [CrossRef]

50. Damaris, R.; Lin, Z.; Yang, P.; He, D. The Rice Alpha-Amylase, Conserved Regulator of Seed Maturation and Germination. *Int. J. Mol. Sci.* **2019**, *20*, 450. [CrossRef]

51. Petková, M.; Švubová, R.; Kyzek, S.; Medvecká, V.; Slováková, L.; Ševčovičová, A.; Gálová, E. The Effects of Cold Atmospheric Pressure Plasma on Germination Parameters, Enzyme Activities and Induction of DNA Damage in Barley. *Int. J. Mol. Sci.* **2021**, *22*, 2833. [CrossRef]

52. Sadhu, S.; Thirumdas, R.; Deshmukh, R.R.; Annapure, U.S. Influence of cold plasma on the enzymatic activity in germinating mung beans (*Vigna radiata*). *LWT* **2017**, *78*, 97–104. [CrossRef]

53. Chen, H.H.; Chang, H.C.; Chen, Y.K.; Hung, C.L.; Lin, S.Y.; Chen, Y.S. An improved process for high nutrition of germinated brown rice production: Low-pressure plasma. *Food Chem.* **2016**, *191*, 120–127. [CrossRef]

54. Izmailov, S.F.; Nikitin, A. V Nitrate Signaling in Plants: Mechanisms of Implementation. *Russ. J. Plant Physiol.* **2020**, *67*, 31–44. [CrossRef]

55. Laurie, S. Antisense SNF1-related (SnRK1) protein kinase gene represses transient activity of an alpha-amylase (*alpha-Amy2*) gene promoter in cultured wheat embryos. *J. Exp. Bot.* **2003**, *54*, 739–747. [CrossRef] [PubMed]

56. Perata, P.; Alpi, A.; Loschiavo, F. Influence of Ethanol on Plant Cells and Tissues. *J. Plant Physiol.* **1986**, *126*, 181–188. [CrossRef]

57. De Azevedo Neto, A.D.; Prisco, J.T.; Enéas-Filho, J.; De Abreu, C.E.B.; Gomes-Filho, E. Effect of salt stress on antioxidative enzymes and lipid peroxidation in leaves and roots of salt-tolerant and salt-sensitive maize genotypes. *Environ. Exp. Bot.* **2006**, *56*, 87–94. [CrossRef]

58. Puač, N.; Škoro, N.; Spasić, K.; Živković, S.; Milutinović, M.; Malović, G.; Petrović, Z.L. Activity of catalase enzyme in *Paulownia tomentosa* seeds during the process of germination after treatments with low pressure plasma and plasma activated water. *Plasma Process. Polym.* **2018**, *15*, 1700082. [CrossRef]

59. Švubová, R.; Slováková, Ľ.; Holubová, Ľ.; Rovňanová, D.; Gálová, E.; Tomeková, J. Evaluation of the Impact of Cold Atmospheric Pressure Plasma on Soybean Seed Germination. *Plants* **2021**, *10*, 177. [CrossRef]

60. Tadege, M.; Dupuis, I.I.; Kuhlemeier, C. Ethanolic fermentation: New functions for an old pathway. *Trends Plant Sci.* **1999**, *4*, 320–325. [CrossRef]

61. Schnarrenberger, C.; Martin, W. Evolution of the enzymes of the citric acid cycle and the glyoxylate cycle of higher plants. A case study of endosymbiotic gene transfer. *Eur. J. Biochem.* **2002**, *269*, 868–883. [CrossRef] [PubMed]

62. Restovic, F.; Espinoza-Corral, R.; Gómez, I.; Vicente-Carbajosa, J.; Jordana, X. An active Mitochondrial Complex II Present in Mature Seeds Contains an Embryo-Specific Iron–Sulfur Subunit Regulated by ABA and bZIP53 and Is Involved in Germination and Seedling Establishment. *Front. Plant Sci.* **2017**, *8*, 277. [CrossRef] [PubMed]

63. Kyzek, S.; Holubová, L.; Medvecká, V.; Tomeková, J.; Gálová, E.; Zahoranová, A. Cold Atmospheric Pressure Plasma Can Induce Adaptive Response in Pea Seeds. *Plasma Chem. Plasma Process.* **2019**, *39*, 475–486. [CrossRef]

64. Tomeková, J.; Kyzek, S.; Medvecká, V.; Gálová, E.; Zahoranová, A. Influence of Cold Atmospheric Pressure Plasma on Pea Seeds: DNA Damage of Seedlings and Optical Diagnostics of Plasma. *Plasma Chem. Plasma Process.* **2020**, *40*, 1571–1584. [CrossRef]

65. Henle, E.S.; Linn, S. Formation, prevention, and repair of DNA damage by iron/hydrogen peroxide. *J. Biol. Chem.* **1997**, *272*, 19095–19098. [CrossRef] [PubMed]

66. Ward, J.F.; Evans, J.W.; Limoli, C.L.; Calabro-Jones, P.M. Radiation and hydrogen peroxide induced free radical damage to DNA. *Br. J. Cancer. Suppl.* **1987**, *8*, 105–112.

67. Suzuki, T.; Inukai, M. Effects of nitrite and nitrate on DNA damage induced by ultraviolet light. *Chem. Res. Toxicol.* **2006**, *19*, 457–462. [CrossRef]

68. Machala, Z.; Chládeková, L.; Pelach, M. Plasma agents in bio-decontamination by dc discharges in atmospheric air. *J. Phys. D Appl. Phys.* **2010**, *43*, 222001. [CrossRef]

69. Machala, Z.; Janda, M.; Hensel, K.; Jedlovský, I.; Leštinská, L.; Foltin, V.; Martišovitš, V.; Morvová, M. Emission spectroscopy of atmospheric pressure plasmas for bio-medical and environmental applications. *J. Mol. Spectrosc.* **2007**, *243*, 194–201. [CrossRef]

70. Tarabová, B.; Lukeš, P.; Janda, M.; Hensel, K.; Šikurová, L.; Machala, Z. Specificity of detection methods of nitrites and ozone in aqueous solutions activated by air plasma. *Plasma Process. Polym.* **2018**, *15*, 1800030. [CrossRef]

71. Abdul-Baki, A.A.; Anderson, J.D. Vigor determination in soybean seed by multiple criteria. *Crop Sci.* **1973**, *13*, 630–633. [CrossRef]

72. Bradford, M.M. A rapid and sensitive method for the quantitation of microgram quantities of protein utilizing the principle of protein-dye binding. *Anal. Biochem.* **1976**, *72*, 248–254. [CrossRef]
73. Matušíková, I.; Salaj, J.; Moravčíková, J.; Mlynárová, L.; Nap, J.-P.; Libantová, J. Tentacles of in vitro-grown round-leaf sundew (*Drosera rotundifolia* L.) show induction of chitinase activity upon mimicking the presence of prey. *Planta* **2005**, *222*, 1020–1027. [CrossRef]
74. Beauchamp, C.; Fridovich, I. Superoxide dismutase: Improved assays and an assay applicable to acrylamide gels. *Anal. Biochem.* **1971**, *44*, 276–287. [CrossRef]
75. Frič, F.; Fuchs, W.H. Veränderungen der Aktivität einiger Enzyme im Weizenblatt in Abhängigkeit von der temperaturlabilen Verträglichkeit fur *Puccinia graminis tritici. J. Phytopathol.* **1970**, *67*, 161–174. [CrossRef]
76. Hodges, D.M.; Andrews, C.J.; Johnson, D.A.; Hamilton, R.I. Antioxidant Enzyme and Compound Responses to Chilling Stress and Their Combining Abilities in Differentially Sensitive Maize Hybrids. *Crop Sci.* **1997**, *37*, 857–863. [CrossRef]
77. Claiborne, A. Catalase Activity. In *CRC Handbook of Methods for Oxygen Radical Research*; Greenwald, R.A., Ed.; CRC Press: Boca Raton, FL, USA, 1985; pp. 283–284.
78. Kumar, D.; Yusuf, M.A.; Singh, P.; Sardar, M.; Sarin, N.B. Histochemical Detection of Superoxide and H_2O_2 Accumulation in *Brassica juncea* Seedlings. *Bio-Protocol* **2014**, *4*, 3–6. [CrossRef]
79. Hartmann, A.; Agurell, E.; Beevers, C.; Brendler-Schwaab, S.; Burlinson, B.; Clay, P.; Collins, A.; Smith, A.; Speit, G.; Thybaud, V.; et al. Recommendations for conducting the in vivo alkaline Comet assay. *Mutagenesis* **2003**, *18*, 45–51. [CrossRef] [PubMed]
80. Collins, A.R. The comet assay for DNA damage and repair: Principles, applications, and limitations. *Mol. Biotechnol.* **2004**, *26*, 249–261. [CrossRef]
81. Gichner, T.; Patková, Z.; Száková, J.; Žnidar, I.; Mukherjee, A. DNA damage in potato plants induced by cadmium, ethyl methanesulphonate and γ-rays. *Environ. Exp. Bot.* **2008**, *62*, 113–119. [CrossRef]

Article

Response of Two Different Wheat Varieties to Glow and Afterglow Oxygen Plasma

Pia Starič [1,*], Silva Grobelnik Mlakar [2] and Ita Junkar [1]

[1] Jožef Stefan Institute, Jamova Cesta 39, 1000 Ljubljana, Slovenia; ita.junkar@ijs.si
[2] Faculty of Agriculture and Life Sciences, University of Maribor, Pivola 10, 2311 Hoče, Slovenia; silva.grobelnik@um.si
* Correspondence: pia.staric@ijs.si

Abstract: Cold plasma technology has received significant attention in agriculture due to its effect on the seeds and plants of important cultivars, such as wheat. Due to climate change, wherein increasing temperatures and droughts are frequent, it is important to consider novel approaches to agricultural production. As increased dormancy levels in wheat are correlated with high temperatures and drought, improving the germination and root growth of wheat seeds could offer new possibilities for seed sowing. The main objective of this study was to evaluate the influence of direct (glow) and indirect (afterglow) radio-frequency (RF) oxygen plasma treatments on the germination of two winter wheat varieties: Apache and Bezostaya 1. The influence of plasma treatment on seed surface morphology was studied using scanning electron microscopy, and it was observed that direct plasma treatment resulted in a high etching and nanostructuring of the seed surface. The effect of plasma treatment on germination was evaluated by measuring the germination rate, counting the number of roots and the length of the root system, and the fresh weight of seedlings. The results of this study indicate that the response of seeds to direct and indirect plasma treatment may be variety-dependent, as differences between the two wheat varieties were observed.

Keywords: cold plasma; nonthermal plasma; wheat; plants; afterglow plasma; glow plasma; SEM; roots; vigor; germination

Citation: Starič, P.; Grobelnik Mlakar, S.; Junkar, I. Response of Two Different Wheat Varieties to Glow and Afterglow Oxygen Plasma. *Plants* **2021**, *10*, 1728. https://doi.org/10.3390/plants10081728

Academic Editors: Vida Mildažienė and Božena Šerá

Received: 23 July 2021
Accepted: 18 August 2021
Published: 20 August 2021

Publisher's Note: MDPI stays neutral with regard to jurisdictional claims in published maps and institutional affiliations.

1. Introduction

Cold plasma technology has a wide range of biological applications, of which decontamination, sterilization, improved seed germination, plant growth, and improved resistance to abiotic stress are of high interest for its application in the agricultural industry [1–8]. Plasma agriculture is a relatively new field of research that offers great potential for various agricultural applications. Plasma treatment techniques could offer an alternative way to treat seeds, which could significantly improve germination as well as reduce or even eliminate the use of environmentally unfriendly chemicals. To control plant diseases and overcome unfavorable growth conditions, developing new technologies to improve the resistance of crops to diseases and various abiotic stresses with a minimal negative impact on the environment is essential. Wheat is one of the most predominant crops, historically and presently, and is economically essential worldwide [9]. Numerous research articles have investigated the effects of cold plasma treatment on improved wheat seed germination, growth, changes in metabolism, and stress resistance [7,8,10–14]. However, the results of these studies are not always consistent and many conflicting results can be found in the literature. The main reason for this is not only the use of different seed varieties, but also the use of different plasma systems, which operate under specific conditions that are usually not well-described. Thus, it is very hard to compare results from different studies. In addition, only a limited number of studies dealt with the effects of different plasma species (ions, atoms, excited species, vacuum ultraviolet radiation, etc.) on the seed surface, which may hold the key to understanding the complex interaction mechanisms taking place between the seed surface and plasma. It is usually difficult to eliminate contributions from different plasma species; however, in the afterglow regime, only neutral molecules and atoms are present. Thus, our work aimed to study the effects of different plasma regimes on two different wheat varieties. Two types of seeds were exposed to glow (direct) and afterglow (indirect) plasma to study changes in the seed surface and their influence on germination.

A few authors have investigated how cold plasma treatment affects different seed varieties of various crops. Volin et al. studied the influence of cold plasma treatment on two pea seed varieties. Their studies revealed no significant differences in the final germination rate between both varieties in response to cold plasma treatment [15]. On the other hand, Ling et al. found that the two oilseed rape cultivars responded differently to the plasma seed treatment. One variety was drought-sensitive (Zhongshuang 7), and the other (Zhongshuang 11) was drought-tolerant. The Zhongshuang 7 variety demonstrated a better germination rate after the plasma treatment compared to untreated seeds. On the other hand, Zhongshuang 11 did not show improved seed germination after the plasma treatment compared to the control. However, both varieties showed greater root and shoot length after the plasma treatment of the seeds [5]. Yodpitak et al. conducted a study on six rice cultivars. Their experiments showed an increase in germination rate, root length, and shoot length in all rice varieties after the plasma treatment of the seeds compared to the control, but the degree of improvement was variety-dependent. Scanning electron microscopy (SEM) analysis showed that the plasma treatment of seeds resulted in a significant etching effect of the seed surface. The seed coat appeared smoother after plasma treatment compared to the rough seed coat of untreated seeds [16]. Similar findings of cold plasma effects on seed coat have also been reported by other researchers [4,8,17,18].

Although several authors have conducted extensive research on the cold plasma treatment of seeds, to the best of our knowledge, none of them have investigated how different wheat varieties respond to plasma treatment, or how glow and afterglow plasma treatment affects seed germination and growth. Previously mentioned studies have investigated the variety-dependent changes in other plant species such as rice, oilseed rape, and pea. Since wheat is an important crop, the objective of our study was to evaluate how two different winter wheat varieties (*Triticum aestivum* L. Apache and Bezostaya 1) respond to radio-frequency (RF) oxygen plasma treatment in two different plasma regimes, the glow and afterglow regime. The effects of plasma treatment on the seeds were evaluated by analyzing the germination rate, root length and the number of roots, shoot height, the fresh weight of seedlings, and seed surface morphology. Our study indicates that the influence of plasma treatment might be variety-dependent, as statistically significant effects were observed in the number of roots, the length of the total root system, and the fresh weight of seedlings.

2. Results

2.1. Glow and Afterglow Plasma Treatment Caused Morphological Changes on the Seed Surface

The main difference between the exposure of seeds to glow and afterglow regimes lies in the interaction of the plasma species with the seed surface [19]. In the afterglow regime, the ions from the plasma are recombined on the wall of the plasma reactor before reaching the seed surface; this results in lower etching and reduces the thermal heating caused by ion bombardment compared to the glow plasma regime [20,21]. Our studies of seed morphology determined from SEM analysis show that the afterglow regime results in much lower etching effects, as virtually no changes in surface morphology were observed. The untreated seed coat exhibited rough net-like structures, while nanostructured morphologies were observed after the exposure of seeds to the plasma glow regime (Figure 1). After 5 s of treatment at 200 W, the surface morphology of the seed was altered and visible remnants of etched debris were observed. The removal of the seed surface coat, presumably a preferential etching of waxes, was even more evident after 30 s of treatment. Longer exposures caused greater etching, resulting in a structured nano-rough seed surface. However, treatment in the afterglow regime at a much higher power input did not change the seed surface morphology compared to the control. Differences between the two wheat varieties, Apache and Bezostaya 1, were also observed. It is already known that the etching rate is highly dependent on the chemical structure of the surface, as the counterparts that are easily etched are removed from the surface much faster, causing nanostructuring. The more pronounced nanostructured surface was observed in Bezostaya 1 after 30 s of treatment (Figure 1), indicating differences in seed coat composition between these two varieties.

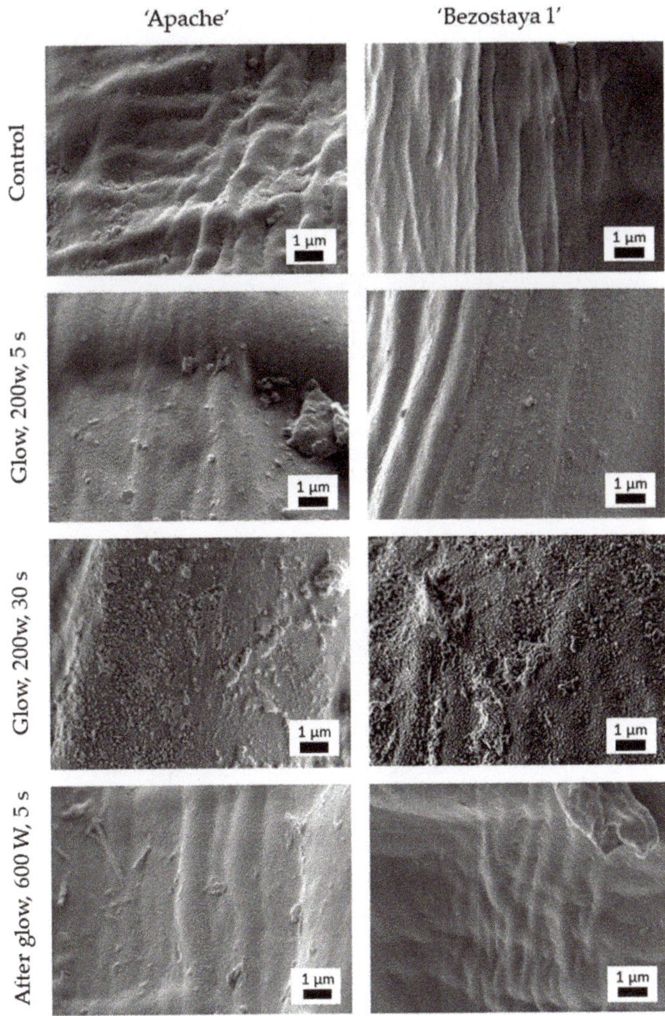

Figure 1. SEM images of the wheat seed coat of the Apache and Bezostaya 1 varieties of non-treated seeds, seeds treated with the glow regime (direct plasma) for 5 and 30 s at a power of 200 W, and seeds treated with the afterglow regime (indirect plasma) for 5 s and power of 600 W.

2.2. Germination Process

As reported by Anjum and Bajwa [22], the calculated parameter S (speed of germination) describes the germination process more accurately than Gt (germination rate) alone. This is also confirmed in our study where oxygen plasma treatment significantly influenced parameter S in both wheat varieties (Figure 2), while no statistically significant effect on the Gt parameter was observed in Apache (Figure 2a). The parameter S (Figure 2b) in Apache is the highest after seed treatment in the afterglow regime at an input power of 600 W for 3 s and is significantly higher compared to the control. The control samples were untreated seeds and seeds exposed to vacuum conditions. No significant differences were observed between these two samples. Interestingly, plasma treatment decreased all germination parameters in Bezostaya 1. The most notable decrease in Gt was observed in seeds exposed to afterglow plasma at 600 W for 3 s, while the negative effect on the parameter describing germination speed was also shown when seeds were exposed to glow plasma at 200 W for a longer time (30 s). These results indicate that the two wheat varieties respond differently to the same plasma treatment conditions. No significant changes in Gt were observed in Apache; however, a notable increase in germination speed (S) was observed in seeds exposed to glow plasma at 600 W for 3 s. On the other hand, seeds from Bezostaya 1 treated with glow and afterglow plasma showed a statistically significant negative trend: a lower germination rate and slower germination than the control. Regarding the plasma treatment conditions, it seems that glow (200 W for 30 s) and afterglow (600 W for 5 s) plasma treatments have similar effects on the Gt and S parameters in seeds of this variety; in both treatment regimes, the Gt and S parameters were similar. Thus, it could be speculated that both treatment conditions have a similar influence on germination and the speed of growth for a particular variety. From the application point of view, the plasma treatment conditions could be optimized to achieve the desired effect in the shortest time possible. The accumulated germination in time of the Apache seeds treated with afterglow plasma for 3 s exhibited faster seed germination on the first day compared to the control and other plasma treatments (Figure 2c). This result supports the statistically significant faster germination (S) in the same plasma treatment. On the other hand, Bezostaya 1 exhibited lower final germination (Gt) and slower germination in all plasma treated seeds except the treatment in glow plasma for 5 s, which is also shown in the accumulated germination in time (Figure 2d).

2.3. Morphological Characteristics of Seedlings

The morphological parameters measured on the germinating wheat plants show that the variety (V) had a statistically significant effect on root number (Rn) and total root system length (Rt). In contrast, individual root length (Ri) and the length of plumule (Pl) were not influenced by factor V (Table 1). Furthermore, the root-to-shoot ratio (R/S) (in fresh weight) was not affected by either V or plasma treatment (T).

(a)

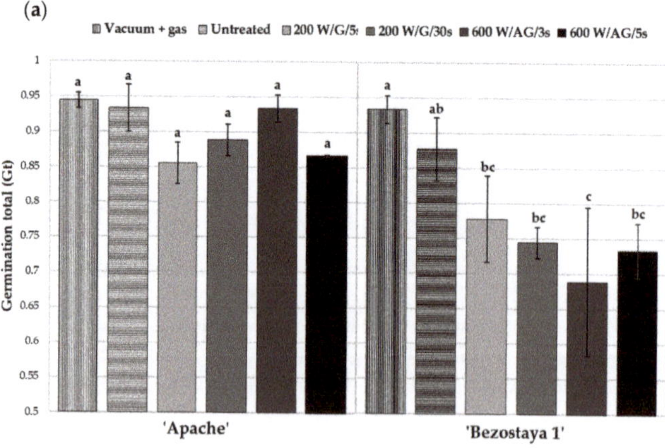

Figure 2. *Cont.*

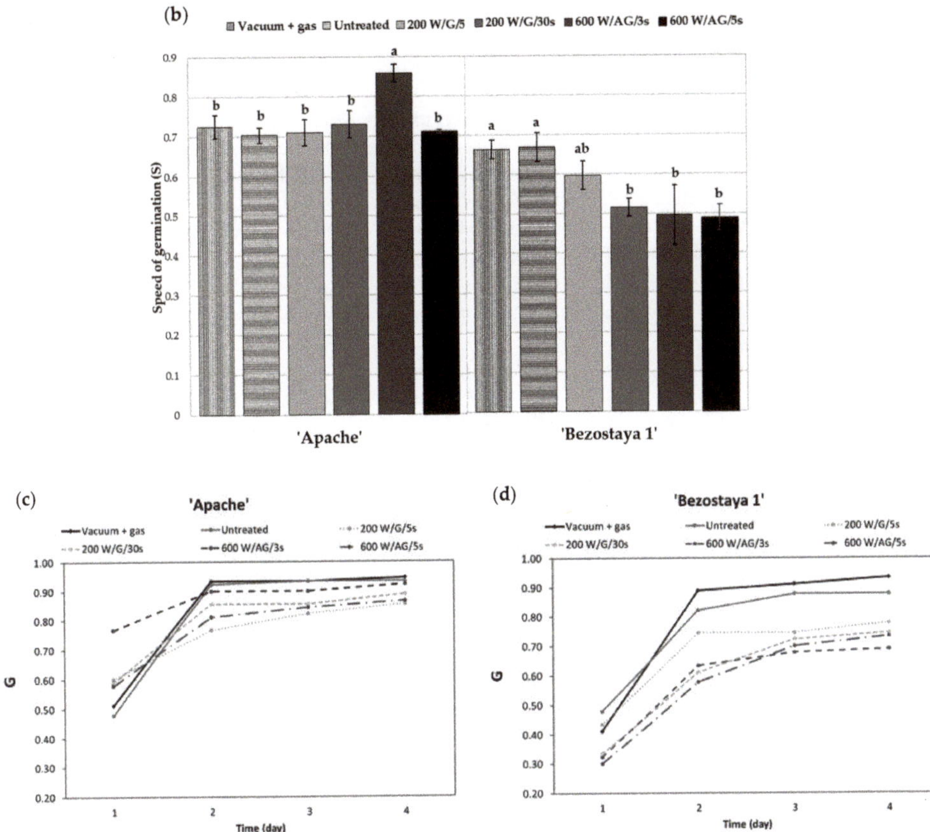

Figure 2. The germination total (**a**), speed of germination (**b**), and accumulated germination in time (**c,d**) of two wheat varieties depending on seed treatment with oxygen plasma at different treatment conditions. "G" in (**c,d**) represents the proportion of germinated seeds. The displayed values are the mean ± SEM of three replications. The different letters (a–c) indicate statistically significant differences among treatments according to Duncan's test ($\alpha = 0.05$).

Table 1. The analysis of variance (F-ratio) of chosen morphological parameters (root number, Rn; length of individual root, Ri; root system length, Rt; plumula length, Pl; seedling fresh weight, Fw; root-to-shoot ratio, R/S and seed vigor index, SV).

Source of Variation	Degree of Freedom	Rn	Ri	Rt	Pl	Fw	R/S	SV
Variety (V)	1	13.18 ***	2.47 ns	16.20 ***	1.30 ns	44.25 ***	0.25 ns	98.36 ***
Treatment (T)	5	2.21 ns	6.41 ***	4.73 ***	4.35 ***	4.59 **	0.87 ns	3.40 *
V × T	5	3.24 **	19.92 ***	5.78 ***	2.18 ns	2.32 ns	0.09 ns	5.11 **
Residuals	22	1.42035	1086.79	13,707.6	1030.5	249.69	0.02	163.81

Mean square of residual; ***, **, * Significant at $p < 0.001$, $p < 0.01$ and $p < 0.05$, respectively; ns = not significant.

Neither the glow nor afterglow plasma treatment of Bezostaya 1 had a statistically significant effect on the number of roots per plant. However, the plasma treatment of Apache seeds exhibited a statistically significant increase in the number of roots after both the glow and afterglow plasma treatments compared to the vacuum control (Figure 3a). However, compared to the control samples, only the glow plasma treatment at 200 W for 5 s increased the root number. The root system length (Figure 3b) of the plasma treated seeds of Bezostaya 1 showed no statistically significant difference compared to the vacuum control. However, the root systems of both glow plasma treated samples, and the samples treated for 5 s with 600 W afterglow plasma, had statistically significantly longer root systems compared to the control. The Apache seeds treated with glow plasma at 200 W for 30 s, on the other hand, showed a statistically significant decrease in root system length compared to the vacuum control, but no significance compared to the control. The treatment of Apache seeds in an afterglow plasma regime at 600 W for 5 s resulted in an increase in the root system length compared to the vacuum control and control samples. In terms of root system length and the number of roots per plant, Apache and Bezostaya 1 responded differently to cold plasma treatment. This could be important, as a large and deep root system may be an advantage when plants grow in moist soil, while a shallower root system may be more beneficial when frequent rainfall during the growing season is the primary source of moisture for the crop plants [23].

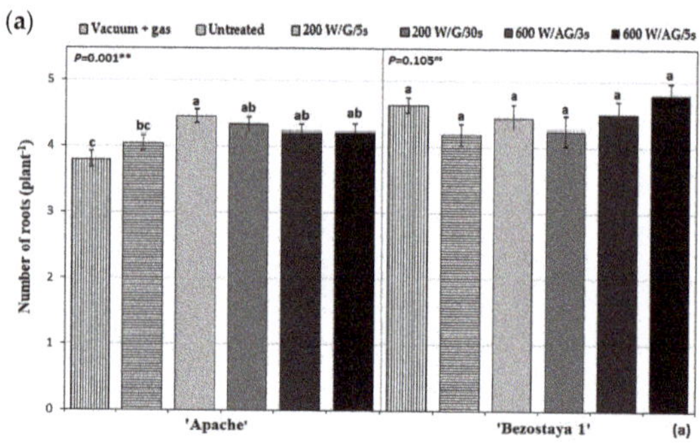

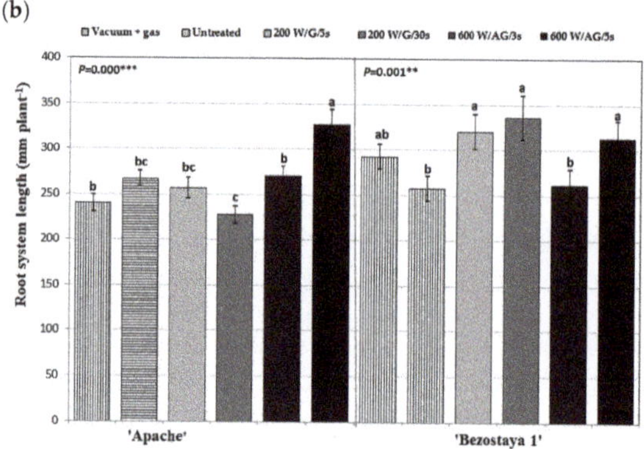

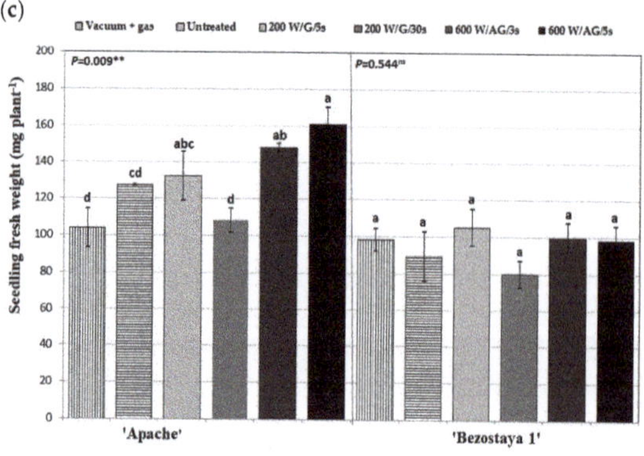

Figure 3. *Cont.*

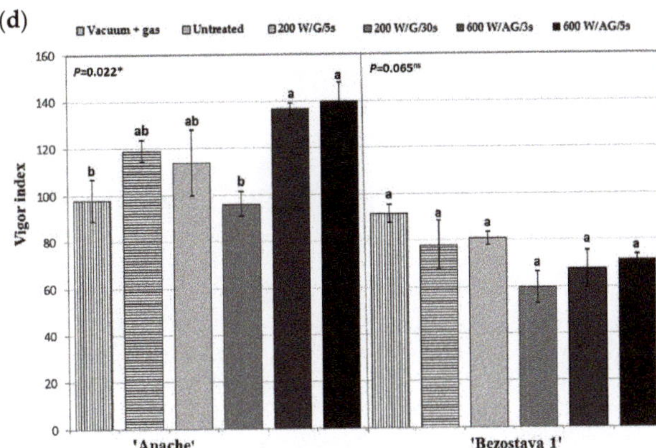

Figure 3. The number of roots per plant (**a**), root system length (**b**), seedling fresh weight (**c**), and seed vigor (**d**) of two wheat varieties depending on seed treatment with oxygen plasma at different treatment conditions. The displayed values are the mean ± SEM of three replications. The different letters (a–c) indicate statistically significant differences among treatments according to Duncan's test ($\alpha = 0.05$). Non-significant differences are marked as "ns".

The fresh weight of the Bezostaya 1 seedlings of cold-plasma-treated seeds did not show statistically significant differences compared to the control or vacuum control (Figure 3c). On the other hand, the Apache seedlings showed an increase in fresh weight when treated with the glow regime (5 s at 200 W), as well as with both afterglow treatment conditions. However, compared to the controls, only the afterglow plasma treatment of seeds (600 W/AG/3 s and 600 W/AG/5 s) exhibited an increase in the fresh weight of the seedlings.

Seed vigor describes the ability of seeds to germinate and establish seedlings rapidly, uniformly, and robustly under various environmental conditions. It is not a single measurable characteristic, but a concept associated with aspects of viable seed performance (e.g., the rate and uniformity of seed germination and parameters describing seedling growth). Seed vigor can describe a complex interaction between seed genetic and environmental components, with genetics being a less understood component of such interactions in the context of producing a robust seed [24]. In the case of Bezostaya 1, cold plasma had no statistically significant effect on its seed vigor index (Figure 3d). However, the seeds of Apache showed an increase in their vigor index when treated with afterglow plasma (600 W for 3 and 5 s).

3. Discussion

It has been shown that each wheat variety has different coping mechanisms and interactions with highly reactive oxygen plasma species. This may be mainly attributed to the different chemical structures of the seed surface, seed size, and genetic diversity within the species [25–27]. The results suggest that Bezostaya 1 is more sensitive to plasma treatment and has poor coping mechanisms for responding to such treatments. In this case, the seed germination rate was lower after plasma treatment, but there were no evident physiological changes in the root system length, the number of roots, or the fresh weight of seedlings following different plasma treatment conditions. Bezostaya 1 was predominantly sown in the 1970s. At that time, it had a high yield potential, but new wheat varieties with higher yields and stress tolerances for various abiotic factors were developed [28]. Apache showed a better responsiveness and adaptability to cold plasma treatment compared to Bezostaya 1. Apache was developed in 1998 and is a newer wheat variety with a higher yield potential compared to Bezostaya 1. Apache also exhibits resistance to water and nitrogen deficits by developing a deep root system under unfavorable conditions [29].

Thus, the ability of each wheat variety to cope with abiotic stress could be a key factor in seed response to cold plasma treatment. By optimizing plasma treatment conditions for specific wheat varieties, the length of the root system can be altered and thus adapted for growth in various environments.

The cold plasma treatment of wheat seeds (Apache and Bezostaya 1) in a glow plasma regime resulted in the nanostructuring of the seed coat, which was not observed in the afterglow plasma regime. This is mainly due to the presence of ions in the glow regime, which cause strong surface etching. In addition, glow plasma contains a higher concentration of relatively aggressive reactive chemical species with a higher energy as well as vacuum ultraviolet radiation (VUV). In contrast, only longer-lived reactive chemical species with a lower energy are present in afterglow plasma [21]. The thermal heating of the samples in the afterglow region is practically negligible, whereas the glow region of plasma can cause significant sample (seeds) heating due to strong ion bombardment [20,21]. Our study indicates that the afterglow treatment of Apache increases the vigor index, root system, fresh weight of seedlings, and the speed of germination compared to the control. Therefore, the application of the afterglow plasma regime could be of interest to further studies on seed processing and the optimization of plasma treatment conditions.

The experiments in this study were carried out on two genetically different winter wheat varieties, which show different response mechanisms and tolerance to the cold plasma treatment of seeds. The results indicate that the response of seeds and seedlings to cold plasma treatment cannot be generalized to seed species based solely on results obtained for one variety. One plant species can have large genetic variations between different varieties [30], which may play a crucial role in seed response to cold plasma treatment and later seedling growth. However, it is important to emphasize that other factors (not only genetic), such as growth location, soil, and weather conditions, also play an important role in seed development and could have an important impact on how seeds respond to plasma treatment.

Our findings in this study present the first steps toward the question of whether or not plasma treatment is variety-dependent. In this study, a laboratory-scale plasma reactor was used to study the influence of the plasma glow and afterglow region on two different wheat varieties. Only smaller amounts of seeds have been tested, as the plasma reactor only allows the treatment of smaller quantities of seeds. Due to the limited number of samples treated in the laboratory plasma reactor, the results and statistical analysis were based on three biological replicates with a relatively small sample size. However, statistically significant results were obtained for specific parameters, indicating that plasma treatment is wheat-variety-dependent. A future experimental setup that enables seed treatment in glow and afterglow plasma, with a larger number of seeds per treatment, would significantly contribute to current discussions of the interaction of cold plasma treatment with seeds. Plasma reactors currently used in research and seed treatment are mostly laboratory-scale plasma reactors, in which only a limited number of seeds can be treated [4,31,32]. For the future application of plasma technologies in agriculture and to obtain meaningful and statistically well-supported results, the development and use of larger, well-defined plasma reactors are necessary.

4. Materials and Methods

4.1. Seed Material

The seeds of winter wheat varieties (*Triticum aestivum* L.) Apache and Bezostaya 1 were obtained from a private collection. Apache is one of the leading modern wheat varieties, while Bezostaya 1 is an old, extensive-type wheat variety, which today is important mainly in wheat breeding [25,26,33]. The tested varieties differed in thousand kernel weight (TKW), grain color, and quality. Apache has a TKW of 44–48 g; it is a high-yielding variety of moderate-to-poor bread-making quality (quality group B2/C) [25]. Bezostaya 1, on the other hand, has a lower TKW (41–44 g) and is an old Russian variety with a lower yield, which is still used in wheat breeding programs as a variety with a high bread-making quality and resistance to low temperatures [26,27]. Seeds of both wheat varieties were produced in the same year in the same field (in Krog near Murska Sobota, Slovenia). Both wheat varieties were sown, treated, harvested, and stored under the same conditions.

4.2. Characteristics of Plasma and Seed Treatment

The plasma treatment was carried out using an in-house designed plasma system, illustrated in Figure 4. Four batches of 90 seeds per variety were exposed to oxygen plasma generated by an RF discharge. The pressure was set to 50 Pa and the input power to 200 W. In this case, seeds were treated in the glow region (direct plasma treatment) for 5 and 30 s. The treatment in the afterglow region (indirect treatment) was carried out at an input power of 600 W for 3 and 5 s. The control samples represented untreated seeds and seeds exposed to vacuum conditions, with the same gas flow as during the plasma treatment.

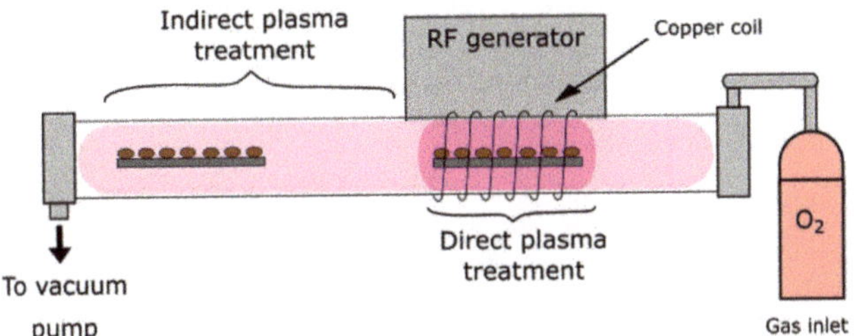

Figure 4. A schematic representation of the plasma system used in the treatment of wheat seeds.

4.3. SEM Imaging and Sample Preparation

For SEM analysis, the seeds were mounted onto aluminum stubs using carbon tape and coated with a thin layer of platinum using a Gatan 682 Precision Etching and Coating System (Gatan Inc. Pleasanton, CA, USA). The samples were imaged by a JSM-7600F Schottky field emission SEM (Jeol Ltd., Tokyo, Japan), and the representative images are shown.

4.4. Seed Germination, Growth Conditions, and Measurements (Parameters Assessed)

The germination experiments were conducted three hours after the plasma treatment. Seeds from each treatment with controls were divided into three equal replicates (each replicate contained 30 seeds) and evenly placed within sterile Petri dishes (10 cm in diameter) lined with filter paper (No. 1 Whatman International, Maidstone, UK). The filter paper was initially moistened with 5 mL of distilled water, and additionally 3 mL of water was added on the third and fourth day of the experiment. The sealed Petri dishes were arranged in a completely randomized design and incubated at $25 \pm 1\,^{\circ}C$ for four days and an additional three days at $20 \pm 1\,^{\circ}C$ (altering 12 h light/12 h darkness cycle).

Germination was recorded every 24 h for four days when untreated seeds reached maximum germination, 100.0% in Apache and 97% in Bezostaya 1. A seed was considered to have germinated when the radicle was visible (~1 mm long). In addition to the total germination (Gt, final germination proportion) parameter, the speed of germination (S) was calculated as recommended by Anjum and Bajwa [22]. Indices Gt and S were calculated as Equations (1) and (2):

$$Gt = G4 = \frac{number\ of\ germinated\ seeds}{30\ seeds}, \tag{1}$$

$$S = \left(\frac{G1}{1}\right) + (G2 - G1) \times \frac{1}{2} + (G3 - G2) \times \frac{1}{3} + (G4 - G3) \times \frac{1}{4}, \tag{2}$$

where G is the proportion of germinated seeds obtained during the first ($G1$) to fourth ($G4$) day of experiment. The maximum value possible for index S is 1.

After seven days of incubation, seedlings were photographed and the lengths of the individual seminal root and shoot were measured with a stereomicroscope (Olympus SZ61, Japan). The fresh weight of the root and shoot were assessed, and the seed vigor index (SV) was calculated according to Equation (3) [27]:

$$SV = \frac{\text{Fresh weight of seedlings (mg)} \times \text{Gt}}{100}. \tag{3}$$

4.5. Statistical Analyses

The obtained results were analyzed using a Statgraphics Centurion XV (Statpoint Technologies, 2005). To improve the normality of distribution and homogeneity of variance (Cochrans C and Bartlett tests), the data describing germination proportion (Gt) was arcsine-transformed. After the multifactor analysis of variance (ANOVA, $p < 0.05$), a one-way ANOVA was carried out to test the effects of plasma treatment within each wheat variety. A comparison of means was performed by Duncan's test ($\alpha = 0.05$). Results are presented as an average of three replications with a standard error of mean ($\pm$SEM).

5. Conclusions

Overall, it can be concluded that not only plasma treatment parameters, but also plant variety and/or seed characteristics, may play a crucial role in optimizing plasma parameters for seed treatment. The physiological responses of plants to the plasma treatment of seeds could be variety-dependent due to different genetic profiles and other factors such as growing location, soil conditions, and the general environmental conditions under which the seeds developed before their plasma treatment. The results of this study serve as a stepping stone for future research in plasma agriculture, as comparisons of varieties of the same plant species could provide important clues for how plasma technology affects seeds and seedlings. It is important to emphasize that results obtained under controlled laboratory conditions may differ from studies conducted in the field where environmental factors are not controlled and could affect plant growth, development, and yield differently.

For the routine use of cold plasma technology in agriculture, it is important to further investigate the exact mechanisms of how treating seeds with cold plasma affects seed germination and seedling growth in relation to specific plant varieties (genotypes) and plasma treatment conditions. This could provide new insights concerning the complex interaction mechanisms of plasma species with the seed surface and biochemical mechanisms underlying the physiological changes in seedlings and plants. It is crucial to investigate which plasma components contribute to the changes in seed physiology and what mechanisms are triggered in seeds by cold plasma treatment. These findings could enable the agricultural industry to fully utilize the potential of plasma technology in the future.

Author Contributions: Conceptualization, S.G.M. and I.J.; methodology, S.G.M. and I.J.; formal analysis, S.G.M. and P.S.; writing—original draft preparation, P.S.; writing—review and editing, P.S., S.G.M., and I.J.; visualization, P.S.; funding acquisition, I.J. All authors have read and agreed to the published version of the manuscript.

Funding: This research was funded by the Slovenian Research Agency and Ministry of Agriculture, Forestry and Food of Slovenia under grant number V4-2001. We also acknowledge the Slovenian Research Agency, (Javna Agencija za Raziskovalno Dejavnost RS–ARRS), young researcher grant number PR-09756, core fund P2-0082 (Thin film structures and plasma surface engineering), and program group P1-02012 (Plant biology).

Institutional Review Board Statement: Not applicable.

Informed Consent Statement: Not applicable.

Data Availability Statement: The data presented in this study are available in this article.

Acknowledgments: We would like to thank Ana Vajsman for their assistance with the execution of experimental work, and Pedro Soares for their help with photo analysis and data extraction.

Conflicts of Interest: The authors declare no conflict of interest.

References

1. Selcuk, M.; Oksuz, L.; Basaran, P. Decontamination of grains and legumes infected with *Aspergillus* spp. and *Penicillum* spp. by cold plasma treatment. *Bioresour. Technol.* **2008**, *99*, 5104–5109. [CrossRef]
2. Niemira, B.A.; Boyd, G.; Sites, J. Cold Plasma Rapid Decontamination of Food Contact Surfaces Contaminated with Salmonella Biofilms. *J. Food Sci.* **2014**, *79*, M917–M922. [CrossRef] [PubMed]
3. Chen, H.H.; Chang, H.C.; Chen, Y.-K.; Hung, C.L.; Lin, S.Y.; Chen, Y.S. An improved process for high nutrition of germinated brown rice production: Low-pressure plasma. *Food Chem.* **2016**, *191*, 120–127. [CrossRef] [PubMed]
4. Gómez-Ramírez, A.; López-Santos, C.; Cantos, M.; García, J.L.; Molina, R.; Cotrino, J.; Espinós, J.P.; González-Elipe, A.R. Surface chemistry and germination improvement of Quinoa seeds subjected to plasma activation. *Sci. Rep.* **2017**, *7*, 5924. [CrossRef]
5. Ling, L.; Jiangang, L.; Minchong, S.; Chunlei, Z.; Yuanhua, D. Cold plasma treatment enhances oilseed rape seed germination under drought stress. *Sci. Rep.* **2015**, *5*, 13033. [CrossRef] [PubMed]
6. Mitra, A.; Li, Y.-F.; Klämpfl, T.G.; Shimizu, T.; Jeon, J.; Morfill, G.E.; Zimmermann, J.L. Inactivation of Surface-Borne Microorganisms and Increased Germination of Seed Specimen by Cold Atmospheric Plasma. *Food Bioprocess. Technol.* **2013**, *7*, 645–653. [CrossRef]
7. Jiang, J.; He, X.; Li, L.; Li, J.; Shao, H.; Xu, Q.; Ye, R.; Dong, Y. Effect of Cold Plasma Treatment on Seed Germination and Growth of Wheat. *Plasma Sci. Technol.* **2014**, *16*, 54–58. [CrossRef]
8. Meng, Y.; Qu, G.; Wang, T.; Sun, Q.; Liang, D.; Hu, S. Enhancement of Germination and Seedling Growth of Wheat Seed Using Dielectric Barrier Discharge Plasma with Various Gas Sources. *Plasma Chem. Plasma Process.* **2017**, *37*, 1105–1119. [CrossRef]
9. Slade, A.J.; Fuerstenberg, S.I.; Loeffler, D.; Steine, M.N.; Facciotti, D. A reverse genetic, nontransgenic approach to wheat crop improvement by TILLING. *Nat. Biotechnol.* **2005**, *23*, 75–81. [CrossRef]
10. Guo, Q.; Wang, Y.; Zhang, H.; Qu, G.; Wang, T.; Sun, Q.; Liang, D. Alleviation of adverse effects of drought stress on wheat seed germination using atmospheric dielectric barrier discharge plasma treatment. *Sci. Rep.* **2017**, *7*, 1–14. [CrossRef]
11. Iranbakhsh, A.; Ardebili, N.O.; Ardebili, Z.O.; Shafaati, M.; Ghoranneviss, M. Non-thermal Plasma Induced Expression of Heat Shock Factor A4A and Improved Wheat (*Triticum aestivum* L.) Growth and Resistance Against Salt Stress. *Plasma Chem. Plasma Process.* **2017**, *38*, 29–44. [CrossRef]
12. Sera, B.; Spatenka, P.; Sery, M.; Vrchotova, N.; Hruskova, I. Influence of Plasma Treatment on Wheat and Oat Germination and Early Growth. *IEEE Trans. Plasma Sci.* **2010**, *38*, 2963–2968. [CrossRef]
13. Roy, N.C.; Hasan, M.M.; Talukder, M.R.; Hossain, M.D.; Chowdhury, A.N. Prospective Applications of Low Frequency Glow Discharge Plasmas on Enhanced Germination, Growth and Yield of Wheat. *Plasma Chem. Plasma Process.* **2017**, *38*, 13–28. [CrossRef]
14. Dobrin, D.; Magureanu, M.; Mandache, N.B.; Ionita, M.-D. The effect of non-thermal plasma treatment on wheat germination and early growth. *Innov. Food Sci. Emerg. Technol.* **2015**, *29*, 255–260. [CrossRef]
15. Volin, J.C.; Denes, F.S.; Young, R.A.; Park, S.M.T. Modification of Seed Germination Performance through Cold Plasma Chemistry Technology. *Crop. Sci.* **2000**, *40*, 1706–1718. [CrossRef]
16. Yodpitak, S.; Mahatheeranont, S.; Boonyawan, D.; Sookwong, P.; Roytrakul, S.; Norkaew, O. Cold plasma treatment to improve germination and enhance the bioactive phytochemical content of germinated brown rice. *Food Chem.* **2019**, *289*, 328–339. [CrossRef]
17. Bafoil, M.; Le Ru, A.; Merbahi, N.; Eichwald, O.; Dunand, C.; Yousfi, M. New insights of low-temperature plasma effects on germination of three genotypes of Arabidopsis thaliana seeds under osmotic and saline stresses. *Sci. Rep.* **2019**, *9*, 1–10. [CrossRef] [PubMed]
18. Bormashenko, E.; Shapira, Y.; Grynyov, R.; Whyman, G.; Bormashenko, Y.; Drori, E. Interaction of cold radiofrequency plasma with seeds of beans (*Phaseolus vulgaris*). *J. Exp. Bot.* **2015**, *66*, 4013–4021. [CrossRef] [PubMed]
19. Bermúdez-Aguirre, D. *Advances in Cold Plasma Applications for Food Safety and Preservation*, 1st ed.; Bermudez-Aguirre, D., Ed.; Academic Press: Cambridge, MA, USA, 2019; ISBN 9780128149218.
20. Han, L.; Patil, S.; Boehm, D.; Milosavljević, V.; Cullen, P.J.; Bourke, P. Mechanisms of Inactivation by High-Voltage Atmospheric Cold Plasma Differ for Escherichia coli and Staphylococcus aureus. *Appl. Environ. Microbiol.* **2016**, *82*, 450–458. [CrossRef] [PubMed]
21. Fridman, G.; Brooks, A.D.; Balasubramanian, M.; Fridman, A.; Gutsol, A.; Vasilets, V.; Ayan, H.; Friedman, G. Comparison of Direct and Indirect Effects of Non-Thermal Atmospheric-Pressure Plasma on Bacteria. *Plasma Process. Polym.* **2007**, *4*, 370–375. [CrossRef]
22. Anjum, T.; Bajwa, R. Importance of Germination Indices in Interpretation of Allelochemical Effects. *Int. J. Agric. Biol.* **2005**, *7*, 417–419.
23. Manschadi, A.M.; Hammer, G.L.; Christopher, J.T.; DeVoil, P. Genotypic variation in seedling root architectural traits and implications for drought adaptation in wheat (*Triticum aestivum* L.). *Plant Soil* **2008**, *303*, 115–129. [CrossRef]
24. Finch-Savage, W.E.; Bassel, G.W. Seed vigour and crop establishment: Extending performance beyond adaptation. *J. Exp. Bot.* **2016**, *67*, 567–591. [CrossRef] [PubMed]

25. Zemljič, A. Agricultural, I. of S. Izbor in Opis Sort Ozimnih Žit Za Setev v Letu 2015/16. Available online: http://www.kgzs-ms.si/wp-content/uploads/2015/09/OPIS-SORT-OZ-ŽIT-2015.pdf (accessed on 23 February 2021).

26. Denčić, S.; Kobiljski, B.; Jestrocić, Z.; Orbović, B.; Pavlović, M. Efekat ruskih sorti Bezostaya 1 i Mirmovska 808 na selekciju pšenice u Institutu za ratarstvo i povrtarstvo u Novom Sadu. *Sel. Semen.* **2009**, *15*, 39–45.

27. Zahoranová, A.; Henselová, M.; Hudecová, D.; Kaliňáková, B.; Kováčik, D.; Medvecká, V.; Černák, M. Effect of Cold Atmospheric Pressure Plasma on the Wheat Seedlings Vigor and on the Inactivation of Microorganisms on the Seeds Surface. *Plasma Chem. Plasma Process.* **2015**, *36*, 397–414. [CrossRef]

28. Keser, M.; Gummadov, N.; Akin, B.; Belen, S.; Mert, Z.; Taner, S.; Topal, A.; Yazar, S.; Morgounov, A.; Sharma, R.C.; et al. Genetic gains in wheat in Turkey: Winter wheat for dryland conditions. *Crop. J.* **2017**, *5*, 533–540. [CrossRef]

29. Postic, F.; Beauchêne, K.; Gouache, D.; Doussan, C. Scanner-Based Minirhizotrons Help to Highlight Relations between Deep Roots and Yield in Various Wheat Cultivars under Combined Water and Nitrogen Deficit Conditions. *Agronomy* **2019**, *9*, 297. [CrossRef]

30. Salem, K.F.M.; Salem, K.; El-Zanaty, A.; Esmail, R. Assessing Wheat (*Triticum aestivum* L.) Genetic Diversity Using Morphological Characters and Microsatellite Markers Maize Breeding and Biotechnological View project Genetic engineering View project Assessing Wheat (*Triticum aestivum* L.) Genetic Diversity Using Morphological Characters and Microsatellite Markers. *World J. Agric. Sci.* **2008**, *4*, 538–544.

31. Junior, C.A.; Vitoriano, J.D.O.; Da Silva, D.L.S.; Farias, M.D.L.; Dantas, N.B.D.L. Water uptake mechanism and germination of Erythrina velutina seeds treated with atmospheric plasma. *Sci. Rep.* **2016**, *6*, srep33722. [CrossRef]

32. Filatova, I.; Azharonok, V.; Shik, A.; Antoniuk, A.; Terletskaya, N. Fungicidal effects of plasma and Radio-Wave pre-treatments on seeds of grain crops and legumes. In *Plasma for Bio-Decontamination, Medicine and Food Security*; Springer: Dordrecht, The Netherlands, 2012; pp. 469–479.

33. Belderok, B.; Mesdag, J.; Donner, D.A. *Bread-Making Quality of Wheat: A Century of Breeding in Europe*, 1st ed.; Belderok, B., Mesdag, J., Eds.; Springer Science and Business Media LLC: Berlin, Germany, 2000.

Article

Non-Thermal Plasma Treatment Influences Shoot Biomass, Flower Production and Nutrition of Gerbera Plants Depending on Substrate Composition and Fertigation Level

Samantha Cannazzaro [1], Silvia Traversari [1,*], Sonia Cacini [1], Sara Di Lonardo [2], Catello Pane [3], Gianluca Burchi [1] and Daniele Massa [1]

[1] CREA Research Center for Vegetable and Ornamental Crops, Council for Agricultural Research and Economics, Via dei Fiori 8, 51012 Pescia, PT, Italy; samantha.cannazzaro@libero.it (S.C.); sonia.cacini@crea.gov.it (S.C.); gianluca.burchi@crea.gov.it (G.B.); daniele.massa@crea.gov.it (D.M.)
[2] Research Institute on Terrestrial Ecosystems (IRET-CNR), Via Madonna del Piano 10, Sesto F.No, 50019 Florence, FI, Italy; sara.dilonardo@cnr.it
[3] CREA Research Center for Vegetable and Ornamental Crops, Council for Agricultural Research and Economics, Via Cavalleggeri 25, 84098 Pontecagnano Faiano, SA, Italy; catello.pane@crea.gov.it
* Correspondence: silvia.traversari@crea.gov.it

Citation: Cannazzaro, S.; Traversari, S.; Cacini, S.; Di Lonardo, S.; Pane, C.; Burchi, G.; Massa, D. Non-Thermal Plasma Treatment Influences Shoot Biomass, Flower Production and Nutrition of Gerbera Plants Depending on Substrate Composition and Fertigation Level. *Plants* **2021**, *10*, 689. https://doi.org/10.3390/plants10040689

Academic Editor: Božena Šerá

Received: 27 February 2021
Accepted: 30 March 2021
Published: 2 April 2021

Publisher's Note: MDPI stays neutral with regard to jurisdictional claims in published maps and institutional affiliations.

Abstract: Non-thermal plasma (NTP) appears a promising strategy for supporting crop protection, increasing yield and quality, and promoting environmental safety through a decrease in chemical use. However, very few NTP applications on containerized crops are reported under operational growing conditions and in combination with eco-friendly growing media and fertigation management. In this work, NTP technology is applied to the nutrient solution used for the production of gerbera plants grown in peat or green compost, as an alternative substrate to peat, and with standard or low fertilization. NTP treatment promotes fresh leaf and flower biomass production in plants grown in peat and nutrient adsorption in those grown in both substrates, except for Fe, while decreasing dry plant matter. However, it causes a decrease in the leaf and flower biomasses of plants grown in compost, showing a substrate-dependent effect under a low fertilization regime. In general, the limitation in compost was probably caused by the high-substrate alkalinization that commonly interferes with gerbera growth. Under low fertilization, a reduction in the photosynthetic capacity further penalizes plant growth in compost. A lower level of fertilization also decreases gerbera quality, highlighting that Ca, Mg, Mn, and Fe could be reduced with respect to standard fertilization.

Keywords: bedding plants; fertilizer saving; floriculture; green compost; nitrogen; peat; reactive oxygen species; reactive nitrogen species

1. Introduction

The treatment of water devoted to agronomic purposes with non-thermal plasma (NTP) is receiving increasing interest following promising results obtained using this technology [1]. NTP is a weakly ionized gas that produces electrons, reactive oxygen species (ROS), reactive nitrogen species (RNS), UV radiation, and a local electric field. It is generated at air temperature using devices as dielectric barrier discharger (DBD) reactors and can be used to treat water, changing its chemical composition [2]. In particular, plasma-activated water (PAW) could present an increase in ROS and RNS and variations in the redox potential and conductivity of the water [3]. The combined effect obtained by the strong oxidant activity of ROS and water acidification, through the generation of nitric acid from RNS, confers an antimicrobial activity to PAW [4], with it thereby representing an environmentally sustainable alternative to chemical sanitizers or pesticides. In fact, plants treated with PAW show a strong inactivation of phytopathogens [5]. Moreover, plasma treatment causes the fixation of air nitrogen, increasing the concentrations of nitrite, nitrous acid, nitrate, and nitric acid, conferring fertilizing proprieties to PAW as well [2].

The use of PAW has been shown to have positive effects on plant yield. Among several examples, the use of PAW generated by DBD increases both seed germination and plant biomass in tomatoes, peppers, and radishes [6]. In fact, the benefit of low concentrations of exogenous ROS and RNS has been already shown; this is probably determined by modulation of plant metabolism and immune response [7,8]. Despite this potential role of NTP for use in agriculture, examples of PAW applications on intensive ornamental and vegetable crop cultivation are still few.

In the ornamental nursery sector, peat moss is widely used as a major component of growing media for potted plant production. However, in recent decades, the horticultural industry has been trying to reduce the use of peat as a soilless substrate due to its environmental unsustainability, related to ecological concerns [9], and its increasing price [10]. Consequently, many new plant substrates and mixtures have been introduced and tested worldwide [11]. A promising alternative growing medium, in line with circular economy concepts, consists of green compost obtained from different organic wastes. This shows potential not only in terms of its positive carbon footprint but also for its low price and suppression of some soil-borne diseases [12]. One of the main problems of compost use can be its high variability of physical and chemical characteristics [13]; however, it may have similar physical features to peat if properly selected [14]. In order to assess the suitability of an alternative substrate for potted plant production, its effects on plant yield and ornamental traits must be assessed during the entire cultivation period, especially on plants that require a specific pH range for their growing medium [15]. As an example, gerbera plants are naturally grown in acidic soils/substrates and show chlorosis symptoms and nutrient deficiency at a pH of around six [16]. The use of different peat-alternative substrates for gerbera production has been the object of several studies [17–19]. The main concern has been the strong influence of substrate composition on the pH and electrical conductivity (EC) of growing media, and consequently on nutrient availability [20]. As an example, the addition of coconut coir and mushroom compost lowered the pH and increased the EC of conventional substrates used for gerbera production (e.g., garden soil, silt, and sand) and consequently could improve plant nutrient uptake [21].

Ornamental potted plants are commonly maintained via protected cultivation fed with a fertigation system that provides nutrients through drip irrigation directly supplied into the active root zone, reducing nutrient release into the environment [22]. Therefore, a reduction of fertilizer concentration might not have negative effects on plants. As an example, in potted gerbera production, a reduction of up to 50% in nutrient concentration does not lead to adverse effects at the end of cultivation when using a mixture of sphagnum peat moss:perlite (4:1) as a substrate [23]. However, a 50% reduction of nutrient supply causes a decrease in growth, and flower number per plant, as well as an increase in bent neck, when using a mixture of soil:compost:sand (2:1:1) [24]. The presence of compost in growing media can indeed support the growth and development of ornamental plants by providing an extra budget of nutrients [25], possibly available to the crop, and for introducing the presence of bioactive organic compounds that may stimulate and improve plant nutrient use efficiency under fertilizer shortage [26]. Therefore, it is strongly recommended that the evaluation of fertigation strength is made in combination with the substrate composition, taking into particular consideration both pH and EC.

The aim of this work is to evaluate the effect of NTP treatment on the suitability of green compost as an alternative substrate to peat for gerbera bedding plant production, and test both substrates using a standard or a low fertilization regime, assessing the opportunity for fertilizer saving. All of these treatments and their combinations are evaluated to assess possible improvements in plant yield, plant nutrition, and ornamental traits, as well as variations in microorganism presence in the rhizosphere, when using NTP to treat the nutrient solution.

2. Results

2.1. Plant Biomass Analyses

Biometric parameters of gerbera plants were simultaneously influenced by the use of NTP-treated solution, the fertilization level, and type of substrate (Figure 1, Table 1). Leaf fresh weight (FW) was higher in plants grown in peat substrates with high fertilization (HF), while the lowest value was measured in plants grown in compost under low fertilization (LF) (Figure 1a, Table 1). NTP treatment increased this parameter in peat while it did not show any effect in compost. A similar trend was measured for the leaf area (Figure 1b, Table 1). As for the leaves, the highest flower FW and number were measured under both untreated or NTP-treated conditions in peat HF plants (on average 42 ± 6.0 g plant^{-1}, and 5 ± 0.5 flowers plant^{-1}, respectively, Figure 1c,d and Table 1). No variation in single flower FW occurred (data not shown). Plant dry weight (DW) (Figure 1e, Table 1) ranked first in peat HF plants in both untreated and NTP-treated plants (on average 19 ± 0.8 g plant^{-1}). Under an LF regime, the NTP treatment influenced the plant biomass in a substrate-dependent way, as was highlighted by a visual evaluation of plant features (Figure 2). Interestingly, NTP-treated solution increased the plant DW in peat while it decreased this parameter in compost. However, NTP did not affect plant biomass under a HF regime in the same compost-based substrate. DW concentration was higher in LF plants and compost, while plants treated with NTP showed a higher degree of tissue hydration.

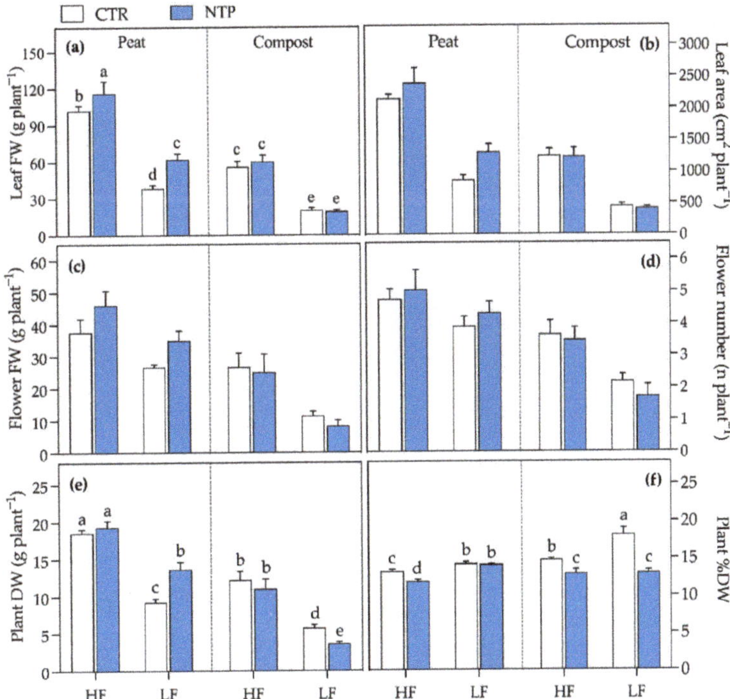

Figure 1. Biometric measurements of gerbera plants grown in peat or compost and fed with high (HF) or low (LF) nutrient solution, untreated (CTR) or NTP-treated (NTP): leaf fresh weight (**a**), leaf area (**b**), flower fresh weight (**c**), flower number (**d**), plant dry weight (**e**), and plant dry weight percentage (**f**). Values represent the means (n = 4) + SD. The presence of different letters represents Tukey's test results for significant triple interaction, while three-way ANOVA p-values are reported in Table 1.

Table 1. The effects of NTP-treatment (NTP), fertilization (F), substrate (S), and their mutual interactions (×) on production, morphometric, and nutrient parameters of gerbera plants (three-way ANOVA results of data reported in Figure 1, Figure 3, and Figure 4).

Parameter	NTP	F	S	NTP × F	NTP × S	F × S	NTP × F × S
Leaf FW	***	***	***	ns	***	***	*
Leaf area	***	***	***	ns	***	***	ns
Flower FW	*	***	***	ns	***	ns	ns
Flower number	ns	***	***	ns	*	**	ns
Plant DW	ns	***	***	*	***	ns	***
Plant % DW	***	***	***	ns	***	ns	***
SLA	***	***	***	ns	***	ns	***
SPAD index	ns	***	***	ns	***	***	ns
Organic nitrogen	ns	***	*	*	*	**	*
Nitrate	***	***	ns	ns	ns	***	ns
Potassium	***	***	ns	ns	ns	***	ns
Calcium	ns	ns	ns	ns	**	*	ns
Magnesium	*	ns	ns	*	*	ns	**
Phosphorus	***	***	**	ns	*	***	ns
Manganese	ns	ns	***	ns	ns	ns	*
Iron	***	ns	***	ns	ns	ns	ns

ns, not significant; * $p \leq 0.05$; ** $p \leq 0.01$; *** $p \leq 0.001$; FW, fresh weight; DW, dry weight; SLA, specific leaf area.

Figure 2. Images of gerbera plants grown in peat or compost and fed with high (HF) or low (LF) nutrient solution, untreated (CTR) or NTP-treated (NTP).

The specific leaf area (SLA) was higher under NTP-HF treatment than in other conditions, while in compost, it was lower for CTR (untreated plants) in combination with LF treatment than in other conditions (Figure 3a, Table 1). Finally, the SPAD index was higher under HF regimes in both peat and compost, while the lowest value was recorded under LF in compost (on average 36 ± 2.2 SPAD units, Figure 3b and Table 1).

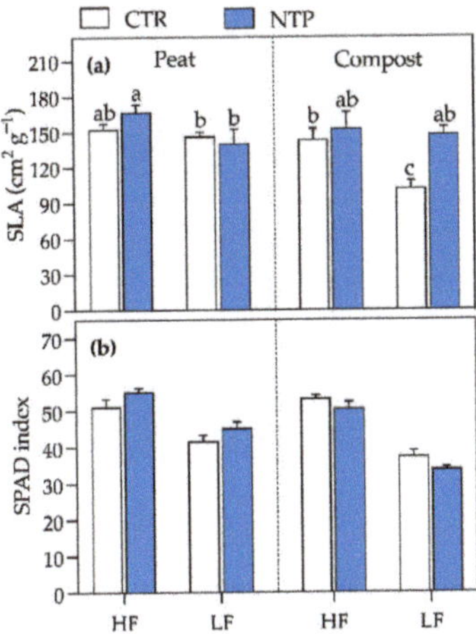

Figure 3. Leaf parameters of gerbera plants grown in peat or compost and fed with high (HF) or low (LF) nutrient solution, untreated (CTR) or NTP-treated (NTP): specific leaf area (SLA) (**a**) and SPAD index (**b**). Values represent the means ($n = 4$) + SD. The presence of different letters represents Tukey's test results for significant triple interaction, while three-way ANOVA *p*-values are reported in Table 1.

2.2. Leaf Elemental Concentrations

The combination of substrate, fertilization level, and use of NTP-treated nutrient solution differently affected leaf nutrient concentration (Figure 4, Table 1). The Ca, Mg, Mn, and Fe contents were not influenced by the LF regime, while the other investigated elements all decreased.

Overall, NTP treatment increased the nitric N, K, P, and Mg, while it decreased the Fe concentration. Interestingly, the P, Mg, and Ca concentrations showed a significant interaction between NTP treatment and substrate type, highlighting a higher absorption of these elements when NTP was applied to compost. The substrate influenced organic N, P, Mn, and Fe assimilation and, in particular, the P and Mn contents were found to be higher in peat, while organic N and Fe were higher in compost treatments. Specifically, organic N was higher under a HF regime, without any difference noted between peat and compost. Meanwhile, it was higher in CTR plants than in NTP ones in peat under an LF regime, showing an additional significant interaction between NTP treatment and substrate type. Mn was higher in the peat substrate than in compost. Fe was lower in peat than in compost and decreased under NTP treatment.

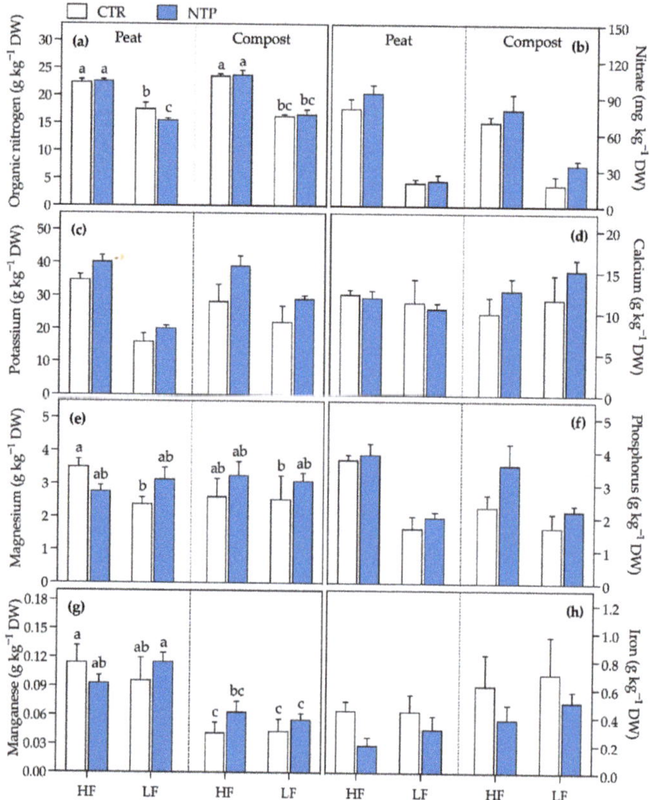

Figure 4. Organic nitrogen (**a**), nitrate (**b**), potassium (**c**), calcium (**d**), magnesium (**e**), phosphorus (**f**), manganese (**g**), and iron (**h**) concentrations within the leaves of gerbera plants grown in peat or compost and fed with a high (HF) or low (LF) nutrient solution, untreated (CTR) or NTP-treated (NTP). Values represent the means (n = 4) + SD. The presence of different letters represents Tukey's test results for significant triple interaction, while three-way ANOVA p-values are reported in Table 1.

2.3. Fungi and Bacteria within the Substrate

The quantification of fungi and bacteria allowed the evaluation of the treatment effect on microorganism population levels within the substrates (Table 2). The bacterial and fungal concentrations showed a specular trend: one group increased as the other decreased. The use of compost strongly modulated the fungal presence, decreasing their density, while the level of fertilization and the use of NTP-treated solution did not show any effect on the amounts of fungal colony forming units (CFUs). Overall, the bacterial concentration was higher in compost and, under an LF regime, the use of NTP-treated solution decreased the bacterial CFU in this substrate.

Table 2. The effects of NTP treatment (NTP), fertilization (F), substrate (S), and their mutual interactions ($\times$) on the colony-forming units (CFU) of fungi and bacteria within the substrate of gerbera plants grown in peat or compost and fed with high (HF) or low (LF) nutrient solution, untreated (CTR) or NTP-treated (NTP). Values represent the means ($n = 3$) $\pm$ SD.

Peat				Compost			
HF		LF		HF		LF	
CTR	NTP	CTR	NTP	CTR	NTP	CTR	NTP
Fungi (CFU $\times 10^3$ g^{-1})							
0.52 ± 0.032	0.40 ± 0.128	0.57 ± 0.131	0.48 ± 0.057	0.10 ± 0.046	0.06 ± 0.009	0.05 ± 0.029	0.14 ± 0.104
Bacteria (CFU $\times 10^3$ g^{-1})							
7.6 ± 2.70	17.6 ± 4.51	18.9 ± 11.4	27.0 ± 13.4	32.8 ± 4.86	42.4 ± 13.60	50.8 ± 6.30	29.9 ± 2.70
ANOVA	NTP	F	S	NTP $\times$ F	NTP $\times$ S	F $\times$ S	NTP $\times$ F $\times$ S
Fungi	ns	ns	***	ns	ns	ns	ns
Bacteria	ns	ns	***	*	ns	ns	ns

ns, not significant; * $p \leq 0.05$; ** $p \leq 0.01$; *** $p \leq 0.001$.

3. Discussion

It is well-known that substrate composition may strongly affect the growth and flowering of gerbera plants, presenting a difficulty with peat replacement [19,21,27]. Therefore, agronomic strategies aimed at improving the performance of peat-free substrates are worth investigating. In this experimental trial, compost decreased plant growth and gerbera plants grown in this substrate produced fewer flowers. This effect was probably due to the high pH in the root zone (around 7.0) monitored through percolate analysis (see methodology). Similarly, Sonneveld and Voogt [28] showed an inhibiting effect of high pH on flower number and weight in gerbera plants. High pH in the root zone may induce many detrimental effects on plant physiology; for example, it could induce nutritional impairments in plants and limit the availability of essential nutrients like P, Fe, and Mn [29]. Compost substrate can indeed affect gerbera yield through a decrease in nutrient availability [18], even if it generally contains an extra budget of nutrients [25]. In these experimental conditions, the nutrient concentrations were similar to those reported by other authors [30] but some nutrients were below the optimal concentrations for greenhouse potted gerbera plants, as reported by [31], almost exclusively in plants grown under an LF regime. P and Mn concentrations were significantly decreased in leaves by compost, which would agree with a reduced biomass accumulation, owing to the role that these elements play in photosynthesis [29,32].

Non-thermal plasma treatment is known to decrease the pH of both the substrate and nutrient solution [33], which can be strategic for the maintenance of an optimal pH level in the root zone, especially in those substrates that show a natural tendency to alkalinization. In the present work, pH was adjusted before irrigation so there would be no differences between the untreated and NTP-treated nutrient solutions. The presence of compost caused an alkalinization of the root zone that was not contained by NTP treatment. Yet, NTP may play a role in plant nutrition by adding nutrient elements, like nitric N [2], or by stimulating the release of nutrients (e.g., K), which are potentially available at high concentrations in some composted [25] and uncomposted [26] organic materials. Accordingly, plants supplemented with NTP-treated solution showed an increase in leaf and flower fresh biomasses, but only in peat. The possible role of NTP in growth promotion has already been documented [34]. However, under an LF regime, the NTP treatment had an opposite effect on plant dry biomass, depending on the tested substrate. Particularly, NTP led to an increase in the DW of gerbera plants grown in peat under nutrient shortage, while it worsened the performance of plants grown in compost. Such a DW reduction in compost-grown plants was caused by a significant decrease in % DW, while no significant differences were observed in plant FW. In a preceding study, we reported a decrease in leaf stomatal conductance in lettuce, under NTP treatment [35]. Stomatal activity regulates photosynthesis through CO_2 uptake and transpiration, controlling water loss; therefore, stomatal closure avoids water loss [36]. A possible reduction in stomatal conductance caused by NTP treatment might increase water reserves, explaining the lack of difference in plant FW. On the other hand, the same reduction in gaseous exchange may have limited CO_2 intake, thus causing an additional stress, in compost-grown plants. Indeed, stomatal closure is induced by ROS [37,38] that might accumulate under NTP treatment; future research in this regard will help to bring about a better understanding of NTP effects. A slight decrease in % DW was also reported in NTP-treated plants under a HF regime in both peat and compost, but it did not affect plant DW, highlighting a general effect of NTP on this parameter.

Manganese plays a crucial role in photosynthetic activity and antioxidant capacity; the deficiency of this element in plants is commonly caused by alkaline soils [32]. Mn was lower in plants grown in compost than in peat, probably due to a high pH that decreased its availability [39] despite its higher concentration in compost (data not shown). Low leaf Mn concentration in compost might contribute to the lower photosynthetic capacity and biomass allocation found in NTP-LF plants, in combination with the higher SLA, a trait ascribed as an indicator of drought resistance due to the higher photosynthetic capacity of plants carrying this feature [40]. Indeed, under an LF regime, plants grown in peat substrates showed a decrease in SLA, both using untreated and NTP-treated nutrient solution, while plants in compost treated with NTP did not present a reduction in this parameter in comparison with HF regimes. Since NTP promotes antioxidant activity in plants due to ROS production [34], it is plausible that an Mn deficiency can occur at the photosynthetic level in these conditions. This finding would be in agreement with the lowest SPAD units measured under such treatment. Therefore, under a LF regime, plants grown in compost suffered a dual source of stress compared to plants in peat.

Despite the postulated role of NTP as a replacement of N fertilizers [33], the leaf organic N of plants grown in peat under an LF regime was lower in those supplemented with NTP-treated solution than in others. However, nitric N was generally increased by NTP treatment. Indeed, nitrates are the main nitrogen form produced when using NTP to treat water [33]. Moreover, nitrates were higher in percolating nutrient solution of NTP-treated plants than in control ones, under a HF regime, supporting an increase of these molecules following NTP application. NTP treatment decreased the Fe concentration in both substrates, probably influencing its availability. However, plants maintained in compost showed an overall increase in Fe in the aerial part, probably supported by the higher Fe concentration within the compost substrate (data not shown), even if an alkaline condition can decrease Fe availability [41]. On the contrary, NTP increased K, P, and Mg concentrations in plants grown on both substrates.

Nutrient solution optimization to meet actual crop requirements has been reported as a key element to limit nutrient losses from soilless cropping, with low environmental impact [42]. In the present work, the level of fertilization affected both shoot and flower biomass, as well as the low fertilization decreased some nutrients (e.g., N, K, and P), suggesting that an important reduction of those elements is not possible for the correct fertigation of gerbera plants. Nevertheless, the Ca, Mg, Mn, and Fe contents in plants were not affected by fertilization strength, highlighting that a possible reduction of these elements within the nutrient solution allowed fertilizer saving, although further studies are necessary to assess the real minimum concentration thresholds. In the case of peat, NTP treatment increased the leaf and flower biomass of plants grown under nutritional stress, which is worth highlighting.

Although the role of NTP in sanitization has been previously elucidated [4], under these experimental conditions, the use of NTP-treated solution did not show clear antimicrobial effects on fungi and bacteria within any substrate. However, a decrease in bacteria occurred under an LF regime in plants grown in compost, the most stressful situation for gerbera plants, likely due to the potential for this microbial group to be carried into the circulating solution exposed to NTP. Fungal concentration, instead, was lower in compost than in the peat substrate. Compost is an impactful conditioner of the telluric environment that enhances microbial biodiversity, reducing the predominance of one or a few groups on the other, which in other specific cases contributes to the effective suppression of soil-borne fungal pathogens [43]. Under the natural pressure of diseases, no biotic attacks occurred during this trial; therefore, the potential role of NTP treatment in contrasting plant diseases remains to be explored.

In conclusion, NTP treatment promoted the fresh biomass production of gerbera plants grown in peat and nutrient uptake in both substrates, except for Fe, while it decreased the plant dry biomass. Interestingly, NTP had a detrimental effect on the biomasses of plants grown in compost under low fertilization, showing under nutrient shortage a substrate-dependent effect. The penalty in compost could be generally associated with high substrate alkalinization and, under low fertilization, a reduction in photosynthetic capacity, probably further decreased plant growth. The low fertilization treatment highlighted that Ca, Mg, Mn, and Fe could be reduced in respect to the standard fertilization used for gerbera production. The results highlighted as the most recommended treatment a combination of the use of peat along with NTP-treated solution containing a high concentration of nutrients, highlighting the suitability of NTP for horticultural practices. On the other hand, the combination of NTP and compost may not be adopted for this type of ornamental species.

4. Materials and Methods

4.1. Plant Material, Treatments and Growing Conditions

Seedlings in 104-hole seed trays (around 40 days old) of *Gerbera jamesonii* (L.) 'Babylon F1' (Sentier, TV, Italy) were transplanted on 20 February 2019 into 1.5 L pots ($\varnothing$ 14 cm) in 50:50 peat:perlite ($n = 320$) or 50:50 compost:perlite ($n = 320$). The peat consisted of a mix of white and black Baltic peat with a fine structure (Hawita Professional Spezial Substrate, HAWITA GmbH, Vechta, Germany), while the compost was a green compost provided by Terflor S.R.L. (BS, Italy). The pots were placed on benches (15 plants m^{-2}) inside the site-pilot greenhouse described by [44], equipped with NTP technology to treat the nutrient solution. This was located at the CREA Research Center for Vegetable and Ornamental Crops in Pescia, Tuscany, Italy (lat. 43°54′ N, long. 10°42′ E). The NTP was produced by a Dielectric Barrier Discharge device (Jonix SRL, Tribano, PD, Italy) ranging from 5–25 kV, therefore, generating 1012–1015 charged molecules cm^{-3}. The nutrient solution was supplied through a drip irrigation system with a flow rate of 2 L h^{-1} pot^{-1}, ensuring a leaching fraction of 15–20%, to minimize possible differences between the root zone conditions of plants within the same treatment. For each substrate, plants were randomly divided into eight blocks of 20 plants: four blocks were irrigated with a reference nutrient solution (high fertilization, HF), while the other four blocks were irrigated with a less concentrated nutrient solution (low fertilization, LF). Specifically, the following nutrient solution was supplied as the HF treatment: N-NO$_3$ 6.4 mmol L^{-1}, N-NH$_4$ 0.8 mmol L^{-1}, P-PO$_4$ 0.8 mmol L^{-1}, K 4 mmol L^{-1}, Ca 1.8 mmol L^{-1}, Mg 0.4 mmol L^{-1}, Na 0.31 mmol L^{-1}, S-SO$_4$ 1.01 mmol L^{-1}, Cl 0.26 mmol L^{-1}, Fe 16 μmol L^{-1}, B 12 μmol L^{-1}, Cu 0.4 μmol L^{-1}, Zn 2 μmol L^{-1}, Mn 2 μmol L^{-1}, and Mo 0.4 μmol L^{-1}. The nutrient solution supplied as the LF treatment was a quarter of the strength of the standard one. The pH of both nutrient solutions was adjusted to 5.7. Moreover, in a block for each HF and LF of both substrates, the nutrient solution was treated with NTP technology. The nutrient solution was prepared by a fertigation unit and stocked in a tank with a capacity of 1 m^3 where 0.5 m^3 of nutrient solution was continuously treated by bubbling NTP-treated air into the solution until a redox potential of roughly 800 mV was achieved.

Eight treatments ($n = 80$, four replicates of 20 plants) were then applied for 11 weeks: 1) standard nutrient solution in peat:perlite substrate (CTR-HF-Peat); 2) NTP-treated standard nutrient solution in peat:perlite substrate (NTP-HF-Peat); 3) low nutrient solution in peat:perlite substrate (CTR-LF-Peat); 4) NTP-treated low nutrient solution in peat:perlite substrate (NTP-LF-Peat); 5) standard nutrient solution in compost:perlite substrate (CTR-

HF-Compost); 6) NTP-treated standard nutrient solution in compost:perlite substrate (NTP-HF-Compost); 7) low nutrient solution in compost:perlite substrate (CTR-LF-Compost); 8) NTP-treated low nutrient solution in compost:perlite substrate (NTP-LF-Compost). In summary, three factors of variability were applied with two levels each, i.e., 1) NTP (or not, in the CTR), 2) nutrient solution concentration (standard HF or LF), and 3) substrate (peat or compost, mixed with perlite).

Climate parameters were monitored by meteorological sensors positioned in the cultivation area and recorded on a five-minute basis. The minimum, mean, and maximum daily global radiation were 2.2, 9.8, and 16.2 MJ m^{-2} day^{-1}, respectively. The minimum, mean, and maximum daily averages of air temperature were 13.6, 17.7, and 22.1 °C, respectively. The air's mean daily relative humidity averaged 56.9%.

The percolating nutrient solution was analyzed about every two weeks to monitor the root zone status and the followed parameters were measured: pH, EC, N-NO$_3$, and P-PO$_4$ (Table 3). N-NO$_3$ was measured by the Model Q46F/82 Auto-Chem Fluoride Monitor equipped with a NO$_3$ electrode (Analytical Technology, Inc., Collegeville, PA, USA). P-PO$_4$ was determined on a nutrient solution using the molybdenum-blue method. Electrical conductivity was two-fold higher in the percolate from a HF regime, in agreement with the higher nutrient concentration (EC of nutrient solution was on average 1400 and 600 µS cm^{-2} for the HF and LF regimes, respectively). Moreover, EC was higher in compost treatments (on average + 250 µS cm^{-2}) than in peat. Generally, NTP treatment caused a slight increase in EC, though this was not always observed. The pH in the percolating nutrient solution showed optimal values in peat (on average 5.7) while it was high in compost (on average 7.0), particularly in composted LF plants (+ 0.5 than in composted HF plants). The P-PO$_4$ in percolating nutrient solution was two-fold higher in HF treatments than in LF ones without any difference in substrates. On the contrary, N-NO$_3$ was higher in peat LF treatment than in compost LF, while the opposite behavior was highlighted under a HF regime.

Table 3. Chemical parameters of the percolating nutrient solution used for the fertigation of gerbera plants grown in peat or compost and fed with high (HF) or low (LF) nutrient solution, untreated (CTR) or NTP-treated (NTP). Nutrient solutions were sampled every two weeks during the trial and these values represent the means $\pm$ SD.

S	F	NTP	EC (µS cm^{-2})	pH	P-PO$_4$ (mmol L^{-1})	N-NO$_3$ (mmol L^{-1})
Peat	HF	CTR	1114 $\pm$ 78.2	5.6 $\pm$ 0.14	0.61 $\pm$ 0.016	3.8 $\pm$ 0.50
		NTP	1305 $\pm$ 101.8	5.6 $\pm$ 0.14	059 $\pm$ 0.010	4.2 $\pm$ 0.30
	LF	CTR	582 $\pm$ 37.5	6.1 $\pm$ 0.17	0.26 $\pm$ 0.008	1.4 $\pm$ 0.29
		NTP	629 $\pm$ 47.6	5.6 $\pm$ 0.25	0.26 $\pm$ 0.016	1.2 $\pm$ 0.33
Compost	HF	CTR	1429 $\pm$ 25.4	6.6 $\pm$ 0.21	0.61 $\pm$ 0.062	4.3 $\pm$ 0.54
		NTP	1630 $\pm$ 21.3	6.9 $\pm$ 0.06	0.58 $\pm$ 0.017	4.5 $\pm$ 0.64
	LF	CTR	833 $\pm$ 57.8	7.4 $\pm$ 0.05	0.25 $\pm$ 0.018	0.8 $\pm$ 0.11
		NTP	752 $\pm$ 68.6	7.0 $\pm$ 0.44	0.25 $\pm$ 0.012	0.8 $\pm$ 0.11

4.2. Plant Sampling and Mineral Element Quantification

The day before plant destructive analysis, SPAD units were measured using a SPAD-502 (Konica Minolta, Chiyoda, Japan) by averaging three measures (basal, median, and apical leaves) on 40 plants per treatment (10 per replicate). On 8 May, 24 plants per treatment (six per replicate) were collected for the final destructive analysis. Therefore, the following measures have been taken as the average value of six plants per replicate. Leaves and flowers were fresh weighted (FW) and, subsequently, oven dried at 65 °C until at a constant dry weight (DW). Specific leaf area (SLA, cm^2 g^{-1}) was also determined as the ratio between plant leaf area, measured through a scanner, and plant leaf DW, using a homogeneous bulk of three leaves from each plant.

Potassium, Ca, Mg, Fe, and Mn were measured by ICP-OES on dried leaf samples (250 mg) following digestion with 5 mL 65% HNO_3 and 2 mL 85% $HClO$ at 210 °C for 2 h. P was determined on the same digested leaves using the molybdenum-blue method. Organic N was determined on leaves and flowers by Kjeldahl distillation after dry matter digestion with H_2SO_4. Nitrates were determined as described by [45], comparing the absorbance at 410 nm against a calibration curve obtained with serial dilutions of a 1000 ppm nitrate standard solution (Merck KGaA, Darmstadt, Germany).

4.3. Quantification of Fungi and Bacteria

The abundance of culturable filamentous fungi and total bacteria were evaluated by the serial ten-fold (10^{-1} to 10^{-7}) dilution method in three replicates, as reported by [43]. Independent samples for each substrate were used. Fungi were counted on PDA (Oxoid, Thermo Fisher Scientific Inc., Waltham, MA, USA) at pH 6, with 150 mg L^{-1} added of nalidixic acid and 150 mg L^{-1} of streptomycin. Total bacteria were counted on a selective medium (glucose 1 g L^{-1}, proteose peptone 3 g L^{-1}, yeast extract 1 g L^{-1}, K_2PO_4 1 g L^{-1}, and agar 15 g L^{-1}), with added actidione 100 mg L^{-1}. Population densities are reported as the colony forming unit (CFU) g^{-1} DW of the substrate.

4.4. Statistics

Data were tested for a normal distribution using the Shapiro–Wilk normality test and were eventually transformed before the ANOVA. Data were analyzed with a three-way ANOVA (NTP, fertilization, and substrate as variables, $p \leq 0.05$) and later a Tukey's posthoc test to assess significant differences. Statistical analyses and graphs were performed with Prism 9 (GraphPad Software, Inc., La Jolla, CA, USA).

Author Contributions: Conceptualization, S.C. (Sonia Cacini) and D.M.; methodology, S.C. (Sonia Cacini), C.P., and D.M.; greenhouse experiment coordination, S.C. (Samantha Cannazzaro); formal analysis, S.C. (Samantha Cannazzaro), S.T., S.C. (Sonia Cacini), S.D.L., and C.P.; data curation, S.T. and D.M.; writing: original draft preparation, S.T. and D.M.; review and editing, all authors; supervision, D.M.; project administration, S.C. (Sonia Cacini), G.B., and D.M.; funding acquisition, G.B. and D.M. All authors have read and agreed to the published version of the manuscript.

Funding: This research was funded by Regione Toscana (Italy) under the call "Bandi POR FESR 2014–2020, Bando 2" for the project "High-Tech House Garden (HT-HG)".

Institutional Review Board Statement: Not applicable.

Informed Consent Statement: Not applicable.

Data Availability Statement: The data presented in this study are available on request from the corresponding author.

Acknowledgments: The authors acknowledge Laboratori ARCHA SRL (Ospedaletto, Pisa, Italy) and SaSe Idraulica SRL (Pistoia, Italy) for their technical support.

Conflicts of Interest: The authors declare no conflict of interest.

References

1. Bradu, C.; Kutasi, K.; Magureanu, M.; Puač, N.; Živković, S. Reactive nitrogen species in plasma-activated water: Generation, chemistry and application in agriculture. *J. Phys. Appl. Phys.* **2020**, *53*, 223001. [CrossRef]
2. Judée, F.; Simon, S.; Bailly, C.; Dufour, T. Plasma-activation of tap water using DBD for agronomy applications: Identification and quantification of long lifetime chemical species and production/consumption mechanisms. *Water Res.* **2018**, *133*, 47–59. [CrossRef] [PubMed]
3. Thirumdas, R.; Kothakota, A.; Annapure, U.; Siliveru, K.; Blundell, R.; Gatt, R.; Valdramidis, V.P. Plasma activated water (PAW): Chemistry, physico-chemical properties, applications in food and agriculture. *Trends Food Sci. Technol.* **2018**, *77*, 21–31. [CrossRef]
4. Oehmigen, K.; Hähnel, M.; Brandenburg, R.; Wilke, C.; Weltmann, K.D.; Von Woedtke, T. The role of acidification for antimicrobial activity of atmospheric pressure plasma in liquids. *Plasma Process. Polym.* **2010**, *7*, 250–257. [CrossRef]

5. Ito, M.; Oh, J.S.; Ohta, T.; Shiratani, M.; Hori, M. Current status and future prospects of agricultural applications using atmospheric-pressure plasma technologies. *Plasma Process. Polym.* **2018**, *15*, 1700073. [CrossRef]

6. Sivachandiran, L.; Khacef, A. Enhanced seed germination and plant growth by atmospheric pressure cold air plasma: Combined effect of seed and water treatment. *RSC Adv.* **2017**, *7*, 1822–1832. [CrossRef]

7. Adhikari, B.; Adhikari, M.; Ghimire, B.; Park, G.; Choi, E.H. Cold atmospheric plasma-activated water irrigation induces defense hormone and gene expression in tomato seedlings. *Sci. Rep.* **2019**, *9*, 1–15. [CrossRef]

8. Kučerová, K.; Henselová, M.; Slováková, Ľ.; Hensel, K. Effects of plasma activated water on wheat: Germination, growth parameters, photosynthetic pigments, soluble protein content, and antioxidant enzymes activity. *Plasma Process. Polym.* **2019**, *16*, 1800131. [CrossRef]

9. Carlile, B.; Coules, A. Towards sustainability in growing media. *Acta Hortic.* **2013**, *1013*, 341–349. [CrossRef]

10. Gruda, N. Sustainable peat alternative growing media. In XXVIII International Horticultural Congress on Science and Horticulture for People (IHC2010), International Symposium on Greenhouse 2010 and Soilless Cultivation. *Acta Hortic.* **2010**, *927*, 973–979.

11. Gruda, N.S. Increasing sustainability of growing media constituents and stand-alone substrates in soilless culture systems. *Agronomy* **2019**, *9*, 298. [CrossRef]

12. Raviv, M. The future of composts as ingredients of growing media. *Acta Hortic.* **2011**, *891*, 19–32. [CrossRef]

13. Massa, D.; Malorgio, F.; Lazzereschi, S.; Carmassi, G.; Prisa, D.; Burchi, G. Evaluation of two green composts for peat substitution in geranium (*Pelargonium zonale* L.) cultivation: Effect on plant growth, quality, nutrition, and photosynthesis. *Sci. Hort.* **2018**, *228*, 213–221. [CrossRef]

14. Restrepo, A.P.; Medina, E.; Pérez-Espinosa, A.; Agulló, E.; Bustamante, M.A.; Mininni, C.; Bernal, M.P.; Moral, R. Substitution of peat in horticultural seedlings: Suitability of digestate-derived compost from cattle manure and maize silage codigestion. *Commun. Soil Sci. Plan.* **2013**, *44*, 668–677. [CrossRef]

15. Larcher, F.; Scariot, V. Assessment of partial peat substitutes for the production of *Camellia japonica*. *HortScience* **2009**, *44*, 312–316. [CrossRef]

16. Savvas, D.; Karagianni, V.; Kotsiras, A.; Demopoulos, V.; Karkamisi, I.; Pakou, P. Interactions between ammonium and pH of the nutrient solution supplied to gerbera (*Gerbera jamesonii*) grown in pumice. *Plant Soil* **2003**, *254*, 393–402. [CrossRef]

17. Barreto, M.S.; Jagtap, K.B. Assessment of substrates for economical production of gerbera (*Gerbera jamesonii* Bolus ex Hooker F.) flowers under protected cultivation. *J. Ornam. Hortic.* **2006**, *9*, 136–138.

18. Caballero, R.; Pajuelo, P.; Ordovás, J.; Carmona, E.; Delgado, A. Evaluation and correction of nutrient availability to *Gerbera jamesonii* H. Bolus in various compost-based growing media. *Sci. Hort.* **2009**, *122*, 244–250. [CrossRef]

19. Khalaj, M.A.; Amiri, M.; Sindhu, S.S. Study on the effect of different growing media on the growth and yield of gerbera (*Gerbera jamesonii* L.). *J. Ornam. Plants* **2011**, *1*, 185–189.

20. Khalaj, M.A.; Suresh Kumar, P.; Roosta, H.R. Evaluation of nutrient uptake and flowering of Gerbera in response of various growing media. *World J. Environ. Biosci.* **2019**, *8*, 12–18.

21. Ahmad, I.; Ahmad, T.; Gulfam, A.; Saleem, M. Growth and flowering of gerbera as influenced by various horticultural substrates. *Pak. J. Bot.* **2012**, *44*, 291–299.

22. Sonneveld, C.; Voogt, W. Nutrient management in substrate systems. In *Plant Nutrition of Greenhouse Crops*; Sonneveld, C., Voogt, W., Eds.; Springer: Dordrecht, The Netherlands, 2009; pp. 277–312.

23. Zheng, Y.; Graham, T.; Richard, S.; Dixon, M. Potted gerbera production in a subirrigation system using low-concentration nutrient solutions. *HortScience* **2004**, *39*, 1283–1286. [CrossRef]

24. Shrikant, M.; Jawaharlal, M. Effect of fertigation level and biostimulants on quality parameters of gerbera (*Gerbera jamesonii* Bolus ex Hooker F.) var. Debora under poly house conditions. *Trends Biosci.* **2014**, *7*, 1134–1137.

25. Massa, D.; Lenzi, A.; Montoneri, E.; Ginepro, M.; Prisa, D.; Burchi, G. Plant response to biowaste soluble hydrolysates in hibiscus grown under limiting nutrient availability. *J. Plant Nutr.* **2018**, *41*, 396–409. [CrossRef]

26. Massa, D.; Bonetti, A.; Cacini, S.; Faraloni, C.; Prisa, D.; Tuccio, L.; Petruccelli, R. Soilless tomato grown under nutritional stress increases green biomass but not yield or quality in presence of biochar as growing medium. *Hortic. Environ. Biotechnol.* **2019**, *60*, 871–881. [CrossRef]

27. Arunesh, A.; Muraleedharan, A.; Sha, K.; Kumar, S.; Joshi, J.L.; Kumar, P.S.; Rajan, E.B. Studies on the effect of different growing media on the growth and flowering of *Gerbera* cv. Goliath. *Plant Arch.* **2020**, *20*, 653–657.

28. Sonneveld, C.; Voogt, W. Effects of pH value and Mn application on yield and nutrient absorption with rockwool grown gerbera. *Acta Hortic.* **1997**, *450*, 139–148. [CrossRef]

29. Barker, A.V.; Pilbeam, D.J. *Handbook of Plant Nutrition*; CRC/Taylor & Francis: Boca Raton, FL, USA, 2007.

30. Mota, P.R.D.A.; Fernandes, D.M.; Ludwig, F. Development and mineral nutrition of gerbera plants as a function of electrical conductivity. *Ornam. Hortic.* **2016**, *22*, 37–42. [CrossRef]

31. Mercurio, G. *Gerbera Cultivation in Greenhouse*; Schreurs: De Kwakel, The Netherlands, 2002; p. 206.

32. Alejandro, S.; Cailliatte, R.; Alcon, C.; Dirick, L.; Domergue, F.; Correia, D.; Castaings, L.; Briat, J.F.; Mari, S.; Curie, C. Intracellular distribution of manganese by the trans-Golgi network transporter NRAMP2 is critical for photosynthesis and cellular redox homeostasis. *Plant Cell* **2017**, *29*, 3068–3084. [CrossRef] [PubMed]

33. Ranieri, P.; Sponsel, N.; Kizer, J.; Rojas-Pierce, M.; Hernández, R.; Gatiboni, L.; Grunden, A.; Stapelmann, K. Plasma agriculture: Review from the perspective of the plant and its ecosystem. *Plasma Processes Polym.* **2020**, e2000162. [CrossRef]

34. Holubová, Ľ.; Kyzek, S.; Ďurovcová, I.; Fabová, J.; Horváthová, E.; Ševčovičová, A.; Gálová, E. Non-Thermal Plasma—A new green priming agent for plants? *Int. J. Mol. Sci.* **2020**, *21*, 9466. [CrossRef]
35. Cannazzaro, S.; Di Lonardo, S.; Cacini, S.; Traversari, S.; Burchi, G.; Pane, C.; Gambineri, F.; Cursi, L.; Massa, D. Opportunities and challenges of using non-thermal plasma treatments in soilless cultures: Experience from greenhouse experiments. In Proceedings of the III International Symposium on Soilless Culture and Hydroponics: Innovation and Advanced Technology for Circular Horticulture, Cyprus, Greece, 19–22 March 2021.
36. Lawson, T.; Vialet-Chabrand, S. Speedy stomata, photosynthesis and plant water use efficiency. *New Phyt.* **2019**, *221*, 93–98. [CrossRef]
37. Medeiros, D.B.; Barros, J.A.; Fernie, A.R.; Araújo, W.L. Eating away at ROS to regulate stomatal opening. *Trends Plant Sci.* **2020**, *25*, 220–223. [CrossRef]
38. Postiglione, A.E.; Muday, G.K. The role of ROS homeostasis in ABA-induced guard cell signaling. *Front. Plant Sci.* **2020**, *11*, 968. [CrossRef]
39. Schmidt, S.B.; Jensen, P.E.; Husted, S. Manganese deficiency in plants: The impact on photosystem II. *Trends Plant Sci.* **2016**, *21*, 622–632. [CrossRef]
40. Rafi, Z.N.; Kazemi, F.; Tehranifar, A. Effects of various irrigation regimes on water use efficiency and visual quality of some ornamental herbaceous plants in the field. *Agr. Water Manage.* **2019**, *212*, 78–87. [CrossRef]
41. Roosta, H.R.; Manzari Tavakkoli, M.; Hamidpour, M. Comparison of different soilless media for growing gerbera under alkalinity stress condition. *J. Plant Nutr.* **2016**, *39*, 1063–1073. [CrossRef]
42. Massa, D.; Magán, J.J.; Montesano, F.F.; Tzortzakis, N. Minimizing water and nutrient losses from soilless cropping in southern Europe. *Agr. Water Manag.* **2020**, *241*, 106395. [CrossRef]
43. Pane, C.; Spaccini, R.; Piccolo, A.; Celano, G.; Zaccardelli, M. Disease suppressiveness of agricultural greenwaste composts as related to chemical and bio-based properties shaped by different on-farm composting methods. *Bio. Control* **2019**, *137*, 104026. [CrossRef]
44. Burchi, G.; Chessa, S.; Gambineri, F.; Kocian, A.; Massa, D.; Milazzo, P.; Rimediotti, L.; Ruggeri, A. Information technology controlled greenhouse: A system architecture. In Proceedings of the IoT Vertical and Topical Summit on Agriculture-Tuscany (IOT Tuscany), Siena, Italy, 8–9 May 2018; pp. 1–6.
45. Cataldo, D.A.; Maroon, M.; Schrader, L.E.; Youngs, V.L. Rapid colorimetric determination of nitrate in plant tissue by nitration of salicylic acid. *Commun. Soil Sci. Plan.* **1975**, *6*, 71–80. [CrossRef]

Review

Biochemical and Physiological Plant Processes Affected by Seed Treatment with Non-Thermal Plasma

Vida Mildaziene [1,*], **Anatolii Ivankov** [1], **Bozena Sera** [2] and **Danas Baniulis** [3]

[1] Faculty of Natural Sciences, Vytautas Magnus University, LT-44404 Kaunas, Lithuania; anatolii.ivankov@vdu.lt

[2] Department of Environmental Ecology and Landscape Management, Faculty of Natural Sciences, Comenius University in Bratislava, 84215 Bratislava, Slovakia; bozena.sera@uniba.sk

[3] Institute of Horticulture, Lithuanian Research Centre for Agriculture and Forestry, LT-54333 Babtai, Lithuania; danas.baniulis@lammc.lt

* Correspondence: vida.mildazienel@vdu.lt; Tel.: +370-610-26-530

Abstract: Among the innovative technologies being elaborated for sustainable agriculture, one of the most rapidly developing fields relies on the positive effects of non-thermal plasma (NTP) treatment on the agronomic performance of plants. A large number of recent publications have indicated that NTP effects are far more persistent and complex than it was supposed before. Knowledge of the molecular basis and the resulting outcomes of seed treatment with NTP is rapidly accumulating and requires to be analyzed and presented in a systematic way. This review focuses on the biochemical and physiological processes in seeds and plants affected by seed treatment with NTP and the resulting impact on plant metabolism, growth, adaptability and productivity. Wide-scale changes evolving at the epigenomic, transcriptomic, proteomic and metabolic levels are triggered by seed irradiation with NTP and contribute to changes in germination, early seedling growth, phytohormone amounts, metabolic and defense enzyme activity, secondary metabolism, photosynthesis, adaptability to biotic and abiotic stress, microbiome composition, and increased plant fitness, productivity and growth on a longer time scale. This review highlights the importance of these novel findings, as well as unresolved issues that remain to be investigated.

Keywords: non-thermal plasma; gene expression; germination; photosynthesis; phytohormones; plant yield; secondary metabolism; seeds; stress signal

Citation: Mildaziene, V.; Ivankov, A.; Sera, B.; Baniulis, D. Biochemical and Physiological Plant Processes Affected by Seed Treatment with Non-Thermal Plasma. *Plants* **2022**, *11*, 856. https://doi.org/10.3390/plants11070856

Academic Editors: Anis Limami and Emerita Paula Jameson

Received: 31 January 2022
Accepted: 17 March 2022
Published: 23 March 2022

Publisher's Note: MDPI stays neutral with regard to jurisdictional claims in published maps and institutional affiliations.

1. Introduction

Research on plasma interaction with seeds is driven by the ever-increasing demand for food and other agricultural products in the context of scarce resources [1]. The development of new environmentally benign technologies for enhancing agricultural production is based on exploiting the natural adaptability of plants, and is essential for reducing unsustainable use of water, nutrients and agricultural chemicals [2]. The pre-sowing seed processing using different methods has been used to improve germination and seedling growth (reviewed in [3,4]). Recently, the use of plasma technologies for seed treatment has attracted increasing interest due to numerous positive effects reported on plant agricultural performance.

The first non-thermal atmospheric pressure plasma (NTP) sources were developed more than three decades ago [5]. Since then, the NTP technology has been used in various industries to modify the physico-chemical properties of treated substances in solid or liquid materials, as well as in the form of microparticles [6]. NTP offers a broad range of industrially interesting applications. The main advantage of plasma compared to other media is its ability to produce active energy-containing species that initiate physical changes and chemical reactions, which otherwise would not occur or proceed with difficulties [7]. Due to its technical and economic advantages, NTP treatment has been increasingly exploited for many practical purposes, such as sterilization, water purification, microfabrication, medicine, and agriculture.

NTP does not cause thermal damage to heat-sensitive biological systems such as living cells and tissues. Therefore, this technology has many applications in biomedical technologies [8–12], for o has shown potential to improve agronomic seed quality by surface decontamination, germination enhancement, and promotion of plant growth, as discussed in numerous reviews published recently [3,13–20]. Intensive research is currently being conducted on the application of NTP in agriculture, forestry, and food industries in many parts of the world. Despite numerous attempts, the molecular mechanisms underlying the effects of seed exposure to NTP remain elusive, and the published reviews do not cover the most recent experimental findings in sufficient detail.

Therefore, we aimed to provide an overview of the existing knowledge on changes in both biochemical and physiological processes, induced by seed treatment with NTP, expecting to aid in the understanding of this issue and to outline the possible development directions. The review consists of an introductory description of plasma as a complex agent that initiates multiple processes by interaction with seeds, followed by sections organized to distinguish the consequent stages in the complex response of plants to stress induced by NTP treatment, starting with the early changes in dry seeds, changes in germinating seeds, and changes observed in growing seedlings and plants. The effects of seed treatment with NTP on DNA methylation, wide-scale changes in gene and protein expression, enzyme activities in the affected seeds and growing plants are considered in the context of the effects observed in plant growth and yield. Particular attention is paid to the importance of seed dormancy, the role of reactive oxygen and nitrogen species and phytohormones, the mobilization of secondary metabolism, increased adaptability to stress, and the effects on the plant-associated microbiome.

2. Definitions of Plasma, Low-Temperature Plasma and Description of Different Types of Devices Used for Seed, Plant or Water Treatment

When sufficient heat is applied, solid material transforms into a liquid first and then, at a higher temperature, into a gas. As the energy supply increases, electrons receive sufficient energy to separate from atoms or molecules of gas and become electrically conductive. In this way, gas undergoes a transition to a partially or completely ionized gas, called the plasma state (physical plasma) [21]. Depending on the type of energy supplied and the amount of energy transferred to the plasma, plasma properties change in terms of electron density or temperature [22].

Physical plasma is the fourth state of matter, in which matter displays a behavior different from that observed in the other three states (solid, liquid, and gas). It is an electrically quasi-neutral gas with chemically reactive species such as electrons, ions, and neutrals [23]. Physical plasma is distinguished into high-temperature plasma (5×10^4–10^6 K) and low-temperature plasma ($\leq 5 \times 10^4$ K), denoted as LTP. LTP is subdivided into thermal plasma and NTP. Thermal plasmas are characterized by a thermodynamic equilibrium among free electrons, ions and neutrals [24].

On the other hand, the energy in NTP is supplied to free electrons only while the overall temperature of ions and neutrals remains significantly lower [5]. NTP contains charged particles (free electrons, ions) and neutral activated species, including gas molecules, free radicals, metastable particles and generated photons (including UV) [24]. The particles are not in a thermodynamic equilibrium; both ions and neutrals are near room temperature, whereas electrons are much hotter. This type of plasma is characterized by a strong thermodynamic non-equilibrium state, high selectivity, low gas temperature, and the presence of reactive chemical species. The temperature of NTP heavy particles (ions, molecules) is low, so they do not damage thermally sensitive materials, but their electrons reach sufficient energy to participate in plasma-chemical reactions. Therefore, NTP does not cause thermal damage to materials, but it is rich in various chemically reactive particles. NTP is used to treat heat-sensitive materials, such as human, animal, and plant tissue, hair, leather, wood, blood, various polymers and proteins [6,25,26].

In the laboratory, NTPs are often generated by electrical discharges in various gases, typically air, oxygen, helium, argon, nitrogen, or their mixtures. Typical discharges differ in electrode arrangement, power sources, gas pressure, and include glow discharge, plasma jets, low-pressure capacitively coupled plasma (CP) discharge, corona discharge, dielectric barrier discharge (DBD), diffuse coplanar surface barrier discharge (DCSBD), etc. (for details, see reviews [19,22,27–30]). In this review, the common abbreviation NTP is used for plasma generated by all types of devices.

Relatively low-level energy (2–5 eV) imparted to electrons initiates dissociation, excitation, and ionization reactions upon collision with gas atoms and molecules at a temperature close to ambient [31]. In the air atmosphere, it leads to the excitation of nitrogen (N_2) molecules, dissociation of molecular oxygen (O_2), and accumulation of ozone (O_3) [32]. An increase in electron energies results in the dissociation of N_2 and the production of nitrogen oxides (NO_x) that inhibit ozone production and subsequently recombine to form several other reactive nitrogen species, including the oxidant and nitrating agent peroxynitrite anion ($ONOO^-$). In the presence of water vapor, a hydroxyl radical ($^{\bullet}OH$) is produced by dissociation of water and by secondary processes involving neutralization of ions or by reactions of excited states of O_2 and N_2 [32].

The main biologically active component of NTP is a complex mixture of reactive oxygen and nitrogen species (ROS and RNS, combined abbreviation—RONS), such as superoxide anion ($O_2^{\bullet-}$), NO, hydrogen peroxide (H_2O_2), $^{\bullet}OH$, or $ONOO^-$ that play important roles as signaling messengers in eukaryotic cells [33] and are involved in the regulation of seed dormancy, germination [33,34], and plant physiology [35].

The unique transfer of chemical reactivity and energy from gaseous plasmas to water occurs in the absence of any other chemicals, but results in a product with a notable transient broad-spectrum biological activity, referred to as plasma-activated water (PAW). These characteristics make PAW a friendly treatment for a wide range of biotechnology applications, including the agriculture and food industry [36]. Although PAW belongs to a broader topic of physical plasma (effects on plant biochemical and physiological processes reviewed recently in [16]), we will not address the issue of PAW in this text.

Thus, NTP is a complex physical stressor that can be applied for seed processing. The variation in plasma sources, specific plasma parameters, and protocols used for seed treatment represents one aspect (physical) of the difficulties in comparing the results obtained by various research groups. In turn, plant seeds are tiny reproductive plant structures that have all the features of complexity typical of biological systems. Therefore, research of interactions between such two systems is an extremely challenging task that requires systematic methods and novel fundamental findings to uncover the most important determinants of both the physical and biological aspects.

3. NTP Effects on Seed Germination and Early Seedling Growth

The first reports on the application of NTP in plant biology were published in 2000 and described the effects on seed germination induced by NTP treatment [37,38]. Dubinov et al. [37] treated oat (*Avena sativa*) and barley (*Hordeum vulgare*) seeds with air glow discharge for several minutes in both continuous and pulsed mode. The continuous mode stimulated seed germination more effectively than the pulsed mode but no changes in early growth of the seedlings were observed. Volin et al. [38] applied longer treatments (2–20 min) of low-pressure radio frequency (RF) rotating plasma in fluorocarbon or nitrogen or carbon-containing compound atmosphere for the treatment of barley (*Hordeum vulgare*), radish (*Raphanus sativus*), pea (*Pisum sativum*), soybean (*Glycine max*), corn (*Zea mays*), and bean (*Phaseolus vulgaris*) seeds, and observed strong negative effects on germination in the majority of cases. A relatively complex work focused on the physiological characteristics of light-induced seed germination affected by NTP was published in 2004 by Živkovič et al. [39]. The authors considered that the stimulation of seed germination of the empress tree (*Paulownia tomentosa*) with NTP could have been explained by three different physical mechanisms: etching, surface functionalization, and deposition of small bioactive molecules. Meiqiang et al. [40] treated tomato seeds with magnetized plasma (arc discharge combined with magnetic field) and observed no effects on germination in vitro. However, seedling emergence in pots was enhanced, and some of the treatment protocols resulted in increased activity of enzymes in seedling tissues (peroxidase in hypocotyls and dehydrogenase in roots), as well as an increase in the number of fruits and fruit biomass per plant. Strong stimulation of Lamb's Quarters (*Chenopodium album* agg.) seed germination after using low-pressure microwave plasma treatment was explained by cracks found on the seed surface (electron microscope scanning), where water could better penetrate seeds [41]. Moreover, the experiment was carried out with dormant Lamb's Quarters seeds, which germinate gradually under natural conditions for many decades, and a threefold increase in seed germination was obtained compared to the control under laboratory conditions, and seedling size was also significantly larger [42].

Since then, numerous studies have been published. The results of a large part of these studies are summarized in Table 1. Considering that germination test results may depend on the used method, only the reports representing in vitro germination tests (tests for *sensu stricto* [43] germination) are presented in Table 1. For brevity, we do not give Latin names of plant species in the following text. The in vitro germination test is typically performed in a Petri dish after seed imbibition on wet filter paper under controllable laboratory conditions (temperature, light, humidity). The number of germinated seeds is periodically counted by the appearance of visible radicles that protrude through the seed coat.

As shown in Table 1, the effects of seed treatment using different NTP devices and protocols have been extensively studied in a wide variety of plants including the main strategic crops (such as wheat, corn, legumes, and oilseeds). Stimulating effects of seed treatment on seed germination in vitro have been demonstrated in most of the publications [37,41,42,44–108]. Most of these reports represent studies limited to the effects on germination and early seedling growth (from 4 days to several weeks), indicating that these two aspects are commonly recognized as the main criteria for estimating plant response to NTP treatment. Neutral [49,64,70,76,92,105,107,109–112] or negative [35,47,72,81,105,110,112–114] effects were reported in a smaller number of studies. This might be explained by the limited chances of publishing negative or neutral results, although such data may also be relevant. Unfortunately, it is difficult to develop an overarching comparative study due to the different methods used to generate plasma and the treatment protocols, i.e., non-standardized treatments.

However, the effects of different equipment on the same seed lots have been compared in a few studies. The germination and early growth of buckwheat were estimated after treatment with low-temperature plasma discharge in air gas generated in 4 types of devices [105]. A positive effect on germination and early growth was observed after applying the gliding arc device, while a strong negative effect was induced by DBD plasma; seed treatment using apparatus with a planar rotating electrode or downstream microwave plasma caused slight inhibition of germination and growth. Such results indicate that treatment protocols must be carefully optimized for each piece of plasma equipment. A recent study compared the effects of three different plasma devices (RF plasma in vacuum, microwave-driven atmospheric-pressure plasma, DBD atmospheric pressure plasma) on corn yield in the field [115]. However, all devices were equally ineffective in their experiment. Air DBD discharge was more effective than the helium plasma jet used to stimulate *Arabidopsis thaliana* seed germination [86].

Different gases can be used for plasma generation (Table 1). Ambient air is used most often, but argon, oxygen, nitrogen, helium, or mixtures of several gases can also be applied. The chemical composition of the generated reactive plasma particles depends on the gaseous phase; therefore, studies comparing the effects of different feeding gases on seed germination are expected to reveal the reactive species responsible for the observed

NTP effects. On the flip side, plasma density and UV radiation parameters, as well as the dose of energy transferred to the sample, also depend on the feeding gas [92,100,116].

Table 1. Effects of different plasma treatment equipment on plant germination in vitro.

Effect	NTP Device	Plant Species (NTP Feeding Gas if Not Air) [References]
Positive	Low-pressure CP	Ajwain [44], bean [45,46], black mulberry [47]; industrial hemp [48], lamb's quarters [41,42], lentil [45], maize [46], mung bean [50], wheat [37,45,49], oilseed rape [52], quinoa [53], red clover [54,55], rice [56], soybean [57], sunflower [58], tomato [40,59]; artichoke (nitrogen) [60], common bean (oxygen) [61], safflower (argon) [62], wheat (helium) [63], wheat (argon) [64].
	DBD plasma	Barley [65], black pine [66], cotton [67], cucumber, [68], green chiretta [69], Norway spruce [70], quinoa [53]; pea [71,72], pepper [68], radish [73–78]; rice [79,80], sunflower [81,82], sweet basil [83], thale cress [84–87], wheat [51,88,89], zinnia [84]; barley (nitrogen + 0.65% air) [90], carrot (argon) [91], coriander (nitrogen) [92], cotton (argon) [67], rice (argon + air) [93], soybean (oxygen, nitrogen) [94,95], soybean (argon) [96], sweet basil (argon + oxygen) [97], wheat (argon/air, argon/oxygen) [98,99].
	Plasma jet	Common bean (helium) [100], *Erythrina velutina* (helium) [101], fenugreek (argon) [102] mung bean (air, oxygen) [103], wheat (nitrogen) [104].
	Gliding arc	Buckwheat (air) [105], garden tree-mallow (nitrogen) [106], industrial hemp (air) [107], maize (air) [108].
Neutral	Low-pressure CP	Blue lupin [49], buckwheat [109], wheat (oxygen) [110].
	Gliding arc	Buckwheat [105], industrial hemp [107].
	DBD plasma	Coriander [92], maize [111], Norway spruce [70], radish [76], Scotish pine [112].
	Plasma jet	Mung bean (helium, nitrogen) [64]
Negative	Low-pressure CP	Barley, radish, pea, soybean, corn, bean (fluorocarbon, nitrogen, carbon-containing compounds) [35], Norway spruce [113], rhododendron [47], buckwheat (oxygen) [114], wheat (oxygen) [110].
	DBD plasma	Buckwheat [105]; Scots pine [112], pea (nitrogen, oxygen) [72], sunflower [81].

DBD plasma treatment stimulated coriander germination only when nitrogen gas was used, but treatments in air and argon gases did not affect germination [92]. The authors found that plasma-generated NO gas also stimulated germination and concluded that NO causes this effect. In contrast, irradiation of radish seeds with DBD plasma using various feeding gases showed that the N_2, He and Ar gases did not promote seedling growth, while plasma irradiation with air, O_2, and NO (10%) + N_2 improved plant growth [74]. Moreover, humid air plasma irradiation was more effective compared to dry oxygen. The authors concluded that the hydroxyl radical $^\bullet$OH and excited oxygen O were the key species responsible for the effect of NTP treatment on the growth of radish seedlings. The effects of atmospheric pressure DBD plasma on wheat germination, seed coat surface changes and permeability were compared using O_2, air, Ar, and N_2 as feeding gases [99]. In this study, O_2 plasma was not effective, but seed treatment for 4 min with air, N_2, and Ar plasma increased the germination potential by 24, 28 and 36%, respectively. These results were supported by a later study [98], in which low-pressure DBD plasma was applied, and the stimulation of wheat germination was stronger using the mixture of Ar/air compared to Ar/O_2 mixture. The effects of NTP on seed germination and growth of mung bean were strongly dependent on the feed gases used to generate plasma in the atmospheric-pressure microplasma array device [103]: air plasma was the most effective; O_2 gas also significantly stimulated germination and seedling growth; seed treatment with N_2 and He plasma did not have an impact. The effects of pea seed treatment with diffuse coplanar surface DBD working at atmospheric pressure were compared in ambient air, O_2 and N_2 atmosphere [72]. However, in this study, positive effects on germination were absent, and the longer duration of all treatments inhibited germination and caused genotoxic effects. In this respect, N_2 plasma was the most effective, while the negative effects of air and O_2 plasma were similar [117].

In summary, the results obtained by different studies on the effects of different feeding gases [74,86,92,98,99] are rather controversial. Despite numerous efforts, the dependence of the observed NTP effects on the physical, chemical and biological components of the NTP interaction with seeds is still far from being understood.

In many studies, the dependence of the exposure time on the effects of the treatment was studied. The duration of plasma treatments used by different authors varied broadly. Some studies observe effects after 0.5 s of treatment [61,87], while others apply treatments for 27 min [67] or even 130 min [62]. This may depend on both the plasma discharge parameters (used equipment) and the response of the plants. An optimal dose has been demonstrated for different plants [57,59,65,66,68,69,79,87,96,99,104,106,116]. That is, certain treatment durations stimulated seed germination, but adverse effects were observed when treatment duration exceeded the optimal value.

The dependence of seed NTP treatment effects on the germination (as well as other effects of NTP) and physiological seed status varies strongly due to a high level of complexity of the biological subject of research. The same treatment protocols can stimulate germination in some plant species but have no effect or even inhibit the germination of other plants. For example, the same duration of low-pressure CP treatment did not affect germination and early growth of blue lupine, but stimulated spring wheat and maize germination [49]; the germination of black mulberry was stimulated by 7 min treatment with a low-pressure CP device, while rhododendron germination was inhibited by the same treatment [47]. The same study [47] reported that the effects of seed treatment with NTP for black mulberry and rhododendron were stronger for freshly harvested seeds compared to the effects observed for the same seeds stored for 6 months. Similar results were obtained when treatment effects were compared in seeds of red clover cultivar 'Vyčiai' stored for different durations of time after harvest—the effects gradually decreased with an increase in seed storage time [54]. However, opposite results were obtained for radish seeds: NTP-induced positive changes in seed germination were stronger in seeds stored for 2 years after harvest compared to seeds harvested one year before the experiment [78].

Moreover, a significant difference in the response to seed treatment with NTP was observed in three different pine species [112] or in different cultivars of pea [37], poppy [116], industrial hemp [105], rapeseed [52,118], brown rice [79], red clover [54], barley [90], wheat [110], and buckwheat [109]. Differences in the effects on germination and growth have been reported even for different genetic families (half-sib) of Norway spruce [70]. In addition, the response to NTP treatment for the same lot of radish [78] or red clover [55] seeds was dependent on the color of the seeds. Such intraspecies differences and the dependence on multiple factors indicate that the response of plant germination and growth to seed treatment with NTP is determined by factors associated with slight genetic differences, peculiarities of seed structure, or physiological seed status.

Numerous studies [57,83,87,93,99,109] have reported that the effects of NTP treatment on germination were followed by similar effects on early seedling growth (from 4 days to 2 or 3 weeks). However, some studies showed that the effects on plant growth for longer periods do not correlate with the effects of NTP on germination. The negative or neutral effects of treatments on seed germination in vitro were followed by improved plant growth on a longer time scale for rhododendron [47], red clover [54], radish [76], common buckwheat [109], Norway spruce [113], and wheat [119]. In contrast, a strong positive effect of NTP on hemp germination was associated with a reduction in female plant growth [48]. Similarly, increased germination was followed by a lower growth rate of pea seedlings [120]. Such findings raise doubts about the validity of in vitro germination tests as a generally accepted criterion for evaluating the effects of the seed treatment.

Additional evidence supporting such doubts comes from comparing the effects obtained by in vitro germination tests with the results of seedling emergence in a substrate or in the field (i.e., under conditions more relevant for agricultural applications). For example, treatment of Norway spruce seeds with low-pressure CP for 2, 5 and 7 min had a strong inhibiting effect on germination in vitro, and this effect increased with the duration of treatment, but only a 5 min treatment inhibited seedling emergence in the substrate [113]. Treatment protocols with the strongest positive effects on *Andrographis paniculata* seed germination in vitro inhibited seedling emergence 7 days after sowing. They were ineffective 30 days after sowing, and the highest percentage of strong seedlings 30 days after sowing was registered for the treatment protocol (5950 V, 10 s) that did not induce changes in the seed vigor index determined in vitro [69]. Although in vitro sunflower germination was stimulated by seed treatment with low-pressure CP for 7 min, such treatment did not influence seedling emergence in the substrate [58]. A decrease in sunflower emergence induced by seed exposure to DBD plasma for 11 min was observed, while in vitro germination was not affected by the same treatment [81]. Treatment of industrial hemp seeds with low-pressure CP for 5 min increased the maximal germination percentage and germination rate in vitro, but germination yield in the field was reduced compared to the control [48]. The in vitro germination test did not reveal the effects of low-pressure CP for two buckwheat cultivars, but the percentage of seedlings that emerged in the field was significantly reduced [109].

The observed differences between NTP effects on in vitro germination kinetics and seedling emergence can be explained by several reasons. Germination in vitro is a measure of germination *sensu stricto* [43], while emergence becomes visible in the later stage of seedling growth. Numerous other factors can be responsible for differences in germination and seedling emergence kinetics, such as water penetration rate, supply of oxygen and light [46]. The presence of various compounds in the substrate and chemical interactions with the soil microbiota may also affect the seed germination rate in the substrate (all these factors are absent when seeds germinate in the Petri dish). However, the reported discrepancy between NTP effects obtained from laboratory germination tests and counting seedling emergence [58,69,81,109,113] contradicts the widespread opinion that the effects on germination can be considered as an informative indicator of plant response to seed treatment. Moreover, such findings demonstrate the importance of longer observations (at least for the entire vegetation season) under conditions used for agricultural plant cultivation.

4. The Mechanisms of NTP Effects on Seed Germination

The effect of NTP treatment on seeds results from the interaction of a physico-chemical stressor and a biological system—a small embryonic plant enclosed in a covering called the seed coat (or testa), usually with some food reserve. Both interacting sides are characterized by a high level of complexity and diversity. NTP is a complex stressor consisting of different components, including an electric discharge, electromagnetic and UV radiation, changes in pressure (in the case of low-pressure NTP) and temperature, the different and unstable composition of reactive chemical species, electrons, and photons. NTP reactor construction, geometry, energies, and treatment exposures vary greatly in different studies. On the other hand, seeds also vary in size, shape, color, external and internal structure, and water content. The structural, physiological, and biochemical properties of seeds strongly depend on the plant species. In addition, seeds can also have considerable intraspecies differences due to genetic polymorphism; in addition, seeds of the same variety and from the same lot are polymorphic by degree of dormancy or other traits [45]. For example, the ability to germinate differs within a population of seeds heterogeneous by size, shape, weight, or color [121].

Nevertheless, all seeds have three main structural parts: seed coat, embryo, and food reserve (endosperm or cotyledons) (Figure 1). The roles of these parts in the interaction with NTP may be quite different, firstly due to various locations and distances from the seed surface, i.e., the external seed layers are easily accessible for such NTP components as reactive particles, photons, and direct discharge energy, while internal structures are shielded. The biological and biochemical properties of the cells that make up these layers are also different. The seed coat is an external protective seed structure, and due to its location, plays a key role in the interaction with both physical and chemical components of NTP. Reactive species generated by plasma discharge are characterized by a short lifetime, which can be further reduced by entering the environment of reduced organic compounds within cells of a living system [122]. The distance covered in plant tissues by the most reactive species, such as the hydroxyl radical OH$^\bullet$ or peroxynitrite ONOO$^-$ (half-life 10^{-9} s) is 1 nm only, for singlet oxygen $^1O_2$ (half-life 10^{-5} s) and superoxide anion $O_2^{\bullet-}$ (half-life 10^{-6} s) this distance is 30 nm [123]. Therefore, it is unlikely that these species could penetrate the seed structures deeper than the external coat surface. Other species, such as hydrogen peroxide H_2O_2 and nitrogen monoxide NO, are characterized by longer half-lives (10^{-3} s and 3–5 s, respectively), can easily diffuse and penetrate membranes (movement distances in order of µm) [123]. These species could be considered as candidates for reaching deeper seed layers when they originate from the outside (plasma source). However, reaching embryonal or aleurone cells for external RONS could only be possible if they are located immediately under a thin, porous, or cracked seed testa (shown by the punctured lines in Figure 1).

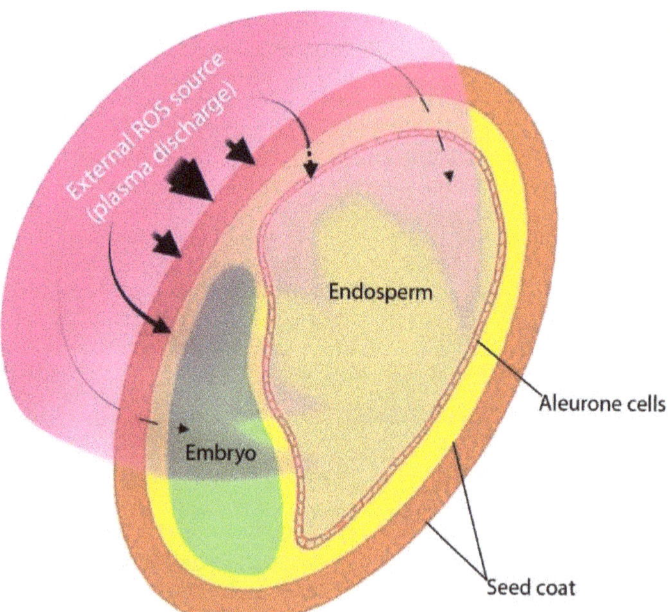

Figure 1. Schematic representation of NTP generated reactive species penetration into structures of the monocot seed.

A model study on the penetration of NTP-generated ROS through the membrane of phospholipid vesicles (cell mimics) has been performed [124]. It was demonstrated that plasma-derived ROS are delivered into cells over a sustained period without compromising cell membrane integrity, but the presence of protein (serum) significantly reduced the transfer efficiency of ROS into the vesicles. However, the seed coat is much more complex and consists of several layers of different cells in most seeds, such as an epidermal, sub-epidermal (parenchymal) cells and a palisade layer, which often contains pigmented cells [48]. Coat cells contain numerous protective secondary metabolites and pigments that function as powerful RONS scavengers [125,126]; therefore, their amount can modulate the penetration and the effects of external RONS. Although seed cells are dehydrated [127], structures of the outer layer tissue can be strongly acidified after an interaction between external NO_2 and NO_3 (generated by NTP) and the remaining water. Seed exposure to external ozone O_3 and hydrogen peroxide H_2O_2 also stimulates germination [128,129], indicating that the interaction of these ROS with the seed surface could contribute to the effects of NTP.

4.1. The Role of NTP-Induced Physical and Chemical Changes in the Seed Coat

In general, the interaction of NTP with different organic surfaces results in chemical modifications, changes in charge and structure [130,131]. The RONS and UV-photons generated by NTP react with the components of the seed surface or penetrate the external layer of the seed coat, inducing significant changes in the elemental composition of the seed coat, while prolonged treatments result in surface etching that is visible in scanning electron microscopy (SEM) pictures. One of the major chemical changes induced by NTP in seed coats of various plants is a decrease in surface carbon content accompanied by an increase in surface oxygen content (an increase in the O:C ratio), while nitrogen and silicon content remain unchanged [51,53,61,68,78,83,132,133]. Additional elements appear on the seed surface (magnesium, calcium) and other elements are found in trace amounts. Using X-ray microfluorescence, Ambrico et al. demonstrated redistribution of P, K, Mg and Zn among the different parts of basil seeds after plasma treatment [83]. Energy-dispersive X-ray analysis performed on radish seeds [78] revealed an increase in the content of C, O, Mg, Al, Si, Cl, K, and Ca after treatment; minerals such as P and S, usually located along the embryo axis [12], were detected on the seed surface after treatment. Furthermore, NTP reduced seed pH [51,134] and increased seed surface saturation with charged oxygen and nitrogen groups, as demonstrated for wheat [51], quinoa [53], beans, and lentil seeds [45]. Analysis of NTP effects by attenuated total reflectance—Fourier transform infrared spectroscopy (ATR-FTIR) was used to characterize the presence of specific chemical groups on the surface of maize [111] and pine seeds [112]. The results indicated the presence of polar nitrogen and oxygen-containing groups and the removal of lipids from the seed surface.

The introduction of polar groups and the removal of hydrophobic substances leads to chemical etching of the seed coat structure and increased hydrophilicity of the seed surface. In numerous studies carried out on seeds of different plants, seed coat erosion after exposure to NTP was observed by structural changes in SEM images, and it was evident that the degree of coat etching increased with the duration of the treatment [45–47,49,51,53,60,62,68,69,71,72,79,86,89,90,92,98,99,102,103,106,110,135]. In some studies, such etching was not observed, possibly due to the shorter duration of NTP treatment [78,85]. These findings led to the conclusion that the stimulating effects on seed germination depend on NTP-induced changes in the physical and chemical structure of the seed coat, and increased hydrophilicity leading to improved wettability and faster penetration of water into the seed after imbibition [45]. Changes in seed surface wettability are measured by water contact angle, and the fact that NTP treatment enhances wettability is well documented (e.g., [45,51,52,57,61,67,72,88,97,100,103,112]). Some studies have also evaluated the water penetration rate into the seed (e.g., [45,57,71,88]) and the results are fully consistent with the statement that water penetration rate is increased in seeds treated with NTP compared to the control.

The correlation of germination rate with changes in seed surface structure after plasma treatment has been demonstrated in numerous studies. However, certain inconsistencies can be observed in the reported relationships between changes in seed wettability and the effects of NTP on germination. For example, wheat germination was not stimulated by NTP despite a very strong increase in wettability [45]. Water contact angle increased with the duration of treatment with coplanar DBD system in pine seeds; however, after 60 s of treatment, a strong inhibition of germination was observed despite the maximal effect on wettability. Similar results were obtained on wheat seeds [51]. Such findings indicate that although the suggested hypothesis that NTP effects on germination can be explained by enhanced wettability and water absorption sounds rational, such an explanation is not sufficient. Water penetration is a key event to initiate germination; however, seed germination is a complex process, controlled by many other determinants besides water.

4.2. The Impact of Seed Dormancy and Phytohormones

Seed formation completes the process of reproduction of the plant vegetation cycle. Many seeds are dormant immediately after maturation; in this state, seeds do not germinate for some time, even when the environmental conditions are favorable for germination [136]. Dormancy is an evolutionarily developed physiological and biochemical mechanism important for plant survival as it prevents premature germination, helps seeds remain viable for a longer time in the soil and under adverse environmental conditions, and favors seed distribution over long distances.

Seed dormancy is divided into exogenous (imposed by the seed coat), endogenous (related to the embryo), and combinational dormancy (caused by exogenous and endogenous reasons) [137]. A more detailed classification system distinguishes five classes of dormancy [43,137]: (1) *physical dormancy* (denoted as PY, is determined by exogenous factors—the seed coat forms a barrier to seed germination); (2) *physiological dormancy* (PD, is an endogenous state of dormancy caused and maintained by phytohormones that inhibit germination); (3) *morphological dormancy* (MD) is also endogenous—seed embryos are not fully developed or immature, although differentiated. Such seeds require time to fully develop, sometimes 4 and more years; (4) combinational dormancy (PY + PD) is determined by internal and external factors, which are a combination of physical and physiological dormancy (germination is restricted by the seed coat and phytohormone-inhibited embryos); non-dormant seeds, ND—dormancy is naturally absent or alleviated.

Physiological dormancy is the most common type among seed-producing plants, characteristic of most (about 80%) plant species [136]. This type of dormancy is under the strict control of phytohormones, a diverse group of biologically active signal molecules comprising a complex network that controls virtually all processes in plants [43,138]. Two antagonistic phytohormones: abscisic acid (ABA) and gibberellins (GA) [43,139,140], are central to the regulation of seed dormancy and germination, but numerous other hormones are involved: auxins, cytokinins, salicylic acid (SA), and jasmonates may all be modulators of effects of ABA and GA [141,142]. In addition, brassinosteroids, ethylene, and NO are also recognized as supplementary regulators of germination [142–144]. ABA is a germination inhibitor, strongly suppressing the synthesis of the germination promoters, the GAs. A high concentration of ABA (and low level of GA) in dormant seeds maintains dormancy, and gradually decreases with seed storage time during natural dormancy alleviation (after-ripening) [139–141], i.e., the transition from a dormant to a germination competent state is associated with an increase in the GA/ABA ratio. Here, it is pertinent to mention, that the biological effects of certain phytohormones in a tightly coordinated hormonal network depend not on their absolute concentrations, but on their ratio with other hormones, since the functions of each phytohormone are modulated by other, antagonistic and synergistic hormones [145]. Seed exposure to various dormancy-breaking agents (such as stratification, scarification or chemical treatments) stimulates germination due to an increase in GA/ABA [136]. The explanations of the effects of NTP on seed germination could be related to different dormancy types, as shown in Table 2.

For example, NTP effects for plant species characterized by physical seed dormancy (e.g., legumes) can depend on NTP-induced modifications on the surface of the seed coat, leading to improved wettability and permeability to water and gases, as well as leakage of germination inhibitors from the imbibed seed. For species belonging to the physiological dormancy type, changes in the seed phytohormone balance should be more important.

Stimulation of germination in seeds with an underdeveloped embryo is hardly possible, since such an embryo needs time to achieve the germination competent state. Despite several repetitive NTP treatments, our efforts to stimulate germination of morphologically dormant seeds of European ash and English yew (both belonging to MD dormancy class), were unsuccessful—seeds did not germinate either in control or treated groups (data not published). The effects on seeds of the combinational dormancy class should be determined by the changes in the seed coat and the amounts of phytohormones. Non-dormant seeds germinate rapidly without stimulation, therefore NTP treatment can be ineffective, while the effects on seedling growth are still possible. In the case of physiological dormancy, freshly harvested seeds are dormant, but their dormancy is gradually alleviated due to after-ripening (related to a progressive increase in GA/ABA, changes in gene expression, and numerous biochemical changes [146]).

Table 2. Possible mechanisms of NTP effects on germination for seeds with different dormancy types.

Dormancy Type	Key Determinant	NTP Effects Due to
Physical, PY	Permeability is limited by the seed coat	Changes in the surface and improved permeability of the seed coat
Physiological, PD	Phytohormone balance (low GA/ABA)	Shift in the balance of phytohormones (GA/ABA increase)
Morphological, MD Morphophysiological, MPD	Under-developed or immature embryo Under-developed embryo and phytohormones	NTP not effective
Combinational: physical and physiological, PY + PD	Germination is limited by the seed coat and inhibited by phytohormones	Combination of the involved factors (both coat and phytohormonal changes)
Non dormant seeds, ND	Seeds are germination competent	Negligible effects on germination

The dependence of the dormancy status on seed storage time could explain the variations in the observed effects of NTP on the same seeds tested at different times after harvesting, which is observed in some studies [47,54].

Several attempts to study the NTP treatment-induced changes in the amounts of phytohormones in dry seeds have been performed using high performance liquid chromatography (HPLC) analysis [55,58,75,81,95,147,148]. Taking into account that HPLC is not sufficient for the quantitation of phytohormones, the validity of the published results must be re-evaluated using a combination of HPLC with mass spectrometry (MS), an adequate method for phytohormone analysis.

In the later study [78], LC/MS analysis was used and it was reported that the stimulation of radish seed germination after treatment with DBD plasma is related to changes in ABA and GA content. Moreover, it was demonstrated that the amounts of phytohormones involved in germination control and the shift in the GA/ABA induced by DBD plasma treatment depend on seed color. For example, in grey radish seeds (2017 harvest) DBD plasma treatment increased in GA/ABA ratio (as recalculated on pmol/g basis from data published in [78]) much stronger (5.3 times) compared to brown seeds (1.3 times). Treatment of grey (but not brown) seeds with a DBD plasma-induced positive effect on maximal germination percentage and seedling growth. The effects of seed treatment with DBD plasma in seeds (2018 harvest) on GA/ABA, germination kinetics were less pronounced in seeds of both colors.

In summary, the reported results [78] provide evidence that at least for some plant species, NTP treatment can stimulate germination due to induced changes in seed phytohormone content, that is, an increase in the GA/ABA ratio. Rapid decrease in ABA content [75] indicates that NTP is an efficient dormancy-breaking agent. This also places phytohormones in the up-stream part of the signal transduction pathway(s) that mediate the effects of NTP on plants.

4.3. Involvement of the Internal Generation of Reactive Oxygen Species in the Effects of NTP on Germination and Plant Growth

Plant cells possess numerous enzymatic systems for the generation of different ROS and RONS species [122,149–152]. These activities comprise an inherent part of the normal metabolism, and the presence of ROS sensing mechanisms determines the key role of ROS in the response of plants to physiological and environmental stimuli through highly complex signal transduction processes [35,151]. The production of reactive species is strongly enhanced under biotic or abiotic stress [35,122]. Both H_2O_2 and NO are also important signaling molecules involved in the regulation of physiological plant processes including seed germination [153–156]. The crosstalk between phytohormones, and ROS regulates seed dormancy and germination specifically [157,158]. Therefore, internal ROS and RNS generation systems should be involved in the biochemical and physiological responses of seeds exposed to NTP treatment.

ROS-induced ROS release has recently been described in plants as the mechanism that mediates long-distance rapid systemic signaling in response to biotic and abiotic stress [159]. That is defined as the production of ROS by one cell that triggers the enhanced production of ROS by a neighboring cell so that a process propagates from one part of the plant to another. This mechanism could be involved in the transfer of NTP-induced signals from the external layers of the seed coat to internal seed structures (endosperm or embryo) (Figure 1). However, it is not easy to obtain experimental evidence for such a hypothesis.

Estimation of the direct effects of NTP treatment on ROS production in intact seeds also represents a methodical challenge. Indirect evidence is provided by several studies that report an increase in the number of paramagnetic centers in seed after NTP treatments detected by EPR spectroscopy [78,113]. Significant enhancement of EPR signal (up to 30% compared with the control) was observed in Norway spruce seeds 20 h after treatment with low-pressure air NTP and radiofrequency electromagnetic field (EMF) [113]. Seed EPR spectra are composed of signals assigned to Fe(III), Mn(II) and stable organic radicals—lipid peroxides, melanin-type pigments and semiquinones originating from oxidized antioxidants located in the seed coat [160,161]. The interaction of antioxidants with reactive species generated by NTP may lead to an increase in EPR signal; however, the finding that EMF treatment also enhanced EPR signal (to a smaller extent compared to NTP) indicated a possible contribution of the internal ROS generating systems in this response of the seed to experienced stress. The results of this study were confirmed on radish seeds, treated with DBD plasma [77]. In addition, it was demonstrated that the increase in EPR signal is higher in grey seeds compared to brown seeds [78].

In several studies, the amounts of different ROS or RONS in the control seeds and those treated with NTP were compared. Low-pressure DBD treatment significantly (up to 3 times) increased H_2O_2 concentration in wheat seeds [89]. Ar/Air plasma induced a larger increase compared to Ar/O_2 (stimulation of germination by Ar/Air treatment was also stronger compared to Ar/O_2). None of these treatments affected the concentration of NO in wheat seeds. However, other authors [51] did not find H_2O_2 in the control or atmospheric DBD plasma-treated wheat seeds, although increased amounts of nitrites and nitrates were detected in the treated seeds. An increased amount of superoxide anion and NO and the intensity of infrared absorption of the hydroxyl group were detected in *Arabidopsis thaliana* seeds treated for short duration (up to 3 min) DBD (air) plasma, related to stimulation of germination [87]. Longer treatments resulted in an increased level of H2O2 and inhibition of germination. Still, it is not clear what part of the increase of RONS amounts found in NTP treated seeds is due to the internal ROS and RONS generating systems. On the other hand, some studies have demonstrated an increase in RONS level in young seedlings growing from NTP treated seeds [72,89,98], and that provides an argument for internal generation. Modulation of H2O2 release from germinating Norway spruce seeds treated with NTP and EMF have also been demonstrated [113].

4.4. Effects of Plasma Treatment on Enzymatic Activities in Dry and Germinating Seeds

Germination, or the appearance of a new plant from a seed, begins with the absorption of water by the dry seed and is completed when the elongating radicle breaks through the seed coat. Hydration of a dry seed is an essential step for seed germination. In addition, numerous external and internal factors, such as seed coat structure, physiological seed condition (dormancy, senescence, etc.), temperature, availability of light, oxygen, stimulators or inhibitors of germination may exert a strong impact on germination kinetics [43]. Biochemical and physiological activities in seeds are activated within minutes of a cell becoming hydrated and oxygenated, even before seed tissues are fully imbibed. At this stage, the rates of numerous metabolic processes such as mitochondrial respiration, selective translation, and degradation of stored mRNAs, DNA repair, synthesis and translation of new mRNAs are increasing with substantial transcription of new genes. These processes lead to the mobilization of reserves required for the rapid growth of embryo cells and their division.

Processes initiated in a dry (not imbibed) seed by NTP treatment may interfere in different ways with the complex molecular machinery which is switched on in the imbibed seed. DNA methylation is a conserved epigenetic modification that is important for gene regulation, genome stability, and plant development; it is also involved in plant responses to biotic and abiotic stress conditions [161]. The dynamics of DNA methylation are significantly impacted by oxidants, such as ROS and NO [162], therefore, at least some of the NTP effects could be related to DNA methylation in dry seeds or in growing seedlings. Until now, NTP-induced epigenetic DNA modification has been reported in a single study, carried out on heat-stressed dry rice seeds [80]. DNA methylation level was modified in the promoter regions of genes encoding enzymes of ABA biosynthesis and degradation, as well as α-amylase genes. Seed treatment with DBD plasma caused significant hypermethylation of the OsNCED5 promoter and hypomethylation of the OsAmy1C and OsAmy3E promoters, and these changes matched their expression patterns. The authors concluded that NTP could facilitate germination by upregulating ABA catabolism genes and downregulating ABA biosynthesis genes in heat-stressed seeds [80]. NTP also restored the expression of α-amylase genes in heat-stressed seeds to the level of control. This enzyme is crucial for starch mobilization in the endosperm during germination [43].

NTP effects on various enzymes in affected dry seeds, resulting in facilitated mobilization of nutrients have been demonstrated in several studies [50,80,87,98] (Table 3).

Table 3. Summary of the published findings on NTP-induced changes in biochemical processes in dry seeds.

NTP Induced Change	Plant Species [Reference]	Implication
Increased number of paramagnetic centers (EPR signal) in seeds	Norway spruce [113], radish [77,78]	Increased production of stable organic radicals indicates the interaction of seed components with ROS (NTP generated NTP or internally produced)
Increased ROS amount in dry and in germinating seeds	Wheat [89], *A. thaliana* [87], Norway spruce [113], soybean [95]	Induced internal RONS production; RONS involved in NTP effects
Change in the balance of phytohormones involved in the control of germination	Radish [78]	NTP effects on germination are related to induced shift in GA/ABA
Gene expression and expression or activities of proteins (including enzymes)	Mung bean [50], rice [56,80], *A. thaliana* [87], spinach [147], soybean [95], wheat [98]	Induced changes in the expression or activities of proteins/enzymes involved in mobilisation of resources for germination and antioxidative defense
DNA methylation	Rice [80]	NTP induces changes in gene expression through changes in DNA methylation.

An increase in the activities of hydrolytic enzymes (amylase, protease, and phytase) along with a decrease in trypsin inhibition activity and phytic acids was reported in dry seeds of mungo bean (*Vigna radiata*) treated with low-pressure plasma [50]. NTP treatment increased the expression of amylolytic enzyme pullulanase in spinach seeds [146]. In wheat seeds [98], activities of superoxide dismutase (SOD) and ascorbate peroxidase (APX) showed no significant changes in response to any of the plasma treatments compared to the non-treated seeds. However, Ar/O_2 (but not Ar/Air) treatment caused a significant increase in the catalase (CAT) activity in the seeds compared to the controls. The effects of DBD plasma on the activities of antioxidant enzymes SOD, CAT and peroxidase (POD) were determined in *Arabidopsis thaliana* seeds and in 7-day-old seedlings [87]. NTP did not affect SOD activity, strongly increased POD and decreased CAT activity in seeds, while the effects observed in seedlings were different: optimal treatment protocols enhanced the activities of all three enzymes.

5. Effects of Plasma Treatment on Biochemical and Physiological Processes in Growing Seedlings and Plants

A large number of studies on the effects induced by seed treatment with NTP on biochemical and physiological processes were carried out on growing seedlings or plants (summarized in Table 4). The body of the reported findings indicates multiple effects of a relatively short seed treatment with NTP on plant traits, detectable at the epigenomic, transcriptomic, proteomic, and metabolic levels and resulting in changes in numerous physiological plant processes.

5.1. Impact on Plant Epigenetics and Protein Expression

Epigenetic changes observed at certain regulatory sites indicate the impact of NTP-induced stress on genome functioning in growing plants. Changes in DNA methylation in the sequences of numerous genes and the up-regulated expression of mRNA of their protein products were reported in 6-day-old soybean sprouts growing from argon DBD plasma treated seeds [96]. The demethylation of cytosine was demonstrated in the regions of five genes—two subunits of ATP synthase, *ATP a1*, *ATP b1*, and *TOR, GRF 5*, and *GRF 6* genes. The authors explain the positive effects of NTP on seedling growth by an increase in the expression of proteins involved in stress response: enzymes important for energy metabolism, antioxidant defense, as well as important regulatory proteins, such as TOR kinase and six GRF proteins.

Several studies using a targeted gene expression analysis provided further insights into plant response to seed treatment with NTP. An experimental study performed on wheat [174] reported that seed treatment with DBD plasma resulted in increased transcription of heat shock factor A4A, which is involved in the plant response to abiotic stressors [200], as well as in an increased expression of POD and phenylalanine ammonia lyase (PAL), a key enzyme for phenylpropanoid biosynthesis [201]. Plasma irradiation upregulated transcription rates of WRKY1 transcription factor were reported for seedlings of industrial hemp [167] and blue sage [166]. WRKY transcription factors have diverse biological functions in plants, but are key players in the plant response to biotic and abiotic stresses [202]. Ghaemi et al. [166] also detected activation of *AREB1*, another transcription factor that regulates ABA signaling involved in stress tolerance [203]. Seed treatment with DBD plasma up-regulated the expression of the *LEA1 and SnRK2 genes* involved in the resistance to drought stress in wheat seedlings [81].

Table 4. Summary of published findings of NTP-induced biochemical changes in growing seedlings and plants.

NTP Induced Change in	Plant Species [Reference]	Implication:
DNA methylation	Soybean [96]	Impact on gene expression through DNA methylation
Gene and protein expression, including proteins involved in photosynthesis, stress response, secondary metabolism	*Arabidopsis* [163,164], bitter melon [165], blue sage [166], industrial hemp [167], maize [168], soybean [94,96], sunflower [58,169,170], tomato [171–173], wheat [81,174,175]	Changed expression and amounts of proteins in growing plants
Enzyme activities	*Arabidopsis* [87], artichoke [60], *A. fridae* [176], green chiretta [69], lemon balm [177], maize [178–180], pea [18], pepper [181], rice [182], soybean [94,96], sweet basil [183], tomato [172,184], wheat [89,98,185,186]	Changes in plant metabolism and antioxidant defense
Amount of phytohormones in plants	Maize [108], pea [71], tomato [173]	
Content of photosynthetic pigments	carrot [91], wheat [63,98,185,187,188], Norway spruce [70], maize [108,180], rice [93,178], soybean [189], spinach [147], tomato [172]	Improved growth due to up-regulated photosynthesis.
Activity or efficiency of photosynthesis	common buckwheat [109], maize [179], rice [182], pea [121], purple coneflower [190], soybean [189], sunflower [81], wheat [191]	
Secondary metabolism	coriander [92], brown rice [56,79], buckwheat [109], industrial hemp [48], maize [180,192], Norway spruce [70], purple coneflower [190], rapeseed [118], red clover [54,148,193], soybean [94], spinach [25], wheat [188,192]	Increased content of secondary metabolites is important for establishment of seedlings, plant fitness, stress resistance, communication with microorganisms.
Communication with microorganisms	*Arabidopsis* [194], sunflower [170], soybean [94], red clover [148]	Changed interactions with pathogens and beneficial microorganisms
Plant growth for the entire vegetation period and production yield	*Arabidopsis* [85], black mulberry [47], common buckwheat [109], garlic [195], industrial hemp [48], maize [179,196,197], Norway spruce [113], purple coneflower [190], peanut [198], red clover [54], rhododendron [47], tomato [40,171,199], wheat [63,191,192,197]	Improved plant growth for longer period of time. Persistent effects show the potential of NTP treatment for Plasma in Agriculture

Li et al. [171] showed that NTP treatment of tomato seeds up-regulated transcription of *9-cis-epoxycarotenoid dioxygenase 1 (NCED1)* that conferred ABA accumulation in a *respiratory burst oxidase homologue 1 (RBOH1)*-dependent manner, leading to an improved tolerance to cold stress in tomato plants. In addition, a higher accumulation of transcripts from ABA signaling pathway genes was observed in 2-day-old *Arabidopsis* seedlings germinated from NTP-treated seeds, although their transcripts were significantly down-regulated after 4 days [163]. It has been proposed that NTP treatment accelerates ABA accumulation in the early growth stages and ABA regulates ROS and Ca^{2+} concentrations to affect the stomatal aperture, which is associated with NTP-induced stimulation of seedling growth. Holubova et al. [168] proposed that the accumulation of the heat shock protein HSP101 and HSP70 encoding gene transcripts is stimulated at the early stage (24 h) of maize seed germination due to the increased demand for the chaperones required to recover the cell proteins damaged by the NTP-treatment. Up-regulation of enzymes involved in the regulation of the cell redox balance, such as catalase (CAT) and superoxide dismutase (SOD), was observed in the roots of wheat seedlings [175] or the leaves of sunflower seedlings grown from NTP-treated seeds [169]. It was suggested that NTP treatment-induced upregulation of ATP synthase plays a stress-mitigating role in soybean and sunflower seedlings [96,169].

Comparative transcriptome analysis of NTP-enhanced early seedling growth in *Arabidopsis thaliana* revealed a differential expression of 218 genes mainly related to pathogen defense or stimulus/stress-response biological processes and involved genes of the MAPK signal transduction pathway or the glutathione, phytohormone or amino acid biosynthesis pathways [164]. Transcriptome analysis of sunflower seedlings revealed the effect of NPT seed treatment on the expression of genes involved in plant growth and development processes such as starch and sucrose metabolism, pentose and glucuronate interconversions, DNA replication, and plant hormone signal transduction [169].

In addition to the stress response signaling pathways, seed treatment has been shown to activate genes involved in the regulation of seedling development. Perez-Piza et al. [94] showed that the accumulation of the expansin gene (*GmEXP1*) transcript involved in root elongation was significantly enhanced in 5-days-old soybean seedlings grown from NTP-treated seeds. The positive effect of NTP on growth was associated with the accumulation of proteins involved in the regulation of cell growth and division, such as growth-regulating factor (*GRF*) and serine/threonine protein kinase *TOR* was observed in the leaves of sunflower seedlings [169]. In the related field of plasma application, an increase in root hair density was associated with PAW-enhanced growth of *Arabidopsis* seedlings, and the relationship of this phenomenon with the function of *COBRA-like 9* involved in root hair development and the cell wall modification enzymes xyloglucan endotransglycosylases/hydrolases *XTH9* and *XTH17* was confirmed by gene expression analysis [204].

Furthermore, several examples of enhanced activity of the plant secondary metabolite synthesis pathway have been described, including up-regulation of four enzymes of the cannabinoid pathway in hemp [167], deacetylvindoline-4-O-acetyltransferase implicated in alkaloid synthesis in pink periwinkle [205] or cinnamoyl-CoA reductases involved in lignin biosynthesis, as well as rosmarinic acid synthase in blue sage [166], and two key enzymes of the carotenoid biosynthesis pathway, phytoene-synthase and phytoene desaturase, in bitter melon (*Momordica charantia*) [165].

Enhanced gene expression of numerous enzymes in tomato seedlings was found after seed treatment with NTP [172]: antioxidant enzymes (POD, CAT, SOD, polyphenol oxidase (PPO), and glutathione transferase (GST)), biosynthetic enzymes (PAL and P450 family enzyme allene oxidase, 12-oxo-phytodienoic acid reductase) and enzymes involved in histone modifications (histone acetyltransferase and histone-lysine N-methyltransferases), enzymes of oxidative signaling (mitogen-activated protein kinase (MAPK) and respiratory burst oxidase (RBOH)).

The results of these studies indicate that plants respond to seed NTP irradiation by a wide scale modulation on the level of genome expression leading to multiple changes in metabolic and physiological processes. Methods that unveil the global balance of gene expression or the accumulation of proteins and metabolites of the cell can provide a comprehensive picture of the biological pathways or processes involved in the intricate response to NTP in a variety of organisms. Proteomics has been used to investigate a complex response of epithelial cells to NTP treatment, to address the perspectives of its medical application in the treatment of skin cancer [206,207].

In plants, an initial proteomics study on common sunflower (*Helianthus annuus*) response to seed treatment with a low-pressure NTP device revealed consistent but low amplitude differences in protein abundance in two-week-old shoots which implies that seed treatment did not trigger a distinct defense response or another stress-induced developmental program, but rather predetermined a subtle modulation of plant metabolic processes associated with enhanced growth [58]. Low-amplitude gene expression changes are characteristic of low-intensity stress (eustress) stimuli such as those described for low-intensity UV-B treatment [208]. Differences in protein abundance in the sunflower were mainly localized to chloroplasts of shoot tissues and were linked to regulation of the photosynthetic activity with no detectable changes in protein abundance in the roots [58]. Furthermore, the effect exerted by the NTP treatment was very similar to that of the EMF treatment leading to the conclusion that the EMF constituent of plasma was likely the cause of the observed plant response. A later study using an ambient atmosphere DBD device revealed similar low-amplitude protein abundance changes in sunflower seedlings but remarkably were mainly localized to the roots [170]. The differentially expressed proteins were involved in amino acid biosynthesis and derived compounds, lipid biosynthesis and protein metabolism, including stress response-related proteins, as well as carbon fixation and energy metabolism. It has been proposed that the discrepancies between the effects of the two plasma types could arise from the presence of atmospheric pressure air in the DBD type device that contributes to an abundance of reactive species during seed treatment, which could be implicated in the direct modulation of root metabolic and stress response processes, as well as induce changes in plant-associated microbiome [170].

It has been presumed that NTP-induced priming of seeds resulting in a long-term effect of stress or disease resistance could be mediated by an epigenetic mechanism of gene expression regulation [176,177]. In addition, this seems a plausible mechanism for the multi-generational effect on plant growth as was reported for *Arabidopsis* and zinnia [84], where the NTP treatment of seeds for two generations resulted in the most prominent growth enhancement of the plants. Recently, it has been suggested [209] that NTP-generated nitric oxide (NO) is involved in the regulation of epigenetic modifications leading to an enhanced proliferation of mammalian stem cells. NO is a well-established epigenetic modifier implicated in the regulation of gene expression and the development of animal [210] and plant [211] cells; however, its role in the NTP-induced plant response remains to be elucidated. Seed treatment performed in the argon atmosphere [96] induced DNA demethylation, therefore a direct role of NTP-produced ROS or RNS in the regulation of DNA methylation should be excluded and suggest the presence of a different mechanism for the NTP-induced epigenetic changes in seeds or implies that the changes occurred at later stages of the seed germination and seedling development.

5.2. Changes in Enzyme Activities

Positive effects of NTP on germination were associated with an enhanced early growth of the seedlings, and numerous reports are available on changes in the amounts of soluble protein [72,96], sugars [81,94,178], proline [178,192], ATP amount [96] or activities of different enzymes, mostly involved in the antioxidant defense, such as SOD, CAT, POD, APX, dehydroascorbate reductase (DHAR), glutathione reductase (GR) (e.g., [18,60,69,87,89,96,98,173,176,179,181–186,189]). In most of these studies, increased activities of antioxidant enzymes in seedlings growing from NTP-treated seeds were observed. These findings are in line with the accepted standpoint that after exposure to biotic and abiotic stressors plants respond to an increase in intracellular ROS levels by raising the level of endogenous antioxidant defense [212]. Thus, increased ROS amounts in NTP treated dry or germinating seeds (Table 3) are followed by an increased expression or activity of enzymes of antioxidant systems in leaves and roots of growing seedlings. An analysis on the extent of enhancement in the activity of enzymes based on different reports was provided in a recent review [17]; the results showed that even a low power and short exposure time NTP treatments can strongly induce cellular antioxidant systems. The largest increase (up to 100% compared to the untreated control) was reported for POD activity, although the observed effects for POD, CAT and SOD remarkably varied. In line with enhanced expression (Section 5.1) and activities of antioxidant enzymes, decreased levels of a lipid peroxidation marker malondialdehyde (MDA) were detected in seedlings of different plants growing from NTP-treated seeds [52,69,89,186,213].

In addition, rice seed treatment with NTP also stimulated enzymes involved in the biosynthesis of secondary metabolites (SMs), such as PAL, polyphenol oxidase (PPO), shikimate dehydrogenase (SKDH), cinnamyl alcohol dehydrogenase (CAD), as well as enzymes of the primary metabolism (sucrose synthase, sucrose phosphate synthase, and acid invertase) under normal and salinity stress conditions [182].

5.3. Effects on the Amount of Phytohormones in Plant Tissues

Phytohormones actively interfere with different signaling pathways, therefore, changes in their amount and balance may be responsible for the multifaceted response to NTP treatments triggered in plants. The expression of genes responsible for the synthesis and degradation of phytohormones and the amounts of phytohormones in seeds is strongly affected by NTP treatments (Table 2). The existing data also reveal significant changes in the amounts of phytohormones in seedlings growing from the treated seeds [108]. It has recently been reported that the positive effects of gliding arc plasma treatment on maize seed germination and seedling growth were related to significant changes in the amounts of phytohormones in the leaves of 14-days old seedlings [108]. The effect of NTP on phytohormones was dependent on the duration of the treatment: short-term (180 s) plasma treatment decreased the levels of stress hormones (ABA, SA, JA and JA isoleucine) as well as active cytokinins, while longer exposures (300 s, 600 s) that had a stronger effect on plant growth, led to their increase. The elevated amount of cytokinins after longer exposure correlated well with an enhanced germination and seedling growth.

The effects of seed exposure to DBD plasma on the content of auxins and cytokinins in 14- and 21-days-old pea seedlings were studied by Stolaric et al. [71]. The results indicated that NTP treatment increased the biosynthesis of some auxins and cytokinins, as well as their catabolites and conjugates. The authors concluded that such changes can lead to improved seedling growth. Tomato seed treatment with NTP resulted in enhanced levels of three cytokinins and reduced ethylene concentrations in leaves exposed to darkness [173].

Reports have been published on NTP treatment-induced changes in amounts of phytohormones in wheat [89] and soybean [95] seedlings. However, only HPLC did not allow for analysis in these studies. The quality of the performed HPLC detection [89,95] was not sufficient for quantitation of phytohormones, in contrast to the LC-MS/MS analysis with internal standards as performed in [71].

5.4. Effect on Photosynthesis

NTP-induced stimulation of plant growth can be explained by stimulated photosynthesis. Numerous studies reported an increase in the amount of photosynthetic pigments in the leaves of seedlings growing from NTP-treated seeds. Such results were obtained for carrot, wheat, maize, rice, spinach, tomato and other plants (Table 4). Thus, positive NTP effects on photosynthesis may be at least in part related to an enhanced content of chlorophyll, which is an important component of the plant photosynthetic system. NTP-induced changes in the expression of numerous proteoforms associated with photosynthetic machinery in sunflower seedlings [58] can be associated with improved photosynthetic indices [81]. Positive effects on the parameters of photosynthetic efficiency were also observed in growing plants of common buckwheat [109], soybean [189], and purple coneflower [190]. Increased activity of photosynthesis in seedlings growing from low-pressure CP treated seeds was observed in wheat [191]. The same was found for DBD plasma on maize [179], while seed exposure to gliding arc plasma did not induce activation of photosynthesis in maize seedlings [108]. In pea seedlings, a decrease in the efficiency of photosynthesis along with retarded growth was reported [120], in line with the statement about a close relationship between the photosynthetic function and plant growth.

5.5. Changes in Amount of Secondary Metabolites

Plants synthesize countless secondary metabolites (SM) of diverse chemical structures and biological activities. SMs, such as terpenes, phenolics, polyketides, and alkaloids, play an important role in plant defense and adaptability to the changing environment [214]. In plants, SM function as signaling compounds, antibiotics, antioxidants, allelochemicals, chelators, pheromones, toxins, differentiation effectors, communication means, etc. In addition, most of the SMs are biologically active substances exerting various effects on the cellular processes in other organisms, therefore SMs are responsible for the medicinal value of plants and have a wide range of practical applications, particularly in pharmaceutics and food industry (as food additives) [215]. It is well established that plant response to biotic and abiotic stressors involves an increased accumulation of SMs [216,217]. Among the many molecular signals involved in the enhancement of SM synthesis in response stress, central roles belong to the phytohormones SA and JA [214], therefore, NTP-induced changes in SMs amounts described below may be caused by NTP-induced changes in the amounts of phytohormones. However, an increase in plant SMs induced by stress experienced in a seed stage was not highlighted. An increase in the amounts of SM in plants growing from NTP treated seeds was documented in numerous studies (Table 4).

An increase in phenolic compounds after seed treatment with NTP was observed for wheat, brown rice, buckwheat, spinach, purple coneflower, red clover, maize, Norway spruce, and coriander (Table 4). In the study of Yodpitak et al. [79] on the six cultivars of brown rice, the concentration of secondary metabolites was periodically measured in extracts of germinating seeds. The maximal increase in concentrations of phenolic compounds, anthocyanins, phytosterols, triterpenoids, and vitamin E was reached one day earlier after NTP treatment compared to the control groups. Low-pressure oxygen CP modulated the amount of glucosinolates in rapeseed seeds, and the effects differed among cultivars: a decrease was found in 'Westar' and 'Kizakinonatane', while an increase was detected in 'Nanashikibu' cultivar [118]. Such findings [79,118] imply that NTP effects on SM synthesis NTP are observed even in germinating seeds.

Treatment of purple coneflower seeds with low-pressure CP for 2, 5 and 7 min resulted in a drastic increase in the content of phenolic acids, vitamin C and radical scavenging activity in the leaves of a widely used medicinal plant [190]. Such NTP effects (increasing biomass, content of biologically active SMs, antioxidant activity) may be relevant to increasing the production of natural products. In the leaves and roots of red clover, low-pressure CP also induced substantial changes in the content of isoflavones. The total amount of the two major isoflavonoids was increased only in 'Sadūnai' cultivar, but in both 'Sadūnai' and 'Vyčiai' cultivars, NTP induced remarkable changes in the ratio of formononetin and biochanin A [54]. The composition of flavonoids in root exudates of red clover plants growing from NTP treated seeds also significantly differed from the control plants, so that the amount of 7,4'-dihydroxyflavone, daidzein, quercetin, and kaempferol was increased more than twice [148]. A strong increase in daidzein and formononetin content in the roots of red clover plants grown from low-pressure CP treated seeds was reported by other authors [193], although isoflavonoid changes in the leaves were different from those obtained in another study [54]. The amounts of isoflavonoids daidzein, genistein, and daidzin were diminished 1.5–1.8 fold compared to the control (while genistin content was not affected) in the roots of soybean plants growing from DBD plasma treated seeds [94]. DBD plasma-induced changes in the amounts of photosynthetic pigments and the total phenolic compounds were different in seven genetic families of the Norway spruce [70].

The amounts of quercetin glycosides and kaempherol glycosides increased in pea seedlings growing from DBD plasma treated seeds [120]. Treatment of industrial hemp seeds with low-pressure CP reduced female plant growth and the amount of cannabidiolic acid (CBDA) in inflorescences, while vacuum treatment significantly increased it [48]. An increase in the amounts of anthocyanins after NTP treatment was reported in the roots of maize seedlings [192] and in wheat leaves [191]. Certain NTP treatment-induced changes were found even in the harvested seeds of common buckwheat: the content of rutin was not affected, but the amount of quercetin was higher in the seeds of 'VB Vokiai' and 'VB Nojai' cultivars, although longer treatment (for 7 min) reduced it in 'VB Nojai' seeds [109].

Thus, numerous studies provide evidence that changes in SM content of seedlings are characteristic of the response of different plant species to seed treatment with NTP (Table 4). Along with the enhanced expression and activities of antioxidant enzymes [18,60,69,87,89,96,98,173,183–186,189], an increased level of SMs may be important for the improvement of plant fitness, adaptability and disease resistance.

5.6. NTP Effects on Plant Adaptability and Stress Resistance

Plants, as sessile organisms, are tied to their habitat and require efficient strategies to avoid or adapt to stress [214]. Depending on nature, plant stressors can be divided into abiotic (drought, heat, salinity, high light, mineral deficiency, low temperature, wounding, ozone, UV-A, UV-B), biotic (insects, pathogens, elicitors, bacteria, fungi, virus), and anthropogenic (herbicides, air pollution, peroxyacyl nitrates, radicals, acid rain, acid fog, heavy metal load) [214–221] With the growing anthropogenic stress load, climate change, and human population growth, innovative ways for making crops more resilient to environmental stressors are highly demanded. To date, several studies have demonstrated that seed treatment can induce adaptive plant responses to various environmental stressors.

Evidence for NTP effects on plant adaptability to abiotic stress (such as chilling, draught, salinity) have been obtained in several studies [52,67,80,89,165,172,174,178,182,222]. NTP-treated cotton seeds were subjected to germination tests that can indicate the ability of the seed to overcome adverse environmental conditions, and the results revealed that NTP treatments enhance germination rates under warm-germination or metabolic-chilling conditions [67]. Wu et al. [178] observed the improved resistance of maize to salt, cold and drought stress at the seedling stage after NTP treatment. Li et al. demonstrated that NTP treatment stimulated oilseed rape seed germination and improved morphometric parameters of seedlings under drought conditions for drought-sensitive and non-sensitive rape cultivars [52]. Sheteiwy et al. [182] found that rice seed treatment with NTP stimulated seedling growth, enhanced the activities of antioxidant enzymes, as well as enzymes involved in SM biosynthesis and primary metabolism, improved the uptake of macro- and micronutrients, resulted in a significant decrease in ROS and MDA contents and helped the plants to recover their cell turgidity under salinity stress. Improved resistance to salinity stress was found in wheat seedlings growing from DBD plasma-treated seeds [174]. Such an effect of NTP was associated with an up-regulated expression of heat shock factor HSFA4A in the roots and increased activities of PAL and POD. The effect of low-pressure CP (air and helium) treatments on the germination and seedling growth of alfalfa seedlings under simulated drought stress conditions was investigated [222]. The authors concluded that NTP treatment had a significant effect on the adaptability of alfalfa seeds in different drought environments since vigor indexes of the treated seeds under different extents of drought stresses were higher compared to untreated controls. Alleviation of the adverse effects of drought stress on wheat germination and seedling growth was induced by DBD treatment, and it was associated with an enhanced ABA synthesis and SOD, CAT, and POD activities, increased amount of proline and reduced ROS content under drought stress [89]. Similar changes along with enhanced tolerance to drought stress was observed in tomato seedlings growing from plasma jet treated seeds [172]. It was found that that treatment of rice seeds with DBD plasma treatment significantly improved the germination of seeds exposed to high-temperature stress during grain filling [80] and these effects are related to changes in epigenetic regulation and expression of genes involved in ABA synthesis and degradation as well as several α-amylase genes.

Several studies have reported an improvement in the adaptive plant response to anthropogenic factors after NTP seeds treatment, such as contamination with toxic chemicals, heavy metals, and nanoparticles. Pre-sowing seed treatment with NTP reduced DNA damage in pea seedlings caused by a toxic concentration of radiomimetic zeocin [223]. Similarly, priming of seeds with NTP seed activated defense-related mechanisms and mitigated toxicity signs of selenium and zinc oxide nanoparticles for lemon balm and bell pepper plants, improving their growth-related characteristics [177,181], similar effects were observed for *Astragalus fridae* seedlings grown in an in vitro culture medium supplemented with silicon nanoparticles [176].

An improved adaptability of plants to biotic stressors was demonstrated by NTP-induced effects on plant resistance to pathogens. It was found that DBD plasma treatment applied to soybean seeds with a high incidence of seed-borne pathogens (*Diaporthe/Phomopsis* complex) increased plant growth and alleviated the negative effects of the disease in seedlings (reduced photosynthetic performance, chlorophyll content, discoloration, retarded growth) [189]. The authors came to the conclusion that the effects of NTP treatments are partially dependent on the removal of pathogens from seeds, however, the impact of NT—induced changes in seedling antioxidant defense potential and content of SM also cannot be excluded.

The effect of pre-sowing treatment of maize, narrow-leaved lupine and winter wheat seeds with low-pressure CP on plant resistance to common diseases during vegetation and crop yield was studied in laboratory and field experiments [192]. Seed treatment suppressed a number of fungal crop diseases such as boil smut in maize, root rot in lupine and winter wheat at different growth stages. The results can be explained by both the decreased level of seed infection and changes in the defensive potential of growing plants, at least in the roots of maize seedlings NTP-induced increase in the content of non-enzymatic antioxidants (proline, anthocyanins as well as total phenolic content) was determined.

Low-pressure CP treatment increased the resistance of tomatoes to bacterial wilt, caused by *Ralstonia solanacearum* with an efficacy of 25% [224]. Such an effect was related to increased production of H_2O_2 and POD, PPO and PAL. NTP treatment increased both germination and plant growth, absorption of calcium and boron compared with the controls.

Thus, improved plant performance and NTP-decreased frequency of infections can be explained by both seed decontamination and mobilization of plant defense mechanisms. Changes in plant communication with beneficial microorganisms also can be involved. In any way, improved adaptability and stress can result in the better establishment of young seedlings, increased growth and yields of plant production.

5.7. Effects on Plant Growth and Productivity

Studies aimed to determine the effects of seed treatment with NTP on plant growth and production yields were performed on annual plants and perennials, and significant positive effects were obtained for *Arabidopsis* [85], black mulberry [47], common buckwheat [109], garlic [195], industrial hemp [48], maize [179,196,197], Norway spruce [113], purple cone-flower [190], peanut [198], red clover [54], rhododendron [47], tomato [40,171,199], wheat [63,191,192,197] (Table 4).

The effects of air DBD plasma irradiation of *Arabidopsis thaliana* seeds on plant growth were studied from the beginning of cultivation to the harvest [85], and growth acceleration in all the growth stages were observed. NTP treatment for 3 min resulted in a shorter harvest period (11%), a significant increase in the total seed weight (56%), one seed weight (12%), and seed number (39%). Low-pressure CP treatment of garlic cloves induced increases in the water uptake and accelerated root growth in a laboratory experiment. The effects were not so obvious in a field experiment, although some trend for increased plant height and dried bulb mass was observed [195]. Compared to the control, plant height, root length and fruit yield (up to 26%) were enhanced and the incidence of disease was decreased in tomato plants growing from DBD plasma treated seeds [199]. Similar effects of NTP treatment on the growth of tomato plants and fruit yield were obtained by Meiqiang et al. [40]. Treatment of tomato seeds with plasma jet discharge resulted in higher shoot length (up to 36%), root length (up to 13%), fresh weight (up to 30%), and increased shoot to root ratio (up to 19%) with respect to control [171]. Positive effects of red clover seed treatment with low-pressure CP on plant biomass gain in the field 5 months after sowing were much stronger (up to 49%) in comparison to the effects (neutral or below 10%) observed in the early growth stages [54]. The field experiment that lasted 1.5-years demonstrated that low-pressure

CP treatment markedly stimulated peanut germination and growth, increased branch number per plant, pod number per plant, compared to the control. The yield was improved by 10% [198].

An NTP-induced increase in wheat growth and yield was reported in several field studies. Seed treatment with low-pressure helium CP increased morphometric plant parameters (plant height, root length and fresh weight, leaf area, etc.) at seedling and booting stages, and the yield of treated wheat was increased by 6% compared to the control [63] Glow discharge (air and air/O_2) plasma treatment stimulated wheat germination, growth in the field and increased yield by 20% over control [197]. A field experiment on wheat seedlings growing from low-pressure CP treated seeds was repeated for two years [191], and the results showed that NTP increased biomass yield (up to 44%), grain yield (up to 35%) and 1000 grain weight (up to 22%).

Glow discharge (Ar + O_2) plasma treatment enhanced the germination of maize seeds, plant growth and development, productivity (1.3%) and improved nutritional composition (moisture, ash, fat, and crude fiber) of leaves, and increased iron and zinc content in grains [180]. Cianioti et al. [196] examined the impact of maize seed treatment with DBD plasma on the germination, physiology, yield and quality characteristics of two maize hybrids with high and low germination capability. It was found that NTP significantly improved the germination and growth of both cultivars. Maize yield increased by 18–25% compared to untreated groups. Low-pressure CP treatment decreased the level of seed infection, stimulated field germination, plant growth and resistance to pathogens during the vegetation period, as a result, the grain yield increased for winter wheat and maize by 2% and for narrow-leaved lupine by 27% compared to control plants. However, in the field study performed in the field by Ahn et al., positive effects of NTP treatment on maize growth were not found [115]. In this study, corn seeds were treated by three types of NTP devices: RF plasma in a vacuum, microwave-driven atmospheric-pressure plasma, DBD atmospheric-pressure plasma, and the effects on the yield of harvested corn observed in six different locations were not significant.

In several cases, NTP effects on plant growth for longer time periods did not coincide with the effects observed on germination in vitro. Treatment of common buckwheat seeds with low-pressure CP did not affect germination in vitro and decreased the percentage of seedlings that emerged under field conditions, NTP treatment strongly improved buckwheat growth and yield, so that the weight of seeds collected per plant for both cultivars was significantly higher (up to 70–97%) compared to the control [109]. Low-pressure CP treatment inhibited germination of the Norway spruce but stimulated plant growth: 17 months after sowing, seedling height was 50–60%, the number of branches was 40–50%, exceeding the same parameters of the control plants. Similar results were obtained for rhododendron—low-pressure CP treatment characterized as distressful based on changes in germination and increased growth of seedlings (stem and root branching, leaf count and surface area) after 13 months [47].

The results of the studies described in this section lead to the conclusion that NTP effects on annual plants persist for the entire vegetation season, and for perennials—at least for several vegetation seasons [47,113,190]. In addition, seed treatment can lead to a significant increase in biomass production and grain yield. Therefore, the application potential of NTP treatments in agriculture is not limited to the effects on germination.

6. NTP-Induced Changes in the Seed Microbiome and Plant-Microbial Interactions

The NTP-generated reactive chemical species and UV can damage microorganisms and the technology finds versatile application as a sterilizing agent used in medical practices and the food processing industry (reviewed by [225–229]). The NTP treatment can be easily applied to most plant seeds or grains due to their small size and low water content. The impact of NTP on the microorganisms residing on a surface or inside seeds can have two-sided implications, which can lead to different applications. On the one hand, the antimicrobial effect of NTP prolongs the shelf life of seeds, it is beneficial for the safety of seed-derived foods, such as sprouts [230–232], and decontamination with NTP could reduce the occurrence of seed-born fungal or bacterial diseases [29,213,230,233–235]. On the other hand, the seeds carry an assembly of microorganisms that are important for the survival and vigor of the germinated seedlings and plants [236–238] and the NTP-mediated inactivation (or activation as suggested by [239]) of this part of the seed microbiome could lead to a long-term effect on plant development, resistance to pathogens and productivity.

Efficient microbial inactivation of seeds by NTP treatment was reported for chickpea [240], alfalfa, onion, radish, cress [241], cucumber, pepper [68], lentil [232], rice [230], buckwheat [114], barley [51] and wheat [235] grains. Incidence of fungal pathogenic strains of *Aspergillus* spp. and *Penicillum* spp. [242], *Rhizoctonia solani* [233], and bacterial pathogen *Xanthomonas campestris* [234] on seeds was largely reduced by NTP treatment in air, sulfur hexafluoride or argon atmosphere. However, filamentous fungi of the genera *Alternaria* and *Epicoccum* proved to be resistant to NTP treatment [114].

NTP treatment was also effective against bacterial spores [243–245]. Efficient inactivation of an indicator strain of spore-forming *Bacillus atrophaeus* was achieved by direct application of NTP [246]. However, *B. atrophaeus* or *Geobacillus stearothermophilus* endospore inactivation on barley and wheat grains was less efficient and required extended treatment [51,247]. This is presumed to result from microorganisms being sheltered by the uneven surface of grains [246,247], and the inactivation efficiency depends on the substrate moisture level and the NTP supply settings that determine the outcome of the reactive species [241].

NTP mediated inactivation of microorganism cells was linked to impairment of cell membrane and wall integrity and damage of integrity and function of intracellular components such as DNA and protein by NTP-generated ROS and RNS produced in air atmosphere [228,248]. When the treatment is carried out in an inert gas atmosphere such as argon, the sterilization effect is proposed to be related to the impact on the microbial membrane by energetic species of electronically excited inert gas ions, metastable particles and atoms [249]. In addition, a negative effect of the NTP-generated UV irradiation on microorganisms has been suggested [249–252] and discussed by [226].

Notably, the NTP effect depends on the dose and composition of generated reactive species, and plasma treatment may enhance the vitality of bacteria and their plant growth-promoting properties [239]. The microbial cell response to sub-lethal NTP doses was addressed in bacteria using proteomic and transcriptomic studies. The response of *Salmonella enteritidis* was associated with an increase in the abundance of proteins related to carbohydrate and nucleotide metabolism, suggesting an enhancement of energy metabolism [253]. Yau et al. [254] linked upregulation of bacterioferritin B protein to NTP-induced oxidative stress response in *Pseudomonas aeruginosa*. Similar activation of the oxidative stress response and DNA repair processes attributed to the concerted action of ROS and UV irradiation were revealed by gene expression analysis in *E. coli* [255] and *Deinococcus radiodurans* [256]. Argon plasma upregulated numerous genes associated with cell wall synthesis and degradation in *E. coli* cells and the response was different from air atmosphere plasma [257,258]. Krewing et al. [259] performed a genome-wide screening in *E. coli* for plasma-protective genes that confer plasma resistance. The study revealed 87 genes, most of which protect against H_2O_2, O_2^- and NO. Upon exposure to low-temperature nitrogen gas plasma of *Bacillus cereus* cells, the transcriptome profile showed a large overlap with profiles obtained from conditions generating reactive oxygen species [260].

Plant-associated microbiota has an immense effect on agro-ecosystem health by supplying nutrients to plants or priming resistance to systemic disease [261]. A reduced microbial diversity is associated with an impairment of the normal state of a plant, often caused by pathogens [262,263]. Therefore, it seems reasonable to investigate the extent of the NTP-induced changes in the composition of microbiota vertically transmitted through seeds and their consequences on plant growth and adaptability. As has been observed for NTP-induced responses in living organisms, NTP antimicrobial properties vary depending on treatment conditions. For instance, Los et al. [51] observed no effect on overall counts of natural microbiota of barley grains upon the NTP treatment that was effective for inactivating bacterial inoculum on the seed surface. However, a significant reduction of the natural seed surface microbiota was detected by Mitra et al. [240]. Recent 16S rRNA gene sequencing-based analysis of the bacterial composition of in vitro germinated *Arabidopsis* seedlings induced by seed treatment with NTP showed a ~4-fold reduction in the number of identified bacterial genera [194]. Furthermore, the analysis of the leaf microbiome of plants germinated from the NTP-treated seeds and grown under greenhouse conditions revealed the effect of NTP on bacterial diversity [194]. A similar effect was observed in 2-week-old seedlings of common sunflower [170], and links between the changes in microbial composition and observed stimulation of root and lateral organ growth were proposed.

However, it remains to be answered whether NTP-induced changes in plant-associated microbiome occurred due to a direct effect of plasma on microorganisms residing on seed surface or inside the seeds or is a consequence of NTP-induced changes in plant and, especially, in root physiology that result in altered interaction with the soil microbiome and colonization by endophytic bacteria. An example of such interaction was obtained in a study of soybean plants by Perez-Piza et al. [94]. It was found that improvement in biometric parameters of soybean seedlings growing from DBD plasma treated seeds was associated with enhanced nodulation and stimulated N-fixation by nodular rhizobacteria. Such findings [94] could be explained by NTP-induced changes in secondary metabolism, since flavonoids released from roots to rhizosphere are recognized as signal molecules promoting the formation of nodules by symbiotic bacteria in the roots of legumes [264]. However, the content of flavonoids in the roots of soybean was decreased [94]. NTP-induced stimulation of root development and enhanced nodulation was reported in red clover roots [148], and increased amounts of flavonoids were detected in root exudates.

In any case, the performed studies [94,148,170,194] have provided evidence that NTP treatment is capable of inducing changes in the plant-associated microbiome that may mediate secondary effects on plant physiology as well as the agroecosystem environment.

7. Perspectives

A large number of recent studies have reported new findings based on epigenomic, transcriptomic, proteomic, and metagenomic approaches. These findings reveal the level of complexity of the molecular mechanisms involved in plant response to stress caused by short-term seed treatment with NTP. Many important aspects at the molecular level (such as DNA methylation, massive changes in gene and protein expression, the contribution of RONS and phytohormones, strong positive effects on plant growth and yield, mobilization of secondary metabolism, increased adaptability to stress, effects on the plant-associated microbiome, etc.) emerged in the last decade. However, despite invaluable progress, a complete structure of wide-scale modulations induced by interaction with NTP at the different hierarchical levels of the plant is far from being understood, and accumulative changes in the biochemical and physiological processes are under-explored.

Taking into account that the effects of seed treatment with NTP are persistent for longer time periods (at least for the entire vegetation season for annuals, Table 4), we suggest the multifaceted effects of NTP on plants should be considered as a multi-step process (Figure 2) which starts from NTP signal perception (stage 1) and early response events in a dry seed (with few exceptions [72,167], imbibed seeds have not been exposed to NTP). Changes induced by NTP in seeds before imbibition (summarized in Table 3) comprise stage 2 of the stress response, and water penetration after seed imbibition induces the further processes (stage 3) resulting in modified kinetics of germination (Table 1) and early seedling growth (effects on seedling growth for several weeks are usually following trends of NTP-induced changes in germination). These changes have an impact on the further processes in the growing plant (at least for the entire season of vegetation in annuals).

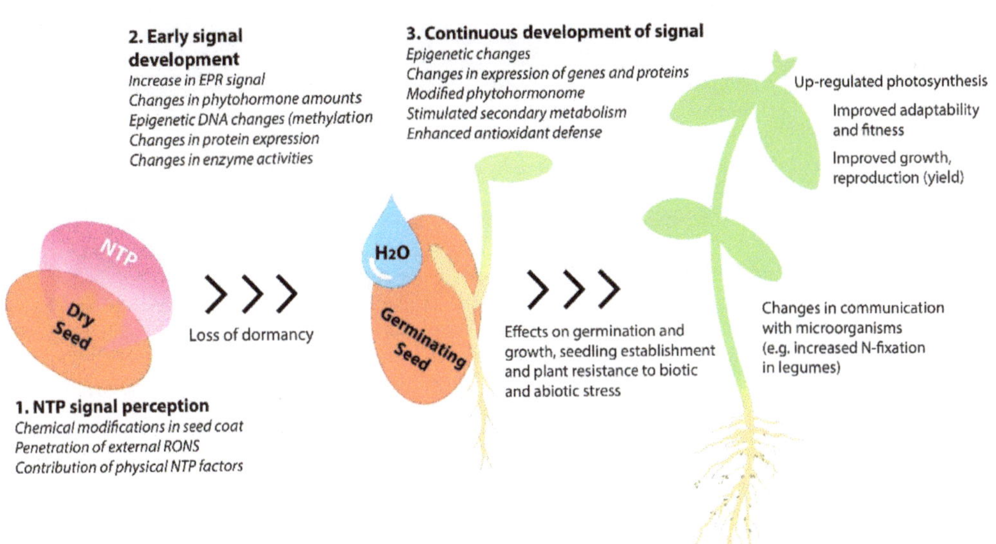

Figure 2. Schematic representation of three stages in the time course of the NTP-induced signal development in seeds and plants.

From a research perspective, one of the most poorly defined parts of the puzzle is the molecular systems in seeds that are responsible for the perception of NTP signals. Receptors for NTP sensing in seeds have not been established; furthermore, more than one receptor can be involved in the perception of different factors constituting a complex NTP signal. The first event in the interaction of seeds with NTP is chemical and structural changes on the seed coat surface (Section 4.1), and NTP receptors are most probably located there. The possible contribution of the perturbations in membrane permeability, the activity of ion channels (e.g., Ca^{2+} channels) or ROS producing enzymes residing in the membranes of cells or cellular walls in the seed coat or in layers of seed structure underneath the coat remain to be determined. In animal cells, Ca^{2+} channels TRPA1 and TRPV1 are involved in the response to atmospheric-pressure NTP [265]. Although TRP channels are not found in plants [266], other ROS-sensitive Ca^{2+} channels function in plant cells [267]. Still, it is not clear if such channels operate in seeds. NTP-induced chemical and physical changes on the seed surface facilitate water penetration or lead to an increase in the EPR signal (Section 4.1). These effects might also be considered among the up-stream factors. ROS-induced ROS release (RIRR), a process in which one cellular compartment or organelle generates or releases ROS, triggering the enhanced production or release of ROS by another compartment or organelle, was first described in animal cells, but later it was discovered in plants [159]. It was supposed that in plants, RIRR is involved in cell-to-cell communication, i.e., enhanced production of ROS by one cell triggers the enhanced production of ROS in a neighboring cell, so that process propagates from one part of the plant to another. It is tempting to speculate about the possible contribution of the external (or NTP-generated) ROS-induced internal ROS release as one of the possible modes for NTP signal perception.

The interaction between receptors and NTP should result in the production of downstream secondary messengers initiating the response to NTP by turning on yet unknown signal transduction pathways. Knowledge of signal perception, as well as signal transduction mechanisms, is crucial for a better understanding of the different outcomes of NTP treatments, a balance between eustress and distress response, or plant species/genotype dependent NTP effects.

A large breakthrough has been made recently in understanding signal development in stages 2 and 3. Evidence of the contribution of epigenetic DNA changes was reported and a large number of studies on changes in gene and protein expression have been published (Sections 4.2 and 5.1). The role of NTP-induced changes in the amounts of phytohormones and RONS production in seeds has been well-documented in agreement with the basic concepts placing cross-talk between phytohormones and ROS at the core of the combinatory plant response to abiotic stress [155,214]. In this review, some rationale for a possible relationship between seed physiology (such as dormancy types) and the effects of NTP on germination was provided (Table 2). However, more detailed information on the role of miRNA, histone modifications or RONS contribution to NTP-induced changes in protein expression through the recruitment of mitogen-activated protein kinases (MAPK) and protein phophorylation, or oxidative modifications of mRNA or posttranslational modifications of proteins (such as carbonylation and nitrosylation) is still not available. The detailed structure or sequence of the involved NTP signal transduction pathways (what is up-stream, what is down-stream?) operating in seeds and in plants is not yet elucidated. Moreover, it can be expected that NTP-triggered signaling cascades are different between different plants or treatment modes.

Multiple changes started in the dry seed develop further and possibly diverge during germination and early seedling growth (Table 3). The down-stream imprint of these changes on the biochemical and physiological processes is observable for the entire vegetation season (or more seasons, for perennials) and manifests by effects on metabolic and protective enzymes, photosynthesis, secondary metabolism, composition of microbiome (Table 4). That results in improved plant growth and reproduction (seed yield), adaptability to stress, increased plant fitness, performance, and better survival chances under unfavorable conditions.

Due to their anhydrobiotic state, seeds are highly resistant to environmental factors [268]. This trait is of key importance for seed longevity, plant reproduction and survival. At the same time, the accumulated knowledge on the outcomes of NTP interaction with seeds reveals that plants have developed mechanisms to respond efficiently to short and rather moderate stress experienced at the seed (embryo) stage. These mechanisms allow seeds to sense environmental changes that could be dangerous for the survival of seedlings and enables plants to respond to such signals by mobilizing internal resources and their defensive potential, leading to improved fitness and competitiveness on a longer time scale (stimulated growth, defense and reproduction). The knowledge of such mechanisms has immense potential for applications in agriculture. However, most results on the effects of seed treatment with NTP are obtained in the laboratory or in small-scale field experiments. For the development of reliable NTP-based agro-biotechnologies, NTP treatment devices for the treatment of large quantities of seeds should be designed and NTP effects on plants should be verified in up-scaled agricultural experiments.

Author Contributions: All authors contributed substantially to the work reported: V.M. conceptualization, V.M., A.I., B.S. and D.B. collected literature, wrote, edited and finalized the manuscript. All authors have read and agreed to the published version of the manuscript.

Funding: This research received no external funding.

Institutional Review Board Statement: Not applicable.

Informed Consent Statement: Not applicable.

Data Availability Statement: Not applicable.

Acknowledgments: The authors thank Vladimir Scholtz for consultations on the physical nature of plasma.

Conflicts of Interest: The authors declare no conflict of interest.

References

1. Rockström, J.; Williams, J.; Daily, G.; Noble, A.; Matthews, N.; Gordon, L.; Wetterstrand, H.; DeClerck, F.; Shah, M.; Steduto, P.; et al. Sustainable intensification of agriculture for human prosperity and global sustainability. *Ambio* **2017**, *46*, 4–17. [CrossRef] [PubMed]
2. Ewel, J.J.; Schreeg, L.A.; Sinclair, T.R. Resources for crop production: Accessing the unavailable. *Trends Plant Sci.* **2019**, *24*, 121–129. [CrossRef] [PubMed]
3. Waskow, A.; Howling, A.; Furno, I. Mechanisms of plasma-seed treatments as a potential seed processing technology. *Front. Phys.* **2021**, *9*, 617345. [CrossRef]
4. Araújo, S.S.; Paparella, S.; Dondi, D.; Bentivoglio, A.; Carbonera, D.; Balestrazzi, A. Physical methods for seed invigoration: Advantages and challenges in seed technology. *Front. Plant Sci.* **2016**, *7*, 646. [CrossRef]
5. Adamovich, I.; Baalrud, S.D.; Bogaerts, A.; Bruggeman, P.J.; Cappelli, M.; Colombo, V.; Czarnecki, U.; Ebert, U.; Eden, J.G.; Favia, P.; et al. The 2017 Plasma Roadmap: Low temperature plasma science and technology. *J. Phys. D Appl. Phys.* **2017**, *50*, 323001. [CrossRef]
6. Weltmann, K.; Kolb, J.F.; Holub, M.; Uhrlandt, D.; Šimek, M.; Ostrikov, K.; Hamaguchi, S.; Cvelbar, U.; Černák, M.; Locke, B.; et al. The future for plasma science and technology. *Plasma Proc. Polym.* **2018**, *16*, e1800118. [CrossRef]
7. Straňák, V.; Tichý, M.; Kříha, V.; Scholtz, V.; Šerá, B.; Špatenka, P. Technological applications of sulfatron produced discharge. *J. Optoelectron. Adv. Mater.* **2007**, *9*, 852–857.
8. Laroussi, M.; Lu, X.; Keidar, M. Perspective: The physics, diagnostics, and applications of atmospheric pressure low temperature plasma sources used in plasma medicine. *J. Appl. Phys.* **2017**, *122*, 020901. [CrossRef]
9. Thirumdas, R.; Kothakota, A.; Annapure, U.; Siliveru, K.; Blundell, R.; Gatt, R.; Valdramidis, V.P. Plasma activated water (PAW): Chemistry, physico-chemical properties, applications in food and agriculture. *Trends Food Sci. Technol.* **2018**, *77*, 21–31. [CrossRef]
10. Liao, X.Y.; Cullen, P.J.; Muhammad, A.I.; Jiang, Z.M.; Ye, X.Q.; Liu, D.H.; Ding, T. Cold plasma-based hurdle interventions: New strategies for improving food safety. *Food Eng. Rev.* **2020**, *12*, 321–332. [CrossRef]
11. Liu, D.W.; Zhang, Y.H.; Xu, M.Y.; Chen, H.X.; Lu, X.P.; Ostrikov, K. Cold atmospheric pressure plasmas in dermatology: Sources, reactive agents, and therapeutic effects. *Plasma Proc. Polym.* **2020**, *17*, e1900218. [CrossRef]
12. Zhu, Y.L.; Li, C.Z.; Cui, H.Y.; Lin, L. Feasibility of cold plasma for the control of biofilms in food industry. *Trends Food Sci. Technol.* **2020**, *99*, 142–151. [CrossRef]
13. Šera, B.; Šery, M. Non-thermal plasma treatment as a new biotechnology in relation to seeds, dry fruits, and grains. *Plasma Sci. Technol.* **2018**, *20*, 044012. [CrossRef]
14. Attri, P.; Ishikawa, K.; Okumura, T.; Koga, K.; Shiratani, M. Plasma agriculture from laboratory to farm: A review. *Processes* **2020**, *8*, 1002. [CrossRef]
15. Staric, P.; Vogel-Mikuš, K.; Mozetic, M.; Junkar, I. Effects of nonthermal plasma on morphology, genetics and physiology of seeds: A review. *Plants* **2020**, *9*, 1736. [CrossRef]
16. Adhikari, B.; Adhikari, M.; Park, G. The effects of plasma on plant growth, development, and sustainability. *Appl. Sci.* **2020**, *10*, 6045. [CrossRef]
17. Song, J.-S.; Kim, S.B.; Ryu, S.; Oh, J.; Kim, D.-S. Emerging plasma technology that alleviates crop stress during the early growth stages of plants: A Review. *Front. Plant Sci.* **2020**, *11*, 988. [CrossRef]
18. Holubová, L.; Kyzek, S.; Durovcová, I.; Fabová, J.; Horváthová, E.; Ševcovicová, A.; Gálová, E. Non-thermal plasma—a new green priming agent for plants? *Int. J. Mol. Sci.* **2020**, *21*, 9466. [CrossRef]
19. Waskow, A.; Avino, F.; Howling, A.; Furno, I. Entering the plasma agriculture field: An attempt to standardize protocols for plasma treatment of seeds. *Plasma Process Polym.* **2021**, *18*, e2100152. [CrossRef]
20. Ranieri, P.; Sponsel, N.; Kizer, J.; Rojas-Pierce, M.; Hernández, R.; Gatiboni, L.; Grunden, A.; Stapelmann, K. Plasma agriculture: Review from the perspective of the plant and its ecosystem. *Plasma Proc. Polym.* **2020**, *18*, e2000162. [CrossRef]
21. Braithwaite, N.S.J. Introduction to gas discharges. *Plasma Sources Sci. Technol.* **2000**, *9*, 517–527.
22. Cheruthazhekatt, S.; Cernak, M.; Slavicek, P.; Havel, J. Gas plasmas and plasma modified materials in medicine. *J. Appl. Biomed.* **2010**, *8*, 55–66. [CrossRef]
23. Nehra, V.; Kumar, A.; Dwivedi, H. Atmospheric non-thermal plasma sources. *Int. J. Eng.* **2008**, *2*, 53–68.
24. Piel, A. *Plasma Physics. An Introduction to Laboratory, Space, and Fusion Plasmas*, 2nd ed.; Springer International Publishing AG: Cham, Switzerland, 2017.
25. Fridman, G. Non-equilibrium plasmas in biology and medicine. In *Biological and Environmental Application of Gas Discharge Plasmas*; Brelles, M.G., Ed.; California State Polytechnic University: Pomona, CA, USA, 2012; pp. 95–184.
26. Misra, N.N.; Schlutter, O.; Cullen, P. Plasma in Food and agriculture. In *Cold Plasma in Food and Agriculture: Fundamentals and Applications*; Misra, N.N., Schlutter, O., Cullen, P.J.M., Eds.; Elsevier: Amsterdam, The Netherlands, 2016; pp. 1–16.
27. Conrads, H.; Schmidt, M. Plasma generation and plasma sources. *Plasma Sources Sci. Technol.* **2000**, *9*, 441–454. [CrossRef]
28. Denes, F.S.; Manolache, S. Macromolecular plasma-chemistry: An emerging field of polymer science. *Prog. Polym. Sci.* **2004**, *29*, 815–885. [CrossRef]
29. Randeniya, L.K.; de Groot, G.J.J.B. Non-Thermal Plasma Treatment of Agricultural Seeds for Stimulation of Germination, Removal of Surface Contamination and Other Benefits: A Review. *Plasma Proc. Polym.* **2015**, *12*, 608–623. [CrossRef]

30. Lu, P.; Cullen, P.J.; Ostrikov, K. Atmospheric Pressure Nonthermal Plasma. In *Cold Plasma in Food and Agriculture: Fundamentals and Applications*; Misra, N.N., Schlutter, O., Culle, P.J.M., Eds.; Elsevier: Amsterdam, The Netherlands, 2016; pp. 83–116.

31. Turner, M. Physics of cold plasma. In *Cold Plasma in Food and Agriculture: Fundamentals and Applications*; Misra, N.N., Schlutter, O., Culle, P.J.M., Eds.; Elsevier: Amsterdam, The Netherlands, 2016; pp. 17–50.

32. Whitehead, J.C. The chemistry of cold plasma. In *Cold Plasma in Food and Agriculture: Fundamentals and Applications*; Misra, N.N., Schlutter, O., Culle, P.J.M., Eds.; Elsevier: Amsterdam, The Netherlands, 2016; pp. 53–81.

33. Droge, W. Free radicals in the physiological control of cell function. *Physiol. Rev.* **2002**, *82*, 47–95. [CrossRef]

34. Jeevan Kumar, S.P.; Rajendra Prasad, S.; Banerjee, R.; Thammineni, C. Seed birth to death: Dual functions of reactive oxygen species in seed physiology. *Ann. Bot.* **2015**, *116*, 663–668. [CrossRef]

35. Del Rio, L.A. ROS and RNS in plant physiology: An overview. *J. Biol. Chem.* **2015**, *66*, 2827–2837. [CrossRef]

36. Zhou, R.; Zhou, R.; Wang, P.; Xian, Y.; Mai-Prochnow, A.; Lu, X.P.; Cullen, P.J.; Ostrikov, K.; Bazaka, K. Plasma-activated water: Generation, origin of reactive species and biological applications. *J. Phys. D Appl. Phys.* **2020**, *53*, 303001. [CrossRef]

37. Dubinov, A.E.; Lazarenko, E.M.; Selemir, V.D. Effect of glow discharge air plasma on grain crops seed. *IEEE Trans. Plasma Sci.* **2000**, *28*, 180–183. [CrossRef]

38. Volin, J.C.; Denes, F.S.; Young, R.A.; Park, S.M.T. Modification of seed germination performance through cold plasma chemistry technology. *Crop Sci.* **2000**, *40*, 1706–1718. [CrossRef]

39. Živkovic, S.; Puac, N.; Giba, Z.; Grubišic, D.; Petrovic, Z.L. The stimulatory effect of non-equilibrium (low temperature) air plasma pretreatment on light-induced germination of *Paulownia tomentosa* seeds. *Seed Sci. Technol.* **2004**, *32*, 693–701. [CrossRef]

40. Meiqiang, Y.; Mingjing, H.; Buzhou, M.; Tengcai, M. Stimulating effects of seed treatment by magnetized plasma on tomato growth and yield. *Plasma Sci. Technol.* **2005**, *7*, 3143–3147. [CrossRef]

41. Šerá, B.; Stranák, V.; Šerý, M.; Tichý, M.; Spatenka, P. Germination of *Chenopodium album* in response to microwave plasma treatment. *Plasma Sci. Technol.* **2008**, *10*, 506–511. [CrossRef]

42. Šerá, B.; Sery, M.; Stranak, V.; Spatenka, P.; Tichy, M. Does cold plasma affect breaking dormancy and seed germination? A study on seeds of Lamb's Quarters (*Chenopodium album* agg.). *Plasma Sci. Technol.* **2009**, *11*, 750–754.

43. Bewley, J.D.; Bradford, K.; Hilhorst, H.; Nonogaki, H. *Seeds. Physiology of Development, Germination and Dormancy*, 3rd ed.; Springer: Berlin/Heidelberg, Germany, 2013; pp. 133–181.

44. Gholami, A.; Safa, N.N.; Khoram, M.; Hadian, J.; Ghomi, H. Effect of low-pressure radio frequency plasma on ajwain seed germination. *Plasma Med.* **2016**, *6*, 389–396. [CrossRef]

45. Bormashenko, E.; Grynyov, R.; Bormashenko, Y.; Drori, E. Cold radiofrequency plasma treatment modifies wettability and germination speed of plant seeds. *Sci. Rep.* **2012**, *2*, 741–748. [CrossRef]

46. Bormashenko, E.; Shapira, Y.; Grynyov, R.; Bormashenko, Y.; Drori, E. Interaction of cold radiofrequency plasma with seeds of beans (Phaseolus vulgaris). *J. Exp. Bot.* **2015**, *66*, 4013–4021. [CrossRef]

47. Mildaziene, V.; Pauzaite, G.; Malakauskiene, A.; Zukiene, R.; Nauciene, Z.; Filatova, I.; Azharonok, V.; Lyushkevich, V. Response of perennial woody plants to seed treatment by electromagnetic field and low-temperature plasma. *Bioelectromagnetics* **2016**, *37*, 536–548. [CrossRef]

48. Ivankov, A.; Nauciene, Z.; Zukiene, R.; Degutyte-Fomins, L.; Malakauskiene, A.; Kraujalis, P.; Venskutonis, P.R.; Filatova, I.; Lyushkevich, V.; Mildaziene, V. Changes in growth and production of non-psychotropic cannabinoids induced by pre-sowing treatment of hemp seeds with cold plasma, vacuum and electromagnetic field. *Appl. Sci.* **2020**, *10*, 8519.

49. Filatova, I.; Azharonok, V.; Lushkevich, V.; Zhukovsky, A.; Gadzhieva, G.; Spasic, K.; Zivkovic, S.; Puac, N.; Lazović, S.; Malović, G.; et al. Plasma Seeds Treatment as a Promising Technique for Seed Germination Improvement. In Proceedings of the 31st International Conference on Phenomena in Ionized Gases, Granada, Spain, 14–19 July 2013; p. PS4-105. Available online: https://www.semanticscholar.org/paper/Plasma-seeds-treatment-as-a-promising-technique-for-Filatova-Azharonok/5e45e5f0a890d8e65a4936292d55173fdddd6c95 (accessed on 15 March 2022).

50. Sadhu, S.; Thirumdas, R.; Deshmukh, R.R.; Annapure, U.S. Influence of cold plasma on the enzymatic activity in germinating mung beans (*Vigna radiate*). *LWT-Food Sci. Technol.* **2017**, *78*, 97–104. [CrossRef]

51. Los, A.; Ziuzina, D.; Boehm, D.; Cullen, P.J.; Bourke, P. Investigation of mechanisms involved in germination enhancement of wheat (*Triticum aestivum*) by cold plasma: Effects on seed surface chemistry and characteristics. *Plasma Proc. Polym.* **2019**, *16*, 1800148. [CrossRef]

52. Li, L.; Li, J.; Shen, M.; Zhang, C.; Dong, Y. Cold plasma treatment enhances oilseed rape seed germination under drought stress. *Sci. Rep.* **2015**, *5*, 13033. [CrossRef]

53. Gómez-Ramírez, A.; López-Santos, C.; Cantos, M.; García, J.L.; Molina, R.; Cotrino, J.; Espinos, J.P.; González-Elipe, A.R. Surface chemistry and germination improvement of Quinoa seeds subjected to plasma activation. *Sci. Rep.* **2017**, *7*, 5924. [CrossRef]

54. Mildaziene, V.; Pauzaite, G.; Nauciene, Z.; Zukiene, R.; Malakauskiene, A.; Norkeviciene, E.; Stukonis, V.; Slepetiene, A.; Olsauskaite, V.; Padarauskas, A.; et al. Effect of seed treatment with cold plasma and electromagnetic field on red clover germination, growth and content of major isoflavones. *J. Phys. D Appl. Phys.* **2020**, *53*, 26.

55. Ivankov, A.; Zukiene, R.; Nauciene, Z.; Degutyte-Fomins, L.; Filatova, I.; Lyushkevich, V.; Mildaziene, V. The effects of red clover seed treatment with cold plasma and electromagnetic field on germination and seedling growth are dependent on seed color. *Appl. Sci.* **2021**, *11*, 4676. [CrossRef]

56. Chen, H.H.; Chang, H.C.; Chen, Y.K.; Hung, C.L.; Lin, S.Y.; Chen, Y.S. An improved process for high nutrition of germinated brown rice production: Low-pressure plasma. *Food Chem.* **2016**, *191*, 120–127. [CrossRef]

57. Li, L.; Jiang, J.; Li, J.; Shen, M.; Xe, X.; Shao, H.; Yuanhua, D. Effects of cold plasma treatment on seed germination and seedling growth of soybean. *Sci. Rep.* **2014**, *4*, 5859–5865. [CrossRef]

58. Mildažienė, V.; Aleknavičiūtė, V.; Žūkienė, R.; Paužaitė, G.; Naučienė, Z.; Filatova, I.; Lyushkevich, V.; Haimi, P.; Tamošiūnė, I.; Baniulis, D. Treatment of Common sunflower (*Helianthus annus* L.) seeds with radio-frequency electromagnetic field and cold plasma induces changes in seed phytohormone balance, seedling development and leaf protein expression. *Sci. Rep.* **2019**, *9*, 6437. [CrossRef]

59. Măgureanu, M.; Sîrbu, R.; Dobrin, D.; Gîdea, M. Stimulation of the germination and early growth of tomato seeds by non-thermal plasma. *Plasma Chem. Plasma Process.* **2018**, *38*, 989–1001. [CrossRef]

60. Hosseini, S.I.; Mohsenimehr, S.; Hadian, J.; Ghorbanpour, M.; Shokri, B. Physico-chemical induced modification of seed germination and early development in artichoke (*Cynara scolymus* L.) using low energy plasma technology. *Phys. Plasmas* **2018**, *25*, 013525. [CrossRef]

61. Recek, N.; Holc, M.; Vesel, A.; Zaplotnik, R.; Gselman, P.; Mozetic, M.; Primc, G. Germination of Phaseolus vulgaris L. seeds after a short treatment with a powerful RF plasma. *Int. J. Mol. Sci.* **2021**, *22*, 6672. [CrossRef] [PubMed]

62. Dhayal, M.; Lee, S.-Y.; Park, S.-U. Using low-pressure plasma for Carthamus tinctorium L. seed surface modification. *Vacuum* **2006**, *80*, 499–506. [CrossRef]

63. Jiang, J.; He, X.; Li, L.; Li, J.; Shao, H.; Xu, Q.; Ye, R.; Dong, Y. Effect of Cold Plasma Treatment on Seed Germination and Growth of Wheat. *Plasma Sci. Technol.* **2014**, *16*, 54–58. [CrossRef]

64. Iqbal, T.; Farooq, M.; Afsheen, S.; Abrar, M.; Yousaf, M.; Ijaz, M. Cold plasma treatment and laser irradiation of *Triticum* spp. seeds for sterilization and germination. *J. Laser Appl.* **2019**, *31*, 042013. [CrossRef]

65. Feizollahi, E.; Iqdiam, B.; Vasanthan, T.; Thilakarathna, M.S.; Roopesh, M.S. Effects of Atmospheric-Pressure Cold Plasma Treatment on Deoxynivalenol Degradation, Quality Parameters, and Germination of Barley Grains. *Appl. Sci.* **2020**, *10*, 3530. [CrossRef]

66. Šerý, M.; Zaharonova, A.; Kerdi, A.; Šerá, B. Seed germination of black pine (*Pinus nigra* Arnold) after diffuse coplanar surface barrier discharge plasma treatment. *IEEE Trans. Plasma Sci.* **2020**, *48*, 939–945. [CrossRef]

67. De Groot, G.J.J.B.; Hundt, A.; Murphy, A.B.; Bange, M.P.; Mai-Prochnow, A. Cold plasma treatment for cotton seed germination improvement. *Sci. Rep.* **2018**, *8*, 14372. [CrossRef]

68. Štepánová, V.; Slavíček, P.; Kelar, J.; Prášil, J.; Smékal, M.; Stupavská, M.; Jurmanová, J.; Cernák, M. Atmospheric pressure plasma treatment of agricultural seeds of cucumber (*Cucumis sativus* L.) and pepper (*Capsicum annuum* L.) with effect on reduction of diseases and germination improvement. *Plasma Process. Polym.* **2018**, *15*, 1700076.

69. Tong, J.; He, R.; Zhang, X.; Zhan, R.; Chen, W.; Yang, S. Effects of atmospheric pressure air plasma pretreatment on the seed germination and early growth of *Andrographis paniculata*. *Plasma Sci. Technol.* **2014**, *16*, 260–266. [CrossRef]

70. Sirgedaitė-Šėžienė, V.; Mildažienė, V.; Žemaitis, P.; Ivankov, A.; Koga, K.; Shiratani, M.; Baliuckas, V. Long-term response of Norway spruce to seed treatment with cold plasma: Dependence of the effects on the genotype. *Plasma Proc. Polym.* **2020**, *18*, 2000159. [CrossRef]

71. Stolárik, T.; Henselová, M.; Martinka, M.; Novák, O.; Zahoranová, A.; Černák, M. Effect of Low-Temperature Plasma on the Structure of Seeds, Growth and Metabolism of Endogenous Phytohormones in Pea (*Pisum sativum* L.). *Plasma Chem. Plasma Proc.* **2015**, *35*, 659–676. [CrossRef]

72. Švubová, R.; Kyzek, S.; Medvecká, V.; Slováková, L.; Gálová, E.; Zahoranová, A. Novel insight at the effect of cold atmospheric pressure plasma on the activity of enzymes essential for the germination of pea (*Pisum sativum* L. cv. Prophet) seeds. *Plasma Chem. Plasma Process.* **2020**, *40*, 1221–1240. [CrossRef]

73. Sarinont, T.; Amano, T.; Kitazaki, S.; Koga, K.; Uchida, G.; Shiratani, M.; Hayashi, N. Growth enhancement effects of radish sprouts: Atmospheric pressure plasma irradiation vs. *heat shock*. *J. Phys. Conf. Ser.* **2014**, *518*, 012017.

74. Sarinont, T.; Amano, T.; Attri, P.; Koga, K.; Hayashi, N.; Shiratani, M. Effects of plasma irradiation using various feeding gases on growth of *Raphanus sativus* L. Arch. *Biochem. Biophys.* **2016**, *605*, 129–140.

75. Degutytė-Fomins, L.; Paužaitė, G.; Žukienė, R.; Mildažienė, V.; Koga, K.; Shiratani, M. Relationship between cold plasma treatment-induced changes in radish seed germination and phytohormone balance. *Jpn. J. Appl. Phys.* **2020**, *59*, SH1001. [CrossRef]

76. Mihai, A.L.; Dobrin, D.; Popa, M.E.; Mihai, A.L.; Dobrin, D.; Măgureanu, M.; Popa, M.E. Positive effect of non-thermal plasma treatment on radish. *Rom. Rep. Phys.* **2014**, *66*, 1110–1117.

77. Koga, K.; Attri, P.; Kamataki, K.; Itakagi, N.; Shiratani, M.; Mildažienė, V. Impact of radish sprouts seeds coat color on the electron paramagnetic resonance signals after plasma treatment. *Jpn. J. Appl. Phys.* **2020**, *59*, SHHF01. [CrossRef]

78. Attri, P.; Ishikawa, K.; Okumura, T.; Koga, K.; Shiratani, M.; Mildaziene, M. Impact of seed color and storage time on the radish seed germination and sprout growth in plasma agriculture. *Sci. Rep.* **2021**, *11*, 2539. [CrossRef]

79. Yodpitak, S.; Mahatheeranont, S.; Boonyawan, D.; Sookwong, P.; Roytrakul, S.; Norkaew, O. Cold plasma treatment to improve germination and enhance the bioactive phytochemical content of germinated brown rice. *Food Chem.* **2019**, *289*, 328–339. [CrossRef] [PubMed]

80. Suriyasak, C.; Hatanaka, K.; Tanaka, H.; Okumura, T.; Yamashita, D.; Attri, P.; Koga, K.; Shiratani, M.; Hamaoka, N.; Ishibashi, Y. Alterations of DNA methylation caused by cold plasma treatment restore delayed germination of heat-stressed rice (*Oryza sativa* L.) Seeds. *ACS Agric. Sci. Technol.* **2021**, *1*, 5–10. [CrossRef]

81. Zukiene, R.; Nauciene, Z.; Januskaitiene, I.; Pauzaite, G.; Mildaziene, V.; Koga, K.; Shiratani, M. DBD plasma treatment induced changes in sunflower seed germination, phytohormone balance, and seedling growth. *Appl. Phys. Express* **2019**, 126003. [CrossRef]

82. Yawirach, S.; Sarapirom, S.; Janpong, K. The effects of dielectric barrier discharge atmospheric air plasma treatment to germination and enhancement growth of sunflower seeds. *J. Phys. Conf. Ser.* **2019**, *1380*, 012148. [CrossRef]

83. Ambrico, P.F.; Šimek, M.; Ambrico, M.; Morano, M.; Prukner, V.; Minafra, A.; Allegretta, I.; Porfido, C.; Senesi, G.S.; Terzano, R. On the air atmospheric pressure plasma treatment effect on the physiology, germination and seedlings of basil seeds. *J. Phys. D Appl. Phys.* **2019**, *53*, 104001. [CrossRef]

84. Sarinont, T.; Amano, T.; Koga, K.; Shiratani, M.; Hayashi, N. Multigeneration effects of plasma irradiation to seeds of *Arabidopsis Thaliana* and Zinnia on their growth. *MRS Online Proc. Libr. Arch.* **2014**, *1723*, 7–11. [CrossRef]

85. Koga, K.; Thapanut, S.; Amano, T.; Seo, H.; Itagaki, N.; Hayashi, N.; Shiratani, M. Simple method of improving harvest by nonthermal air plasma irradiation of seeds of *Arabidopsis thaliana* (L.). *Appl. Phys. Express* **2016**, *9*, 016201. [CrossRef]

86. Bafoil, M.; Jemmat, A.; Martinez, Y.; Merbahi, N.; Eichwald, O.; Dunand, C.; Yousfi, M. Effects of low temperature plasmas and plasma activated waters on *Arabidopsis thaliana* germination and growth. *PLoS ONE* **2018**, *13*, e0195512. [CrossRef]

87. Cui, D.; Yin, Y.; Wang, J.; Wang, Z.; Ding, H.; Ma, R.; Jiao, Z. Research on the physio-biochemical mechanism of non-thermal plasma regulated seed germination and early seedling development in Arabidopsis. *Front. Plant Sci.* **2019**, *10*, 1322. [CrossRef]

88. Zahoranová, A.; Henselová, M.; Hudecová, D.; Kaliňáková, B.; Kováčik, D.; Medvecká, V.; Černák, M. Effect of cold atmospheric pressure plasma on the wheat seedlings vigor and on the inactivation of microorganisms on the seeds surface. *Plasma Chem. Plasma Process.* **2015**, *36*, 397–414. [CrossRef]

89. Guo, Q.; Wang, Y.; Zhang, H.; Qu, G.; Wang, T.; Sun, Q.; Liang, D. Alleviation of adverse effects of drought stress on wheat seed germination using atmospheric dielectric barrier discharge plasma treatment. *Sci. Rep.* **2017**, *7*, 16680. [CrossRef] [PubMed]

90. Park, Y.; Oh, K.S.; Oh, J.; Seok, D.C.; Kim, S.B.; Yoo, S.J.; Lee, M.-J. The biological effects of surface dielectric barrier discharge on seed germination and plant growth with barley. *Plasma Process. Polym.* **2016**, *15*, 1600056. [CrossRef]

91. Guragain, R.P.; Baniya, H.B.; Pradhan, S.P.; Dhungana, S.; Chhetri, G.K.; Sedhai, B.; Basnet, N.; Panta, G.P.; Joshi, U.M.; Pandey, B.P.; et al. Impact of non-thermal plasma treatment on the seed germination and seedling development of carrot (*Daucus carota sativus* L.). *J. Phys. Commun.* **2021**, *5*, 125011.

92. Ji, S.H.; Kim, T.; Panngom, K.; Hong, Y.J.; Pengkit, A.; Park, D.H.; Kang, M.H.; Lee, S.H.; Im, J.S.; Kim, J.S.; et al. Assessment of the effects of nitrogen plasma and plasma-generated nitric oxide on early development of *Coriandum sativum*. *Plasma Proc. Polym.* **2015**, *12*, 1164–1173. [CrossRef]

93. Billah, M.; Karmakar, S.; Mina, F.B.; Haque, M.N.; Hasan, M.F.; Acharjee, U.K. Investigation of mechanisms involved in seed germination enhancement, enzymatic activity and seedling growth of rice (*Oryza Sativa* L.) using LPDBD (Ar+Air) plasma. *Arch. Biochem. Biophys.* **2021**, *698*, 108726. [CrossRef] [PubMed]

94. Pérez-Pizá, M.C.; Cejas, E.; Zilli, C.; Prevosto, L.; Mancinelli, B.; Santa-Cruz, D.; Yannarelli, G.; Balestrasse, K. Enhancement of soybean nodulation by seed treatment with non–thermal plasmas. *Sci. Rep.* **2020**, *10*, 4917.

95. Pérez-Pizá, M.C.; Prevosto, L.; Zilli, C.; Cejas, E.; Kelly, H.; Balestrasse, K. Effects of non–thermal plasmas on seed-borne Diaporthe/Phomopsis complex and germination parameters of soybean seeds. *Innov. Food Sci. Emerg.* **2018**, *49*, 82–91. [CrossRef]

96. Zhang, J.J.; Jo, J.O.; Huynh, D.L.; Mongre, R.K.; Ghosh, M.; Singh, A.K.; Lee, S.B.; Mok, Y.S.; Hyuk, P.; Jeong, D.K. Growth-inducing effects of argon plasma on soybean sprouts via the regulation of demethylation levels of energy metabolism-related genes. *Sci. Rep.* **2017**, *7*, 41917.

97. Singh, R.; Kishor, R.; Singh, V.; Singh, V.; Prasad, P.; Aulakh, N.S.; Tiwari, U.K.; Kumar, B. Radio-frequency (RF) room temperature plasma treatment of sweet basil seeds (*Ocimum basilicum* L.) for germination potential enhancement by immaculation. *J. Appl. Res. Med. Aromat. Plants* **2022**, *26*, 100350. [CrossRef]

98. Rahman, M.M.; Sajib, S.A.; Rahi, M.S.; Tahura, S.; Roy, N.C.; Parvez, S.; Reza, M.A.; Talukder, M.R.; Kabir, A.H. Mechanisms and signaling associated with LPDBD plasma mediated growth improvement in wheat. *Sci. Rep.* **2018**, *8*, 10498. [CrossRef]

99. Meng, Y.; Qu, G.; Wang, T.; Sun, Q.; Liang, D.; Hu, S. Enhancement of germination and seedling growth of wheat seed using dielectric barrier discharge plasma with various gas sources. *Plasma Chem. Plasma Proc.* **2017**, *37*, 1105–1119. [CrossRef]

100. Volkov, A.G.; Hairston, J.S.; Marshall, J.; Bookal, A.; Dholichand, A.; Patel, D. Plasma seeds: Cold plasma accelerates *Phaseolus vulgaris* seed imbibition, germination, and speed of seedling growth. *Plasma Med.* **2020**, *10*, 139–158.

101. Alves, C., Jr.; de Oliveira Vitoriano, J.; da Silva, D.L.; de Lima Farias, M.; de Lima Dantas, N.B. Water uptake mechanism and germination of *Erythrina velutina* seeds treated with atmospheric plasma. *Sci. Rep.* **2016**, *6*, 33722. [CrossRef]

102. Fadhlalmawla, S.A.; Mohamed, A.A.H.; Almarashi, J.Q.M.; Boutraa, T. The impact of cold atmospheric pressure plasma jet on seed germination and seedlings growth of fenugreek (*Trigonella foenum-graecum*). *Plasma Sci. Technol.* **2019**, *21*, 105503. [CrossRef]

103. Zhou, R.; Zhou, R.; Zhang, X.; Zhuang, J.; Yang, S.; Bazaka, K.; Ostrikov, K. Effects of atmospheric-pressure N_2, He, air, and O_2 microplasmas on mung bean seed germination and seedling growth. *Sci. Rep.* **2016**, *6*, 32603. [CrossRef]

104. Lotfy, K.; Al-Harbi, N.A.; Abd El-Raheem, H. Cold atmospheric pressure nitrogen plasma jet for enhancement germination of wheat seeds. *Plasma Chem. Plasma Process.* **2019**, *39*, 897–912. [CrossRef]

105. Šerá, B.; Gajdovâ, I.; Černák, M.; Gavril, B.; Hnatiuc, E.; Kováčik, D.; Kříha, V.; Sláma, J.; Šerý, M.; Špatenka, P. How various plasma sources may affect seed germination and growth. In Proceedings of the 2012 13th International Conference on Optimization of Electrical and Electronic Equipment (OPTIM), Brasov, Romania, 24–26 May 2012; pp. 1365–1370.

106. Pawlat, J.; Starek, A.; Sujak, A.; Kwiatkowski, M.; Terebun, P.; Budzen, M. Effects of atmospheric pressure plasma generated in GlidArc reactor on *Lavatera thuringiaca* L. seeds' germination. *Plasma Process. Polym.* **2018**, *15*, 1700064.

107. Sera, B.; Sery, M.; Gavril, B.; Gajdova, I. Seed Germination and Early Growth Responses to Seed Pre-Treatment by Non-Thermal Plasma in Hemp Cultivars (*Cannabis sativa* L.). *Plasma Chem. Plasma Process.* **2017**, *37*, 207–221.

108. Šerá, B.; Vanková, R.; Rohácek, K.; Šerý, M. Gliding arc plasma treatment of maize (*Zea mays* L.) grains promotes seed germination and early growth, affecting hormone pools, but not significantly photosynthetic parameters. *Agronomy* **2021**, *11*, 2066. [CrossRef]

109. Ivankov, A.; Naučienė, Z.; Degutytė-Fomins, L.; Žūkienė, R.; Januškaitienė, I.; Malakauskienė, A.; Jakstas, V.; Ivanauskas, L.; Romanovskaja, D.; Slepetiene, A.; et al. Changes in agricultural performance of common buckwheat induced by seed treatment with cold plasma and electromagnetic Field. *Appl. Sci.* **2021**, *11*, 4391. [CrossRef]

110. Staric, P.; Grobelnik Mlakar, S.; Junkar, I. Response of Two Different Wheat Varieties to Glow and Afterglow Oxygen Plasma. *Plants* **2021**, *10*, 1728. [CrossRef] [PubMed]

111. Zahoranová, A.; Hoppanová, L.; Šimoncicová, J.; Tuceková, Z.; Medvecká, V.; Hudecová, D.; Kalináková, B.; Kovácik, D.; Cernák, M. Effect of Cold Atmospheric Pressure Plasma on Maize Seeds: Enhancement of Seedlings Growth and Surface Microorganisms Inactivation. *Plasma Chem. Plasma Process.* **2018**, *38*, 969–988.

112. Šerá, B.; Šerý, M.; Zahoranová, A.; Tomekova, J. Germination improvement of three pine species (*Pinus*) after diffuse coplanar surface barrier discharge plasma treatment. *Plasma Chem. Plasma Process.* **2021**, *41*, 211–226. [CrossRef]

113. Paužaitė, G.; Malakauskienė, A.; Naučienė, Z.; Žūkienė, R.; Filatova, I.; Lyushkevich, V.; Azarko, I.; Mildažienė, V. Changes in Norway spruce germination and growth induced by pre-sowing seed treatment with cold plasma and electromagnetic field: Short-term versus long-term effects. *Plasma Process. Polym.* **2018**, *15*, 1700068. [CrossRef]

114. Mravlje, J.; Regvar, M.; Staric, P.; Mozetic, M.; Vogel-Mikuš, K. Cold plasma affects germination and fungal community structure of buckwheat seeds. *Plants* **2021**, *10*, 85. [CrossRef]

115. Ahn, C.; Gill, J.; Ruzic, D.N. Growth of plasma-treated corn seeds under realistic conditions. *Sci. Rep.* **2019**, *9*, 4355. [CrossRef]

116. Šerá, B.; Gajdová, I.; Šerý, M.; Špatenka, P. New physicochemical treatment method of poppy seeds for agriculture and food industries. *Plasma Sci. Technol.* **2013**, *15*, 935–938.

117. Tomeková, J.; Kyzek, S.; Medvecká, V.; Gálová, E.; Zahoranová, A. Influence of cold atmospheric pressure plasma on pea seeds: DNA damage of seedlings and optical diagnostics of plasma. *Plasma Chem. Plasma Process.* **2020**, *40*, 1571–1584. [CrossRef]

118. Maruyama-Nakashita, A.; Ishibashi, Y.; Yamamoto, K.; Liu Zhang, L.; Morikawa-Ichinose, T.; Sun-Ju Kim, S.-J.; Hayashi, N. Oxygen plasma modulates glucosinolate levels without affecting lipid contents and composition in *Brassica napus* seeds. *Biosci. Biotechnol. Biochem.* **2021**, *85*, 2434–2441. [CrossRef]

119. Iranbakhsh, A.; Ghoranneviss, M.; Oraghi Ardebili, Z.; Oraghi Ardebili, N.; Hesami Tackallou, S.; Nikmaram, H. Non-thermal plasma modified growth and physiology in *Triticum aestivum* via generated signaling molecules and UV radiation. *Biol. Plantarum* **2016**, *61*, 702–708. [CrossRef]

120. Bußler, S.; Herppich, W.B.; Neugart, S.; Schreiner, M.; Ehlbeck, J.; Rohn, S.; Schlüter, O. Impact of cold atmospheric pressure plasma on physiology and flavonol glycoside profile of peas (*Pisum sativum* "Salamanca"). *Food Res. Int.* **2015**, *76*, 132–141. [CrossRef]

121. Matilla, A.; Matilla, A.; Gallardo, M.; Puga-Hermida, M.I. Structural, physiological and molecular aspects of heterogeneity in seeds: A review. *Seed Sci. Res.* **2005**, *15*, 63–76. [CrossRef]

122. Bhattacharjee, S. ROS and oxidative stress: Origin and implication. In *Reactive Oxygen Species in Plant Biology*; Springer Nature India Private Limited: Dheli, India, 2019. [CrossRef]

123. Wahid, A.; Farooq, M.; Siddique, K.H.M. Implications of Oxidative Stress for Crop Growth and Productivity. In *Handbook of Plant and Crop Physiology*, 3rd ed.; Pessarakli, M., Ed.; Taylor and Francis Group: Boca Raton, FL, USA, 2014; pp. 549–556. [CrossRef]

124. Hong, S.-H.; Szili, E.J.; Jenkins, A.T.A.; Short, R.D. Ionized gas (plasma) delivery of reactive oxygen species (ROS) into artificial cells. *J. Phys. D Appl. Phys.* **2014**, *47*, 362001. [CrossRef]

125. Lepiniec, L.; Debeaujon, I.; Routaboul, J.-M.; Baudry, A.; Pourcel, L.; Nesi, N.; Caboche, M. Genetics and biochemistry of seed flavonoids. *Ann. Rev. Plant Biochem.* **2006**, *57*, 405–430. [CrossRef]

126. Zhou, W.; Chen, F.; Luo, X.; Dai, Y.; Yang, Y.; Zheng, C.; Yang, W.; Shu, K. A matter of life and death: Molecular, physiological and environmental regulation of seed longevity. *Plant Cell Environ.* **2020**, *43*, 293–302. [CrossRef] [PubMed]

127. Leprince, O.; Pellizzaro, A.; Berriri, S.; Buitink, J. Late seed maturation: Drying without dying. *J. Exp. Bot.* **2017**, *68*, 827–841. [CrossRef] [PubMed]

128. Wojtyla, Ł.; Lechowska, K.; Kubala, S.; Garnczarska, M. Different modes of hydrogen peroxide action during seed germination. *Front. Plant Sci.* **2016**, *7*, 66. [CrossRef] [PubMed]

129. Pandiselvam, R.; Mayookha, V.P.; Kothakota, A.; Sharmila, L.; Ramesh, S.V.; Bharathi, C.P.; Gomathy, K.; Srikanth, V. Impact of ozone treatment on seed germination–a systematic review. *Ozone: Sci. Eng* **2020**, *42*, 331–346. [CrossRef]

130. Neyts, E.C. Plasma-surface interactions in plasma catalysis. *Plasma Chem Plasma P* **2016**, *36*, 185–212. [CrossRef]

131. Bazaka, K.; Jacob, M.V.; Crawford, R.J.; Ivanova, E.P. Plasma-assisted surface modification of organic biopolymers to prevent bacterial attachment. *Acta Biomater.* **2011**, *7*, 2015–2028. [CrossRef]

132. Homola, T.; Prukner, V.; Artemenko, A.; Hanuš, J.; Kylián, O.; Šimek, M. Direct treatment of pepper (*Capsicum annuum* L.) and melon (*Cucumis melo*) seeds by amplitude-modulated dielectric barrier discharge in air. *J. Appl. Phys.* **2021**, *129*, 193303. [CrossRef]

133. Molina, R.; López-Santos, C.; Gómez-Ramírez, A.; Vílchez, A.; Espinós, J.P.; González-Elipe, A.R. Influence of irrigation conditions in the germination of plasma treated nasturtium seeds. *Sci. Rep.* **2018**, *8*, 16442. [CrossRef] [PubMed]

134. Lee, K.H.; Kim, H.-J.; Woo, K.S.; Jo, C.; Kim, J.-K.; Kim, S.H.; Park, H.Y.; Oh, S.-K.; Kim, W.H. Evaluation of cold plasma treatments for improved microbial and physicochemical qualities of brown rice. *LWT-Food Sci. Technol.* **2016**, *73*, 442–447. [CrossRef]

135. Shapira, Y.; Multanen, V.; Whyman, G.; Bormashenko, Y.; Chaniel, G.; Barkay, Z.; Bormashenko, E. Plasma Treatment Switches the Regime of Wetting and Floating of Pepper Seeds. *Colloid Surf. B* **2017**, *157*, 417–423. [CrossRef]

136. Baskin, J.M.; Baskin, C.C. A classification system for seed dormancy. *Seed Sci. Res.* **2004**, *14*, 1–16. [CrossRef]

137. Nikolaeva, M.G. Factors controlling the seed dormancy pattern. In *The Physiology and Biochemistry of Seed Dormancy and Germination*; Khan, A.A., Ed.; North-Holland Publishing Co.: Amsterdam, The Netherlands, 1977; pp. 51–74.

138. Baskin, C.C.; Baskin, J.M. *Seeds: Ecology, Biogeography and Evolution of Dormancy and Germination*, 2nd ed.; Academic Press: New York, NY, USA, 2014; pp. 37–77.

139. Graeber, K.; Nakabayashi, K.; Miatton, E.; Leubner-Metzger, G.; Soppe, W.J.J. Molecular mechanisms of seed dormancy. *Plant Cell Environ.* **2012**, *35*, 1769–1786. [CrossRef] [PubMed]

140. Yan, A.; Chen, Z. The pivotal role of abscisic acid signaling during transition from seed maturation to germination. *Plant Cell Rep.* **2017**, *36*, 689–703. [CrossRef]

141. Shu, K.; Liu, X.D.; Xie, Q.; He, Z.H. Two faces of one seed: Hormonal regulation of dormancy and germination. *Mol. Plant.* **2016**, *9*, 34–45. [CrossRef]

142. Seo, M.; Jikumaru, Y.; Kamiya, Y. Profiling of hormones and related metabolites in seed dormancy and germination studies. *Methods Mol. Biol.* **2011**, *773*, 99–111. [CrossRef]

143. Arc, E.; Sechet, J.; Corbineau, F.; Rajjou, L.; Marion-Poll, A. ABA crosstalk with ethylene and nitric oxide in seed dormancy and germination. *Front. Plant Sci.* **2013**, *4*, 63. [CrossRef]

144. Hu, Y.; Yu, D. BRASSINOSTEROID INSENSITIVE2 interacts with ABSCISIC ACID INSENSITIVE5 to mediate the antagonism of brassinosteroids to abscisic acid during seed germination in Arabidopsis. *Plant Cell* **2014**, *26*, 4394–4408. [CrossRef] [PubMed]

145. Kohli, A.; Sreenivasulu, N.; Lakshmanan, P.; Kumar, P.P. The phytohormone crosstalk paradigm takes center stage in understanding how plants respond to abiotic stresses. *Plant Cell Rep.* **2013**, *32*, 945–957. [CrossRef] [PubMed]

146. Hou, L.; Wang, M.; Wang, H.; Zhang, W.-H.; Mao, P. Physiological and proteomic analyses for seed dormancy and release in the perennial grass of *Leymus chinensis*. *Environ. Exp. Bot.* **2019**, *162*, 95–102. [CrossRef]

147. Ji, S.H.; Choi, K.H.; Pengkit, A.; Im, J.S.; Kim, J.S.; Kim, Y.H.; Park, Y.; Hong, E.J.; Jung, S.K.; Choi, E.H.; et al. Effects of high voltage nanosecond pulsed plasma and micro DBD plasma on seed germination, growth development and physiological activities in spinach. *Arch. Biochem. Biophys.* **2016**, *605*, 117–128. [CrossRef]

148. Mildažienė, V.; Ivankov, A.; Paužaitė, G.; Naučienė, Z.; Žūkienė, R.; Degutytė-Fomins, L.; Pukalskas, A.; Venskutonis, P.R.; Filatova, I.; Lyuskevich, V. Seed treatment with cold plasma and electromagnetic field induces changes in red clover root growth dynamics, flavonoid exudation, and activates nodulation. *Plasma Proc. Polym.* **2020**, *18*, 2000160.

149. Hansen, C.C.; Nelson, D.R.; Møller, B.L.; Werck-Reichhart, D. Plant cytochrome P450 plasticity and evolution. *Molecular Plant* **2021**, *14*, 1244–1265. [CrossRef] [PubMed]

150. Mittler, R. ROS are good. *Trends Plant Sci.* **2017**, *22*, 11–19. [CrossRef]

151. Waszczak, C.; Carmody, M.; Kangasjärvi, J. Reactive Oxygen Species in Plant Signaling. *Ann. Rev. Plant Biol.* **2018**, *69*, 209–236. [CrossRef]

152. Skalak, J.; Nicolas, K.L.; Vankova, R.; Hejatko, J. Signal integration in plant abiotic stress responses via multistep phosphorelay signaling. *Front. Plant Sci.* **2021**, *12*, 644823. [CrossRef]

153. Arasimowicz, M.; Floryszak-Wieczorek, J. Nitric oxide as a bioactive signalling molecule in plant stress responses. *Plant Sci.* **2007**, *172*, 876–887.

154. Santisree, P.; Adimulam, S.S.; Sharma, K.; Bhatnagar-Mathur, P.; Sharma, K.K. Insights into the Nitric Oxide Mediated Stress Tolerance in Plants. In *Plant Signaling Molecules*; Khan, M.I.R., Reddy, P.S., Ferrante, A., Khan, N.A., Eds.; Elsevier: Amsterdam, The Netherlands, 2019; pp. 385–406.

155. Bailly, C. The signalling role of ROS in the regulation of seed germination and dormancy. *Biochem. J.* **2019**, *476*, 3019–3032. [CrossRef] [PubMed]

156. Wang, P.; Zhu, J.-K.; Lang, Z. Nitric oxide suppresses the inhibitory effect of abscisic acid on seed germination by S-nitrosylation of SnRK2 proteins. *Plant Sign. Behav.* **2015**, *10*, e1031939. [CrossRef]

157. Bahin, E.; Bailly, C.; Sotta, B.; Kranner, I.; Corbineau, F.; Leymarie, J. Crosstalk between reactive oxygen species and hormonal signalling pathways regulates grain dormancy in barley. *Plant Cell Environ.* **2011**, *34*, 980–993. [CrossRef] [PubMed]

158. Oracz, K.; Karpinski, S. Phytohormones signaling pathways and ROS involvement in seed germination. *Front. Plant Sci.* **2016**, *7*, 864. [CrossRef]

159. Zandalinas, S.I.; Mittler, R. ROS-induced ROS release in plant and animal cells. *Free Rad. Biol. Med.* **2017**, *122*, 21–27. [CrossRef]

160. Khan, M.M.; Hendry, G.A.F.; Atherton, N.M.; Vertucci-Walters, C.W. Free radical accumulation and lipid peroxidation in testas of rapidly aged soybean seeds: A light-promoted process. *Seed Sci. Res.* **1996**, *6*, 101–107. [CrossRef]

161. Hepburn, H.A.; Goodman, B.A.; McPhail, D.B.; Matthews, S.; Powell, A.A. An Evaluation of EPR Measurements of the Organic Free Radical Content of Individual Seeds in the Non-destructive Testing of Seed Viability. *J. Exp. Bot.* **1986**, *37*, 1675–1684. [CrossRef]

162. Zhang, H.; Lang, Z.; Zhu, J.-K. Dynamics and function of DNA methylation in plants. *Nat. Rev. Mol. Cell Biol.* **2018**, *19*, 489–506. [CrossRef]

163. Wang, J.; Cui, D.; Wang, L.; Du, M.; Yin, Y.; Ma, R.; Sun, H.; Jiao, Z. Atmospheric pressure plasma treatment induces abscisic acid production, reduces stomatal aperture and improves seedling growth in *Arabidopsis thaliana*. *Plant Biol.* **2021**, *23*, 564–573. [CrossRef]

164. Cui, D.J.; Yin, Y.; Li, H.D.; Hu, X.X.; Zhuang, J.; Ma, R.N.; Jiao, Z. Comparative transcriptome analysis of atmospheric pressure cold plasma enhanced early seedling growth in *Arabidopsis thaliana*. *Plasma Sci. Technol.* **2021**, *23*, 085502. [CrossRef]

165. Seddighinia, F.S.; Iranbakhsh, A.; Ardebili, Z.O.; Soleimanpour, S. Seed-priming with cold plasma and supplementation of nutrient solution with carbon nanotube enhanced carotenoid contents and the expression of psy and pds in Bitter melon (*Momordica charantia*). *J. Appl. Bot. Food Qual.* **2021**, *94*, 7–14. [CrossRef]

166. Ghaemi, M.; Majd, A.; Iranbakhsh, A. Transcriptional responses following seed priming with cold plasma and electromagnetic field in *Salvia nemorosa* L. *J. Theor. Appl. Phys.* **2020**, *14*, 323–328. [CrossRef]

167. Iranbakhsh, A.; Ardebili, Z.O.; Molaei, H.; Ardebili, N.O.; Amini, M. Cold plasma up-regulated expressions of WRKY1 transcription factor and genes involved in biosynthesis of cannabinoids in hemp (*Cannabis sativa* L.). *Plasma Chem Plasma Process.* **2020**, *40*, 527–537. [CrossRef]

168. Holubova, L.; Svubova, R.; Slovakova, L.; Bokor, B.; Krockova, V.C.; Rencko, J.; Uhrin, F.; Medvecka, V.; Zahoranova, A.; Galova, E. Cold atmospheric pressure plasma treatment of maize grains-induction of growth, enzyme activities and heat shock proteins. *Int. J. Mol. Sci.* **2021**, *22*, 8509. [CrossRef] [PubMed]

169. Han, B.; Yu, N.N.; Zheng, W.; Zhang, L.N.; Liu, Y.; Yu, J.B.; Zhang, Y.Q.; Park, G.; Sun, H.N.; Kwon, T. Effect of non-thermal plasma (NTP) on common sunflower (*Helianthus annus* L.) seed growth via upregulation of antioxidant activity and energy metabolism-related gene expression. *Plant Growth Regul.* **2021**, *95*, 271–281. [CrossRef]

170. Tamosiune, I.; Gelvonauskiene, D.; Haimi, P.; Mildaziene, V.; Koga, K.; Shiratani, M.; Baniulis, D. Cold plasma treatment of sunflower seeds modulates plant-associated microbiome and stimulates root and lateral organ growth. *Front. Plant Sci.* **2020**, *11*, 1347. [CrossRef]

171. Li, K.; Zhong, C.S.; Shi, Q.H.; Bi, H.G.; Gong, B. Cold plasma seed treatment improves chilling resistance of tomato plants through hydrogen peroxide and abscisic acid signaling pathway. *Free Rad. Biol. Med.* **2021**, *172*, 286–297. [CrossRef]

172. Adhikari, B.; Adhikari, M.; Ghimire, B.; Adhikari, B.C.; Park, G.; Choi, E.H. Cold plasma seed priming modulates growth, redox homeostasis and stress response by inducing reactive species in tomato (*Solanum lycopersicum*). *Free Radical Bio. Med.* **2020**, *156*, 57–69. [CrossRef]

173. Li, K.; Zhang, L.; Shao, C.; Zhong, C.; Cao, B.; Shi, Q.; Gong, B. Utilising cold plasma seed treatment technologies to delay cotyledon senescence in tomato seedlings. *Sci. Horticult.* **2021**, *281*, 109911. [CrossRef]

174. Iranbakhsh, A.; Ardebili, N.O.; Ardebili, Z.O.; Shafaati, M.; Ghoranneviss, M. Non-thermal plasma Induced expression of heat shock factor A4A and improved wheat (*Triticum aestivum* L.) growth and resistance against salt stress. *Plasma Chem. Plasma Process.* **2018**, *38*, 29–44. [CrossRef]

175. Hasan, M.; Sohan, M.; Sajib, S.A.; Hossain, M.; Miah, M.; Maruf, M.; Khalid-Bin-Ferdaus, K.M.; Kabir, A.H.; Talukder, M.R.; Rashid, M.M.; et al. The effect of low-pressure dielectric barrier discharge (LPDBD) plasma in boosting fermination, growth, and nutritional properties in wheat. *Plasma Chem. Plasma Process.* **2022**, *42*, 339–362. [CrossRef]

176. Moghanloo, M.; Iranbakhsh, A.; Ebadi, M.; Satari, T.N.; Ardebili, Z.O. Seed priming with cold plasma and supplementation of culture medium with silicon nanoparticle modified growth, physiology, and anatomy in *Astragalus Fridae* as an endangered species. *Acta Physiol. Plant* **2019**, *41*, 54. [CrossRef]

177. Babajani, A.; Iranbakhsh, A.; Ardebili, Z.O.; Eslami, B. Seed priming with non-thermal plasma modified plant reactions to selenium or zinc oxide nanoparticles: Cold plasma as a novel emerging tool for plant science. *Plasma Chem. Plasma Prosess.* **2019**, *39*, 21–34. [CrossRef]

178. Wu, Z.H.; Chi, L.H.; Bian, S.F.; Xu, K.Z. Effects of plasma treatment on maize seeding resistance. *J. Maize Sci.* **2007**, *15*, 111–113.

179. Henselová, M.; Slováková, L'.; Martinka, M.; Zahoranová, A. Growth, anatomy and enzyme activity changes in maize roots induced by treatment of seeds with low-temperature plasma. *Biologia* **2012**, *67*, 490–497. [CrossRef]

180. Karmakar, S.; Billah, M.; Hasan, M.; Sohan, S.R.; Hossain, M.F.; Hoque, K.M.F.; Kabir, A.H.; Rashid, M.M.; Talukder, M.R.; Reza, M.A. Impact of LFGD (Ar+O2) plasma on seed surface, germination, plant growth, productivity and nutritional composition of maize (*Zea mays* L.). *Heliyon* **2021**, *7*, e06458. [CrossRef]

181. Iranbakhsh, A.; Ardebili, Z.O.; Ardebili, N.O.; Ghoranneviss, M.; Safari, N. Cold plasma relieved toxicity signs of nano zinc oxide in *Capsicum Annuum Cayenne* via modifying growth, differentiation, and physiology. *Acta Physiol. Plant* **2018**, *40*, 154. [CrossRef]

182. Sheteiwy, M.S.; An, J.; Yin, M.; Jia, X.; Guan, Y.; He, F.; Hu, J. Cold plasma treatment and exogenous salicylic acid priming enhances salinity tolerance of *Oryza sativa* seedlings. *Protoplasma* **2019**, *256*, 79–99. [CrossRef]

183. Singh, R.; Prasad, P.; Mohan, R.; Verma, M.K.; Kumar, B. Radiofrequency cold plasma treatment enhances seed germination and seedling growth in variety CIM-Saumya of sweet basil (*Ocimum basilicum* L.). *J. Appl. Res. Med. Arom. Plants* **2019**, *12*, 78–81. [CrossRef]

184. Zhang, B.; Li, R.; Yan, J. Study on activation and improvement of crop seeds by the application of plasma treating seeds equipment. *Arch. Biochem. Biophys.* **2018**, *655*, 37–42. [CrossRef]

185. Kabir, A.H.; Rahman, M.M.; Das, U.; Sarkar, U.; Roy, N.C.; Reza, M.A.; Talukder, M.R.; Uddin, M.A. Reduction of cadmium toxicity in wheat through plasma technology. *PLoS ONE* **2019**, *14*, e0214509. [CrossRef]

186. Li, Y.; Wang, T.; Meng, Y.; Qu, G.; Sun, Q.; Liang, D.; Hu, S. Air atmospheric dielectric barrier discharge plasma induced germination and growth enhancement of wheat seed. *Plasma Chem. Plasma Process.* **2017**, *37*, 1621–1634. [CrossRef]

187. Saravana Kumar, R.M.; Wang, Y.; Zhang, X.; Cheng, H.; Sun, L.; He, S.; Hao, F. Redox components: Key regulators of epigenetic modifications in plants. *Int. J. Mol. Sci.* **2020**, *21*, 1419. [CrossRef]

188. Sera, B.; Spatenka, P.; Sery, M.; Vrchotova, N.; Hruskova, I. Influence of plasma treatment on wheat and oat germination and early growth. *IEEE Transact. Plasma Sci.* **2010**, *38*, 2963–2968. [CrossRef]

189. Pérez-Pizá, M.C.; Prevosto, L.; Grijalba, P.E.; Zilli, C.G.; Cejas, E.; Mancinelli, B.; Balestrasse, K.B. Improvement of growth and yield of soybean plants through the application of non-thermal plasmas to seeds with different health status. *Heliyon* **2019**, *54*, e01495. [CrossRef]

190. Mildažienė, V.; Paužaitė, G.; Naučienė, Z.; Malakauskiene, A.; Žūkienė, R.; Januškaitienė, I.; Jakštas, V.; Ivanauskas, L.; Filatova, I.; Lyushkevich, V. Pre-sowing seed treatment with cold plasma and des electromagnetic field increases secondary metabolite content in purple coneflower (*Echinacea purpurea*) leaves. *Plasma Process. Polym.* **2018**, *14*, 1700059.

191. Saberi, M.; Modarres-Sanavy, S.A.M.; Zare, R.; Ghomi, H. Amelioration of photosynthesis and quality of wheat under non-thermal radio frequency plasma treatment. *Sci. Rep.* **2018**, *8*, 11655. [CrossRef]

192. Filatova, I.; Lyushkevich, V.; Goncharik, S.; Zhukovsky, A.; Krupenko, N.; Kalatskaja, J. The effect of low-pressure plasma treatment of seeds on the plant resistance to pathogens pathogens and crop yields. *J. Phys. D Appl. Phys.* **2020**, *53*, 244001. [CrossRef]

193. Nedved, H.; Kalatskaja, J.; Kopylova, N.; Herasimovich, K.; Rybinskaya, E.; Vusik, N.; Filatova, I.; Lyushkevich, V.; Laman, N. Short-term and long-term effects of plasma and radio wave seed treatments on red clover plants. *IOP Conf. Ser. Earth Environ. Sci.* **2021**, *937*, 022137. [CrossRef]

194. Tamosiune, I.; Gelvonauskiene, D.; Ragauskaite, L.; Koga, K.; Shiratani, M.; Baniulis, D. Cold plasma treatment of *Arabidopsis thaliana* (L.) seeds modulates plant-associated microbiome composition. *Appl. Phys. Express* **2020**, *13*, 076001. [CrossRef]

195. Holc, M.; Primc, G.; Iskra, J.; Titan, P.; Kovač, J.; Mozetič, M.; Junkar, I. Effect of oxygen plasma on sprout and root growth, surface morphology and yield of garlic. *Plants* **2019**, *8*, 462. [CrossRef]

196. Cianioti, S.; Katsenios, N.; Efthimiadou, A.; Stergiou, P.; Xanthou, Z.; Giannoglou, M.; Dimitrakellis, P.; Gogolides, E.; Katsaros, G. Pre-sowing treatment of maize seeds by cold atmospheric plasma and pulsed electromagnetic fields: Effect on plant and kernels characteristics. *Aust. J. Crop Sci.* **2021**, *15*, 251–259. [CrossRef]

197. Roy, N.C.; Hasan, M.M.; Talukder, M.R.; Hossain, M.D.; Chowdhury, A.N. Prospective applications of low frequency glow discharge plasmas on enhanced germination, growth and yield of wheat. *Plasma Chem. Plasma Process.* **2018**, *38*, 13–28. [CrossRef]

198. Li, L.; Li, J.; Shen, M.; Hou, J.; Shao, H.; Dong, Y.; Jiang, J. Improving seed germination and peanut yields by cold plasma treatment. *Plasma Sci. Technol.* **2016**, *18*, 1027–1033. [CrossRef]

199. Zhou, Z.; Huang, Y.; Yang, S.; Chen, W. Introduction of a new atmospheric pressure plasma device and application on tomato seeds. *Agricult. Sci.* **2011**, *2*, 23–27. [CrossRef]

200. Yoon, Y.; Seo, D.H.; Shin, H.; Kim, H.J.; Kim, C.M.; Jang, G. The role of stress-responsive transcription factors in modulating abiotic stress tolerance in plants. *Agronomy* **2020**, *10*, 788. [CrossRef]

201. Heldt, H.-W.; Piechulla, B. Phenylpropanoids comprise a multitude of plant-specialized metabolites and cell wall components. In *Plant Biochemistry*, 5th ed.; Academic Press: London, UK, 2021; pp. 411–429.

202. Bai, Y.; Sunarti, S.; Kissoudis, C.; Visser, R.G.F.; van der Linden, C.G. The role of tomato WRKY genes in plant responses to combined abiotic and biotic stresses. *Front. Plant Sci.* **2018**, *9*, 801. [CrossRef]

203. Yoshida, T.; Fujita, Y.; Sayama, H.; Kidokoro, S.; Maruyama, K.; Mizoi, J.; Shinozaki, K.; Yamaguchi-Shinozaki, K. AREB1, AREB2, and ABF3 are master transcription factors that cooperatively regulate ABRE-dependent ABA signaling involved in drought stress tolerance and require ABA for full activation. *Plant J.* **2010**, *61*, 672–685. [CrossRef]

204. Lee, Y.K.; Lim, J.; Hong, E.J.; Kim, S.B. Plasma-activated water regulates root hairs and cotyledon size dependent on cell elongation in *Nicotiana tabacum* L. *Plant Biotechnol. Rep.* **2020**, *14*, 663–672. [CrossRef]

205. Ghasempour, M.; Iranbakhsh, A.; Ebadi, M.; Oraghi Ardebili, Z. Seed priming with cold plasma improved seedling performance, secondary metabolism, and expression of deacetylvindoline O-acetyltransferase gene in *Catharanthus roseus*. *Contrib. Plasm. Phys.* **2020**, *60*, e201900159. [CrossRef]

206. Dezest, M.; Chavatte, L.; Bourdens, M.; Quinton, D.; Camus, M.; Garrigues, L.; Descargues, P.; Arbault, S.; Burlet-Schiltz, O.; Casteilla, L.; et al. Mechanistic insights into the impact of cold atmospheric pressure plasma on human epithelial cell lines. *Sci. Rep.* **2017**, *7*, 41163. [CrossRef] [PubMed]

207. Scharf, C.; Eymann, C.; Emicke, P.; Bernhardt, J.Ã.; Wilhelm, M.; Görries, F.; Winter, J.; von Woedtke, T.; Darm, K.; Daeschlein, G.; et al. Improved wound healing of airway epithelial cells is mediated by cold atmospheric plasma: A time course-related proteome Analysis. *Oxid. Med. Cell. Longev.* **2019**, *2019*, 7071536. [CrossRef] [PubMed]

208. Hideg, Ã.V.; Jansen, M.A.K.; Strid, Ã.K. UV-B exposure, ROS, and stress: Inseparable companions or loosely linked associates? *Trends Plant Sci.* **2013**, *18*, 107–115. [CrossRef] [PubMed]

209. Park, J.; Suh, D.; Tang, T.; Lee, H.J.; Roe, J.S.; Kim, G.C.; Han, S.; Song, K. Non-thermal atmospheric pressure plasma induces epigenetic modifications that activate the expression of various cytokines and growth factors in human mesoderm-derived stem cells. *Free Rad. Biol. Med.* **2020**, *148*, 108–122. [CrossRef] [PubMed]

210. Socco, S.; Bovee, R.C.; Palczewski, M.B.; Hickok, J.R.; Thomas, D.D. Epigenetics: The third pillar of nitric oxide signaling. *Pharmacol. Res.* **2017**, *121*, 52–58. [CrossRef]

211. Lindermayr, C.; Rudolf, E.E.; Durner, J.Ã.; Groth, M. Interactions between metabolism and chromatin in plant models. *Mol. Metab.* **2020**, *38*, 100951. [CrossRef]

212. Sharma, P.; Jha, A.B.; Dubey, R.S.; Pessarakli, M. Reactive oxygen species, oxidative damage, and antioxidative defense mechanism in plants under stressful conditions. *J. Bot.* **2012**, *2012*, 217037. [CrossRef]

213. Adhikari, B.; Pangomm, K.; Veerana, M.; Mitra, S.; Park, G. Plant disease control by non-thermal atmospheric-pressure plasma. *Front. Plant Sci.* **2020**, *11*, 77. [CrossRef]

214. Lacchini, E.; Goossens, A. Combinatorial control of plant specialized metabolism: Mechanisms, functions, and consequences. *Ann. Rev. Cell Dev. Bi.* **2020**, *36*, 291–313. [CrossRef]

215. Wink, M. Secondary metabolites, the role in plant diversification of. In *Encyclopedia of Evolutionary Biology*, 1st ed.; Kliman, R., Ed.; Elsevier Inc.: Oxford, UK, 2016; Volume 4, pp. 1–9.

216. Selmar, D.; Kleinwächter, M. Stress enhances the synthesis of secondary plant products: The impact of stress-related over-reduction on the accumulation of natural products. *Plant Cell Physiol.* **2013**, *54*, 817.

217. Das, K.; Roychoudhury, A. Reactive oxygen species (ROS) and response of antioxidants as ROS-scavengers during environmental stress in plants. *Front. Env. Sci.* **2014**, *2*, 53. [CrossRef]

218. Sewelam, N.; Kazan, K.; Schenk, P.M. Global plant stress signaling: Reactive oxygen species at the cross-road. *Front. Plant Sci.* **2016**, *7*, 187. [CrossRef] [PubMed]

219. Seleiman, M.F.; Al-Suhaibani, N.; Ali, N.; Akmal, M.; Alotaibi, M.; Refay, Y.; Dindaroglu, T.; Abdul-Wajid, H.H.; Battaglia, M.L. Drought stress impacts on plants and different approaches to alleviate its adverse effects. *Plants* **2021**, *10*, 259. [CrossRef] [PubMed]

220. Isayenkov, S.V.; Maathuis, F.J.M. Plant salinity stress: Many unanswered questions remain. *Front. Plant Sci.* **2019**, *10*, 80. [CrossRef]

221. Dubey, S.; Shri, M.; Gupta, A.; Rani, V.; Chakrabarty, D. Toxicity and detoxification of heavy metals during plant growth and metabolism. *Environ. Chem. Lett.* **2018**, *16*, 1169–1192. [CrossRef]

222. Feng, J.; Wang, D.; Shao, C.; Zhang, L.; Tang, X. Effects of cold plasma treatment on alfalfa seed growth under simulated drought stress. *Plasma Sci. Technol.* **2018**, *20*, 035505. [CrossRef]

223. Kyzek, S.; Holubová, L'.; Medvecká, V.; Tomeková, J.; Gálová, E.; Zahoranová, A. Cold atmospheric pressure plasma can induce adaptive response in pea seeds. *Plasma Chem Plasma Processing* **2019**, *39*, 475–486. [CrossRef]

224. Jiang, J.; Lu, Y.; Li, J.; Li, L.; He, X.; Shao, H.; Dong, Y. Effect of seed treatment by cold plasma on the resistance of tomato to *Ralstonia solanacearum* (Bacterial Wilt). *PLoS ONE* **2014**, *9*, e97753. [CrossRef]

225. De Geyter, N.; Morent, R. Nonthermal plasma sterilization of living and nonliving surfaces. *Annu. Rev. Biomed. Eng.* **2012**, *14*, 255–274. [CrossRef]

226. Patil, S.; Bourke, P.; Cullen, P.J. Principles of nonthermal plasma decontamination. In *Cold Plasma in Food and Agriculture: Fundamentals and Applications*; Misra, N.N., Schlutter, O., Culle, P.J., Eds.; Elsevier: Amsterdam, The Netherlands, 2016; pp. 143–177.

227. Bourke, P.; Ziuzina, D.; Han, L.; Cullen, P.J.; Gilmore, B.F. Microbiological interactions with cold plasma. *J. Appl. Microbiol.* **2017**, *123*, 308–324. [CrossRef]

228. Niedzwiedz, I.; Wasko, A.; Pawlat, J.; Polak-Berecka, M. The state of research on antimicrobial activity of cold plasma. *Pol. J. Microbiol.* **2019**, *68*, 153–164. [CrossRef] [PubMed]

229. Van, I.J.; Smet, C.; Tiwari, B.; Greiner, R.; Ojha, S.; Stulic, V.; Vukušić, T.; Režek Jambrak, A. State of the art of nonthermal and thermal processing for inactivation of micro-organisms. *J. Appl. Microbiol.* **2018**, *125*, 16–35. [CrossRef]

230. Khamsen, N.; Onwimol, D.; Teerakawanich, N.; Dechanupaprittha, S.; Kanokbannakorn, W.; Hongesombut, K.; Srisonphan, S. Rice (*Oryza sativa* L.) seed sterilization and germination enhancement via atmospheric hybrid nonthermal discharge plasma. *ACS Appl. Mater. Interfaces* **2016**, *8*, 19268–19275. [CrossRef] [PubMed]

231. Simoncicova, J.; Krystofova, S.; Medvecka, V.; Durisova, K.; Kalinakova, B. Technical applications of plasma treatments: Current state and perspectives. *Appl. Microbiol. Biotechnol.* **2019**, *103*, 5117–5129. [CrossRef]

232. Waskow, A.; Betschart, J.; Butscher, D.; Oberbossel, G.; Kloti, D.; Buttner-Mainik, A.; Adamcik, J.; von Rohr, P.R.; Schuppler, M. Characterization of efficiency and mechanisms of cold atmospheric pressure plasma decontamination of seeds for sprout production. *Front. Microbiol.* **2018**, *9*, 3164. [CrossRef]

233. Nishioka, T.; Takai, Y.; Kawaradani, M.; Okada, K.; Tanimoto, H.; Misawa, T.; Kusakari, S. Seed disinfection effect of atmospheric pressure plasma and low pressure plasma on *Rhizoctonia solani*. *Biocontrol. Sci.* **2014**, *19*, 99–102. [CrossRef]

234. Nishioka, T.; Takai, Y.; Mishima, T.; Kawaradani, M.; Tanimoto, H.; Okada, K.; Misawa, T.; Kusakari, S. Low-pressure plasma application for the inactivation of the seed-borne pathogen *Xanthomonas campestris*. *Biocontrol. Sci.* **2016**, *21*, 37–43. [CrossRef]

235. Kordas, L.; Pusz, W.; Czapka, T.; Kacprzyk, R. The effect of low-temperature plasma on fungus colonization of winter wheat grain and seed quality. *Pol. J. Environ. Stud.* **2015**, *24*, 433–438.

236. Barret, M.; Briand, M.; Bonneau, S.; Preveaux, A.; Valiere, S.; Bouchez, O.; Hunault, G.; Simoneau, P.; Jacques, M.-A. Emergence shapes the structure of the seed microbiota. *Appl. Environ. Microbiol.* **2015**, *81*, 1257. [CrossRef]
237. Klaedtke, S.; Jacques, M.A.; Raggi, L.; Preveaux, A.; Bonneau, S.; Negri, V.; Chable, V.; Barret, M. Terroir is a key driver of seed-associated microbial assemblages. *Environ. Microbiol.* **2016**, *18*, 1792–1804. [CrossRef]
238. Nelson, E.B. The seed microbiome: Origins, interactions, and impacts. *Plant Soil.* **2018**, *422*, 7–34. [CrossRef]
239. Ji, S.H.; Kim, J.S.; Lee, C.H.; Seo, H.S.; Chun, S.C.; Oh, J.; Choi, E.-H.; Park, G. Enhancement of vitality and activity of a plant growth-promoting bacteria (PGPB) by atmospheric pressure non-thermal plasma. *Sci. Rep.* **2019**, *9*, 1044. [CrossRef] [PubMed]
240. Mitra, A.; Li, Y.F.; Klämpfl, T.G.; Shimizu, T.; Jeon, J.; Morfill, G.E.; Zimmermann, J.L. Inactivation of surface-borne microorganisms and increased germination of seed specimen by cold atmospheric plasma. *Food Bioprocess Technol.* **2014**, *7*, 645–653. [CrossRef]
241. Butscher, D.; Van, L.H.; Waskow, A.; Rudolf von Rohr, P.; Schuppler, M. Plasma inactivation of microorganisms on sprout seeds in a dielectric barrier discharge. *Int. J. Food. Microbiol.* **2016**, *238*, 222–232. [CrossRef]
242. Selcuk, M.; Oksuz, L.; Basaran, P. Decontamination of grains and legumes infected with *Aspergillus* spp. and *Penicillum* spp. by cold plasma treatment. *Bioresour. Technol.* **2008**, *99*, 5104–5109. [CrossRef]
243. Hong, Y.F.; Kang, J.G.; Lee, H.Y.; Uhm, H.S.; Moon, E.; Park, Y.H. Sterilization effect of atmospheric plasma on *Escherichia coli* and *Bacillus subtilis* endospores. *Lett. Appl. Microbiol.* **2009**, *48*, 33–37. [CrossRef]
244. Reineke, K.; Langer, K.; Hertwig, C.; Ehlbeck, J.; Schluter, O. The impact of different process gas compositions on the inactivation effect of an atmospheric pressure plasma jet on *Bacillus* spores. *Innov. Food Sci. Emerg.* **2015**, *30*, 112–118. [CrossRef]
245. Liao, X.; Muhammad, A.I.; Chen, S.; Hu, Y.; Ye, X.; Liu, D.; Ding, T. Bacterial spore inactivation induced by cold plasma. *Crit. Rev. Food Sci. Nutr.* **2019**, *59*, 2562–2572. [CrossRef]
246. Schnabel, U.; Niquet, R.; Krohmann, U.; Winter, J.; Schluter, O.; Weltmann, K.D.; Ehlbeck, J. Decontamination of microbiologically contaminated specimen by direct and indirect plasma treatment. *Plasma Process Polym.* **2012**, *9*, 569–575. [CrossRef]
247. Butscher, D.; Zimmermann, D.; Schuppler, M.; Rudolf von Rohr, P. Plasma inactivation of bacterial endospores on wheat grains and polymeric model substrates in a dielectric barrier discharge. *Food Control* **2016**, *60*, 636–645. [CrossRef]
248. Han, L.; Patil, S.; Boehm, D.; Milosavljevic, V.; Cullen, P.J.; Bourke, P. Mechanisms of inactivation by high-voltage atmospheric cold plasma differ for *Escherichia coli* and *Staphylococcus aureus*. *Appl. Environ. Microbiol.* **2016**, *82*, 450–458. [CrossRef] [PubMed]
249. Yu, Q.S.; Huang, C.; Hsieh, F.H.; Huff, H.; Duan, Y. Bacterial inactivation using a low-temperature atmospheric plasma brush sustained with argon gas. *J. Biomed. Mater. Res.* **2007**, *80B*, 211–219. [CrossRef]
250. Boudam, M.K.; Moisan, M.; Saoudi, B.; Popovici, C.; Gherardi, N.; Massines, F. Bacterial spore inactivation by atmospheric-pressure plasmas in the presence or absence of UV photons as obtained with the same gas mixture. *J. Phys. D Appl. Phys.* **2006**, *39*, 3494–3507. [CrossRef]
251. Moisan, M.; Barbeau, J.; Moreau, S.; Pelletier, J.; Tabrizian, M.; Yahia, L. Low-temperature sterilization using gas plasmas: A review of the experiments and an analysis of the inactivation mechanisms. *Int. J. Pharm.* **2001**, *226*, 1–21. [CrossRef] [PubMed]
252. Moreau, S.; Moisan, M.; Tabrizian, M.; Barbeau, J.; Pelletier, J.; Ricard, A.; Yahia, L. Using the flowing afterglow of a plasma to inactivate *Bacillus subtilis* spores: Influence of the operating conditions. *J. Appl. Phys.* **2000**, *88*, 1166–1174. [CrossRef]
253. Ritter, A.C.; Santi, L.; Vannini, L.; Beys-da-Silva, W.O.; Gozzi, G.; Yates, J., 3rd; Ragni, L.; Brandelli, A. Comparative proteomic analysis of foodborne *Salmonella Enteritidis* SE86 subjected to cold plasma treatment. *Food Microbiol.* **2018**, *76*, 310–318. [CrossRef] [PubMed]
254. Yau, K.P.S.; Murphy, A.B.; Zhong, L.; Mai-Prochnow, A. Cold plasma effect on the proteome of *Pseudomonas aeruginosa*-Role for bacterioferritin. *PLoS ONE* **2018**, *13*, e0206530. [CrossRef]
255. Sharma, A.; Collins, G.; Pruden, A. Differential gene expression in *Escherichia coli* following exposure to nonthermal atmospheric pressure plasma. *J. Appl. Microbiol.* **2009**, *107*, 1440–1449. [CrossRef]
256. Roth, S.; Feichtinger, J.; Hertel, C. Response of *Deinococcus radiodurans* to low-pressure low-temperature plasma sterilization processes. *J. Appl. Microbiol.* **2010**, *109*, 1521–1530. [CrossRef]
257. Winter, T.; Bernhardt, J.Ä.; Winter, J.Ä.; Mader, U.; Schluter, R.; Weltmann, K.D.; Hecker, M.; Kusch, H. Common versus noble *Bacillus subtilis* differentially responds to air and argon gas plasma. *Proteomics* **2013**, *13*, 2608–2621. [CrossRef]
258. Winter, T.; Winter, J.Ä.; Polak, M.; Kusch, K.; Mader, U.; Sietmann, R.; Ehlbeck, J.; van Hijum, S.; Weltmann, K.D.; Hecker, M.; et al. Characterization of the global impact of low temperature gas plasma on vegetative microorganisms. *Proteomics* **2011**, *11*, 3518–3530. [CrossRef] [PubMed]
259. Krewing, M.; Jarzina, F.; Dirks, T.; Schubert, B.; Benedikt, J.; Lackmann, J.W.; Bandow, J.E. Plasma-sensitive *Escherichia coli* mutants reveal plasma resistance mechanisms. *J. R. Soc. Interface* **2019**, *16*, 20180846. [CrossRef] [PubMed]
260. Mols, M.; Mastwijk, H.; Nierop Groot, M.; Abee, T. Physiological and transcriptional response of *Bacillus cereus* treated with low-temperature nitrogen gas plasma. *J. Appl. Microbiol.* **2013**, *115*, 689–702. [CrossRef] [PubMed]
261. Mendes, R.; Garbeva, P.; Raaijmakers, J.M. The rhizosphere microbiome: Significance of plant beneficial, plant pathogenic, and human pathogenic microorganisms. *FEMS Microbiol. Rev.* **2013**, *37*, 634–663. [CrossRef] [PubMed]
262. Berg, G.; Koberl, M.; Rybakova, D.; Muller, H.; Grosch, R.; Smalla, K. Plant microbial diversity is suggested as the key to future biocontrol and health trends. *FEMS Microbiol. Ecol.* **2017**, *93*, fix050. [CrossRef]
263. Sessitsch, A.; Mitter, B. 21st century agriculture: Integration of plant microbiomes for improved crop production and food security. *Microb. Biotechnol.* **2015**, *8*, 32–33. [CrossRef]

264. Ferguson, B.J.; Mens, C.; Hastwell, A.H.; Zhang, M.; Su, H.; Jones, C.H.; Chu, X.; Gresshoff, P.M. Legume nodulation: The host controls the party. *Plant Cell Environ.* **2019**, *42*, 41–51. [CrossRef]

265. Kawase, M.; Chen, W.; Kawaguchi, K.; Nyasha, M.R.; Sasaki, S.; Hatakeyama, H.; Kaneko, T.; Kanzaki, M. TRPA1 and TRPV1 channels participate in atmospheric-pressure plasma-induced $[Ca^{2+}]_i$ response. *Sci. Rep.* **2020**, *10*, 9687. [CrossRef]

266. Himmel, N.J.; Cox, D.N. Transient receptor potential channels: Current perspectives on evolution, structure, function and nomenclature. *Proc. R. Soc. B* **2020**, *287*, 20201309. [CrossRef]

267. Demidchik, V.; Shabala, S.; Isayenkov, S.; Cuin, T.A.; Pottosin, I. Calcium transport across plant membranes: Mechanisms and functions. *New Phytologist.* **2018**, *220*, 49–69. [CrossRef]

268. Sano, N.; Rajjou, L.; North, H.M.; Debeaujon, I.; Marion-Poll, A.; Seo, M. Staying alive: Molecular Aspects of Seed Longevity. *Plant Cell Physiol.* **2015**, *57*, 660–674. [CrossRef] [PubMed]

 *plants*

Review

Effects of Non-Thermal Plasma Treatment on Seed Germination and Early Growth of Leguminous Plants—A Review

Božena Šerá [1,*], Vladimír Scholtz [2], Jana Jirešová [2], Josef Khun [2], Jaroslav Julák [3] and Michal Šerý [4]

1 Department of Environmental Ecology and Landscape Management, Faculty of Natural Sciences, Comenius University in Bratislava, Ilkovičova 6, 842 15 Bratislava, Slovakia
2 Department of Physics and Measurements, University of Chemistry and Technology Prague, Technická 5, 166 28 Prague, Czech Republic; vladimir.scholtz@vscht.cz (V.S.); jana.jiresova@vscht.cz (J.J.); josef.khun@vscht.cz (J.K.)
3 Institute of Immunology and Microbiology, First Faculty of Medicine, Charles University and General University Hospital in Prague, Studničkova 7, 128 00 Prague, Czech Republic; jaroslav.julak@lf1.cuni.cz
4 Faculty of Education, Department of Physics, University of South Bohemia, Jeronýmova 10, 371 15 České Budějovice, Czech Republic; kyklop@pf.jcu.cz
* Correspondence: bozena.sera@uniba.sk

Citation: Šerá, B.; Scholtz, V.; Jirešová, J.; Khun, J.; Julák, J.; Šerý, M. Effects of Non-Thermal Plasma Treatment on Seed Germination and Early Growth of Leguminous Plants—A Review. *Plants* **2021**, *10*, 1616. https://doi.org/10.3390/plants10081616

Academic Editor: Maurizio Cocucci

Received: 16 June 2021
Accepted: 3 August 2021
Published: 6 August 2021

Publisher's Note: MDPI stays neutral with regard to jurisdictional claims in published maps and institutional affiliations.

Abstract: The legumes (*Fabaceae* family) are the second most important agricultural crop, both in terms of harvested area and total production. They are an important source of vegetable proteins and oils for human consumption. Non-thermal plasma (NTP) treatment is a new and effective method in surface microbial inactivation and seed stimulation useable in the agricultural and food industries. This review summarizes current information about characteristics of legume seeds and adult plants after NTP treatment in relation to the seed germination and seedling initial growth, surface microbial decontamination, seed wettability and metabolic activity in different plant growth stages. The information about 19 plant species in relation to the NTP treatment is summarized. Some important plant species as soybean (*Glycine max*), bean (*Phaseolus vulgaris*), mung bean (*Vigna radiata*), black gram (*V. mungo*), pea (*Pisum sativum*), lentil (*Lens culinaris*), peanut (*Arachis hypogaea*), alfalfa (*Medicago sativa*), and chickpea (*Cicer aruetinum*) are discussed. Likevise, some less common plant species i.g. blue lupine (*Lupinus angustifolius*), Egyptian clover (*Trifolium alexandrinum*), fenugreek (*Trigonella foenum-graecum*), and mimosa (*Mimosa pudica, M. caesalpiniafolia*) are mentioned too. Possible promising trends in the use of plasma as a seed pre-packaging technique, a reduction in phytotoxic diseases transmitted by seeds and the effect on reducing dormancy of hard seeds are also pointed out.

Keywords: *Fabaceae*; legumes; low temperature plasma; plasma treatment; seed; seedling

1. Introduction

Plasma, called also the fourth state of matter, is a partially or fully ionized gas. A distinction is made between thermal (high temperature, equilibrium) and non-thermal (cold, low temperature, non-equilibrium) plasma. The thermal plasma reaches the temperatures of thousands of Kelvins and occurs in the Sun, lighting, electric sparks, tokamaks, etc. and it is therefore not applicable in biological applications. On the other hand, the non-thermal plasma (NTP), also called low-temperature or cold plasma, occurs at nearly ambient temperature and the high kinetic energy is stored in electrons only. Its biological and also medical applications are very wide and include, among others, disinfection processes, acceleration of blood coagulation and improved wound and infection healing, dental applications or cancer therapy. These are summarized in numerous reviews, such as [1–6], or in the comprehensive books of Shintani and Sakudo [7] and Metelmann et al. [8].

The use of non-thermal plasmas in agriculture or plant biology has also been widely reported in the last few years. The topics, related to the decontamination of seeds, modification of surface properties, metabolomic pathways, and enzymatic activity, enhancing seed germination and the initial growth, are summarized e.g., in [9–16]. Plant disease control [17] or mycotoxin degradation [18,19] were also reported. The nature of chemical reactions in NTP is rather complex, see e.g., [20–23].

NTP may be easily generated in various electric discharges, among which the most commonly used ones are corona discharges, plasma jets (called also plasma needles, plasma torches or plasma pens), dielectric barrier discharges, gliding arcs and microwave discharges. For a general description of plasma sources, see e.g., [24–27]. In addition, the described effects are not constrained to the direct NTP treatment, but on a lower scale are also mediated by the effects of plasma-activated water (PAW) or air, i.e., the water exposed to NTP prior to the application to desired objects. The described effects can persist for many months after exposure, mainly due to the presence of stable reactive oxygen and nitrogen particles as described e.g., in [28–30].

The *Fabaceae* family (*Leguminosae,* legumes) are a large group of flowering plants with a worldwide distribution. With some 20,000 species, the *Fabaceae* are the third largest family of higher plants. Members of this family are dominant species in some ecosystems (e.g., *Acacia* sp. in parts of Africa and Australia). They are ecologically important for their ability to symbiotically fix nitrogen [31]. The roots of many *Leguminosae* host distinct and specific symbiotic nitrogen-fixing bacteria (*Rhizobium* sp.). They are sources of oils, timber, gums, dyes, and insecticides [32]. The legumes are second only to cereal crops in agricultural importance based on area harvested and total production [33]. As typical examples, the following species may be mentioned: soybeans (*Glycine max*), peanut (*Arachis hypogaea*), common bean (*Phaseolus vulgaris*), lentil (*Lens culinaris*), pea (*Pisum sativum*); flavouring plants, such as carob (*Ceratonia siliqua*); fodder and soil rotation plants, such as alfalfa (*Medicago sativa*) and clovers (*Trifolium* sp.).

The use of NTP has already been addressed by the authors in relation to wheat (*Triticum aestivum*) [10] and seeds that can be used as raw seed [34]. In this communication, we aim to provide an overview of the rapidly growing amount of knowledge gained by scientific teams in testing the effect of NTP on plant seeds. This manuscript is a continuation to this review trend. We chose seeds of legume plants because they are readily available, are large and therefore easy to work with, and so they can also serve as a model for studying other organisms. In scientific databases, the number of articles dealing with germination and initial growth after plasma application is increasing. The aim of this work was to search, analyze and synthesize contents of scientific articles dealing with the effect of NTP plasma on legumes. Figure 1 summarizes main essentials of our review.

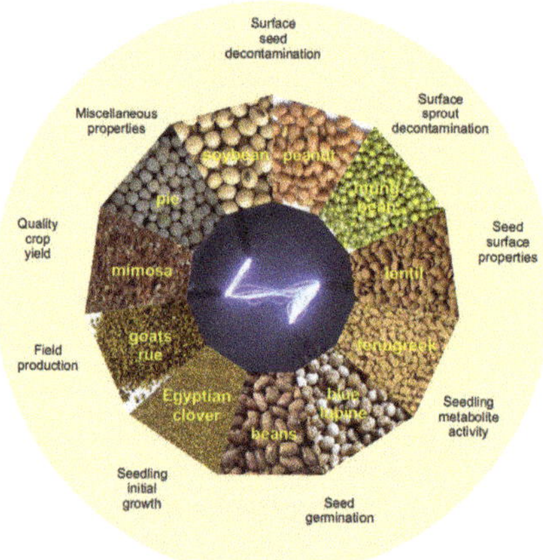

Figure 1. A schematic overview of *Fabaceae* species for which the effect of non-thermal plasma (NTP) has been followed.

2. Surface Seed and Sprout Decontamination

Both microbial and toxin decontamination of seed surface and plant sprout were included. Runtzel et al. [35] reported the effective fungal inactivation of *Aspergillus parasiticus* and *Penicillium* sp. on the surface of common bean after 10–30 min exposure of dielectric barrier discharge (DBD). Selcuk et al. [36] reported the inactivation of pathogenic fungi—*Aspergillus* sp. and *Penicillium* sp.—by NTP in a SF_6 atmosphere on artificially contaminated seeds of common bean, chickpea, lentil and soybean without affecting the cooking time and other food qualities. A significant reduction of 3-log was achieved within 15 min.

Mitra et al. [37] showed a significant reduction of the initial natural microbial load on the chickpea seed surface of 4.5 ± 0.02 log colony-forming unit (CFU) by 1 and 2-log after 2 and 5 min of NTP treatment. The reduction of *Alternaria* sp., *Mucor* sp., *Fusarium* sp., *Penicillium* sp., *Stemphylium* sp., *Cladosporium* sp. on chickpea and fenugreek was reported in [38], however, without detailed specifications.

A significant reduction of the seed-borne microbial contamination on pea was observed by Khatami and Ahmadinia [39], where the amount of microorganisms decreased by ca 3-log from initial 5.5 to 2.5 log $CFU/mL/cm^2$ (the units are not explained in the original paper) after 60 s of exposure. Peanut decontamination was reported by Basaran et al. [40], where 1-log and 5-log reductions of *Aspergillus parasiticus* after 5 min treatment in air or a SF_6 atmosphere were observed, respectively.

The decontamination of several leguminous species (blue lupine, goat-rue, honey clover, soybean, and pea) has been studied [41,42], where the plasma treatment contributed to better fungicidal effect against of *Fusarium* sp., *Alternaria* sp., and *Stemphilium* sp. on seeds. On the other hand, a possible reduction of toxin production on peanut kernels was reported [43], where a significant reduction of aflatoxin levels without any negative sensory effect was reported.

The decontamination of soybean seeds contaminated with bacteria using PAW was reported by Lee et al. [44]. PAW reduced the overall 4.3-log CFU/mL amount of aerobic microbes and 7.0-log CFU/mL of artificially inoculated *Salmonella* Typhimurium within 5 min and 2 min, respectively. Two following works reported the decontamination of sprouts, both by PAW only. Schnabel et al. [45] contaminated mung beans sprouts with the bacteria *Escherichia coli*, *Pseudomonas fluorescens*, *Pseudomonas marginalis*, *Pectobacterium carotovorum*, and *Listeria innocua*. The experimental results showed a reduction from 2.5-log to 3.5-log of bacteria and better growth of the mung bean sprouts, while untreated samples became strongly glassy and cell liquor was released, no influence of treated samples was observed. Similar results with mung bean were reported also by Xiang et al. [46], where reductions of 2.3 to 2.8-log were observed in aerobic bacteria, yeasts and moulds.

3. Effects on Seed Surface Properties

The applications of NTP or PAW also affect the properties of samples. These changes appeared to be beneficial and are listed below, and may be divided into seed surface properties and seed internal content properties. The primary effect on the surface is a decrease of the surface energy leading to better wettability or higher hydrophilicity, as measured by the contact angle of water droplets. This change in the wetting properties of seeds is at least partially due to oxidation of their surface by NTP.

Bormashenko et al. [47,48] reported a contact angle decrease in common beans and their markedly accelerated water absorption after tens of seconds of cold radiofrequency plasma treatment. The treatment leads to hydrophilization of the cotyledon and tissues constituting the seed coat when they are exposed to plasma separately. On the contrary, when the entire seed is exposed to plasma treatment, only the external surface of the common bean is hydrophilized by the cold plasma.

Similar results and a possible explanation were presented by Runtzel et al. [35] who observed on common beans scanning electron microscope (SEM), where both the testa and cotyledon structures showed disruption effects on their cell membranes. The inner surface topography of the cotyledon of chickpea was analyzed by Mitra et al. [37], who observed significant changes in the roughness, leading to a change of membrane permeability. The related conductivity alone increased by more than 100%.

Shapira et al. [49] observed that on lentil seeds this effect is irreversible and that it is not related directly to the electrical charge. Da Silva et al. [50] analyzed the wettability and imbibition of *Mimosa caesalpiniafolia* seeds. The wettability and imbibition were found to be directly related to the treatment duration, probably caused by the chemical alternation of the seeds' lipid layers. After its complete modification, the increase of wettability saturates, as was also observed after 9 min of exposure. The chemical alternation is probably mainly caused by oxidative processes, as reported on mung beans [51]: the higher effect was obtained in an air atmosphere, while the effect was negligible in He or N_2.

All these statements are in agreement with the results of [52,53], where the authors observed pea seeds. SEM and Fourier transform infrared (FTIR) surface analyses showed small changes in the surface layer caused by the oxidation of lipids and polysaccharides (the consequences are mentioned in the original work). Moreover, the result of performed geno-toxicological tests also confirmed that the level of DNA damage is minimal. A significant increase in water imbibition was also reported for soybean seeds [54,55].

Surface modification was also reported after treatment by PAW. Fan et al. [56] reported a water absorption rate increase from 65% for control samples to 75% for treated mung beans. Sajib et al. [57] reported a similar lipid or wax coat alteration of black gram due to the interactions of $NO_2{}^-$ and H_2O_2 with wax. Zhou et al. [58] confirmed by SEM that the seed coat of mung bean is chapped and that it improves the water and nutrients absorption, which is a condition that could enhance the germination rate of mung bean and promote the growth of hypocotyls and radicles.

4. Seed Germination

The NTP treatment can have positive effect on seed germination apparent in an acceleration of germination, an increased germination rate and the breaking of seed dormancy.

Tang et al. [59] found that NTP stimulation significantly increased the germination rate and vigor of alfalfa seeds after 20 s of treatment. Bormashenko et al. [47,48] have found an acceleration of seed germination and germination rate for common beans. Rundzel et al. [35] also found similar effects, improving the germination speed and increasing hypocotyl and radicle length in common beans, after 5 min of treatment exposure. However, after longer exposures (20–30 min) saturation and negative effects occur. Fenugreek was studied by Fadhlalmawla et al. [60], where enhancements of the seed germination rate by 7 and 4 times and growth parameters could occur, probably due to the etching of the seed surface stimulated by the plasma streamers and high electric field. Vejrazka et al. [38] confirmed inhibitory effects on fenugreek, when the highest reduction of germination (over 40%) was recorded after 50 s of NTP treatment.

Mitra et al. [37] also reported increases in the germination percentage, the speed of germination, the shoot and root length, the seedling dry weight, and the vigor index in chickpea, however only for short exposure times. These effects saturate and decrease after 2 min of treatment. Very similar effects were described for mung bean seeds [51,61], pea seeds [39,42,52,62], lentil seeds [47] and peanuts [63], where also a yield improvement by 10% was reported, soybean seeds [54,64] and for many leguminous species, like blue lupine, goat-rue, honey clover, and soybean [41,42]. For red clover seeds, according to Mindaziene et al. [65] this effect is correlated with changes in the phytohormone content, where the amount of abscisic acid decreased and gibberellin/abscisic acid ratio increased.

Moreover, the following two works should be mentioned in more details. Tomekova et al. [66], treated pea seeds with diffuse coplanar surface barrier discharge (DCSBD) plasma operating in air, nitrogen, oxygen and mixtures. Aside from an improvement of germination and growth, DNA damage was also detected. This damage increased with increasing amount of nitrogen, due to the much more intensive UV radiation, and also with increasing treatment time. It was concluded that ambient air seems to be the most suitable atmosphere because of the combination of the plasma chemical composition with water vapour. Svubova et al. [53] declared that the main positive effect on pea seeds, i.e., the overall activation of lytic enzymes in seedlings, is caused by the DCSBD generated in air and nitrogen atmospheres. Increased concentrations of radicals in young seedlings and activation of antioxidant enzymes suggest that low NTP doses act as low stressors, which paradoxically have a stimulating effect on the germination, growth and development of seedlings. Small changes in the surface layer caused by oxidation of lipids and polysaccharides, altering the hydrophilicity and thereby increasing imbibition, were also observed. The DNA damage was minimal after short treatment times. It seems that the positive effect is caused by low doses of NTP stress.

A rapid increase of germination percentage from 5% to 50% after 3 min of exposure was reported for *Mimosa caesalpiniafolia* seeds [50]. This effect may be classified as an overcoming of dormancy due to the effects of NTP on the seed surface. After longer times, this effect decreased. It is generally known that the seeds of the *Fabaceae* family are often dormant. These seeds do not germinate immediately after ripening, but rather they often need an additional stimulus. These seeds germinate well after the disruption of their hard impermeable seed coats. NTP treatments are likely to erode the surface of the seeds so that seeds can absorb water better and begin to germinate faster. Thus, NTP treatment can help dormant seeds with a hard seed coat to break dormancy.

The improvement of seed germination caused by PAW treatment was reported by the following works: for mung beans [56,58,67], where the increase of total phenolic and flavonoid contents was also reported; and for soybean seeds in Lo Potro et al. [68]. The following four works reporting the effects of PAW are mentioned in more details.

In [44], the authors also observed an increase in ascorbate, asparagine and γ-aminobutyric acid (GABA), and followed the development of cotyledon and hypocotyl in germinating soybean seed of soybean from the 1st to the 4th day of cultivation. In Zhou et al. [58], the authors applied PAW on mung beans and reported that the high concentration of ROS might contribute to the chapping of seed coat, which improved its absorption of water and nutrients. The activities of superoxide dismutase (SOD), malondialdehyde (MDA), typical phytohormones affecting the growth of plants indole acetic acid (IAA) and abscisic acid (ABA) contents in mung bean seedlings were followed. It was demonstrated that PAW could reduce membrane lipid peroxidation damage by increasing antioxidant enzyme activities, thus significantly reducing the accumulation of MDA. On the other hand, reactive nitrogen species (RNS, nitrogen oxide NO_X, HNO_X) were partially responsible for the acidification of the solution. In [69] authors used PAW and plasma activated liquid fertilizer on lentils. While the PAW caused germination rates as high as 80% against 30% for control ones, the plasma activation of the liquid fertilizer unifies two effects: an early stage boost (probably due to the fertilizer) and an enhancement of growth rate (probably due to the plasma-activated liquid).

Compared to the previous ones, the authors of [70] modified the germination characteristics of pea, common bean, and soybean by plasma coating of the seed surface with macromolecules. To delay germination, two different hydrophobic source gases were utilized: carbon tetrafluoride (CF4) and octadecafluorodecalin (ODFD). Seeds of pea (*P. sativum* cv. Little Marvel and cv. Alaska) treated with CF_4 displayed a significant delay in germination. Similarly, plasma treatment with ODFD delayed germination in soybean, and common bean seeds. The degree of delay was dependent on the amount of coating applied.

In the recent study of Svubova et al. [71], the effects of cold atmospheric pressure plasma exposure on seed germination of soybean was defined. Seed treatment with lower doses of plasma generated in ambient air and oxygen significantly increased the activity of succinate dehydrogenase (a Krebs cycle enzyme), proving the switching of the germinating seed metabolism from anoxygenic to oxygenic. A positive effect on seed germination was documented, while the seed and seedling vigour were also positively affected.

5. Seedling Initial Growth

Affecting of initial seedling growth is closely related to the previously mentioned effects on germination. Thus, the appropriate NTP exposure or the use of PAW causes an initial growth improvement.

The authors of [39] reported an increase of the length of shoots and roots after 30 s and 60 s of plasma treatment with the optimum being observed at 30 s for pea and zucchini (*Cucurbita pepo, Cucurbitaceae* family) seedlings. The improvement of root and shoot length, dry weight, and the vigor, together with changes in the production of endogenous hormones (auxins and cytokinins and their catabolites and conjugates) was also reported by Stolárik et al. [52]. In a study by Bußler et al. [62] they determined the effects on the flavonol glycoside profile, considering the impact on their metabolic activity in different growth stages. In 15 day-old seedlings, the concentration of flavonoid glycosides was dose-dependently decreased after two NTP treatments compared to none or three treatments. The photosynthetic efficiency of treated pea sprouts and seedlings declined, indicating a negative effect of NTP treatment on plant metabolism.

For soybean, in [55] the authors report the enhancement of seedling growth and that DBD treatment incremented 1.6 fold the nitrogenase activity in nodules, while leghaemoglobin content was increased two times, indicating that mutualistic bacteria in the nodes fixed nitrogen more actively than the control. Accordingly, the nitrogen content increased by 64% and 23%, respectively, in nodules and the aerial part of plants. In [54], the authors reported a root weight increase by 27% in seed soybean after NTP treatment and that the soluble sugar and protein contents were 16% and 25% higher than those of the control. Soybean seedling growth improvement was reported also in Zhang et al. [72]. Improvements were reported also for black gram [73], where the seed germination rate, seedling growth, total chlorophyll content, total soluble protein and sugar concentrations increased by 13%, 37%, 37%, 53% and 51%, respectively. Similar results were found for red clover [65].

Interesting results concerning the germination and seedling growth under simulated drought stress conditions were reported in the following two works. In [74], the authors found that appropriate NTP alfalfa seed treatment led to increased germination, and the seedlings presented good adaptability to different drought conditions (the consequences are mentioned in the original work). Higher doses had the opposite effect. In Fadhlalmawla et al. [60], the authors showed for fenugreek seeds that plasma treatment affected the growth of seedlings, measured as root and shoot length and fresh and dry weight of root and shoot, but the effect on seedling growth was not consistent. The changes in red clover plant's internal processes, the beneficial root nodulation and their communication with microorganisms were reported [65,75], where the NTP stress change the amounts of flavonoids important for communication with nitrogen fixing strains of rhizobacteria on the roots of red clover.

Positive effects of PAW were reported in the following works. In Judée et al. [76], the daily irrigation of lentils by PAW led to increases of seedlings length by 34% and 128% after 3 days and 6 days, respectively. In [68], the authors irrigated soybean seeds in soil by PAW. Faster growth and taller soybeans plants were observed, and average stem length values increased from 10 cm to 17 cm. Similar results for black gram were reported in Sajib et al. [57]. For mung beans, the works [56,58] reported positive effects with the optimum of PAW contents, related with the time of water activation and the plasma atmosphere. The optimal activation time was 15 s, as a longer activation led to a decrease of the positive effects. The best positive effects were observed for PAW prepared in air in comparison with that prepared in O_2, He, and N_2.

6. Seedling Metabolite Activity Affection

The NTP and PAW also affect the properties of the inner contents of seeds or plants related to change in metabolite activity. Bußler et al. [62] studied the effect on pea seedlings' (*P. sativum* 'Salamanca') flavonol glycoside profiles after DBD treatment, while considering the impact on their metabolic activity in different growth stages. Non-acylated and monoacylated triglycerides of quercetin and kaempferol dominated the flavonol glycoside profile, quercetin-3-O-p-coumaroyl-triglucoside being the main flavonoid glycoside. In 15 day-old pea seedlings, the concentration of flavonoid glycosides was dose-dependently decreased in DBD-treated samples. Furthermore, the photosynthetic efficiency of treated pea sprouts and seedlings declined, potentially indicating a negative effect of DBD treatment on plant metabolism.

The oxidative stress caused by NTP and resulting metabolic responses are presented in the following four works. Ebrahimibasabi et al. [77] observed increasing activities of fenugreek catalase by 24%, glutathione peroxidase by 53%, ascorbate peroxidase by 86%. Gebramical et al. [78] reported a decrease in unsaturated fatty acid and moisture content and increased saturated fatty acids, peroxide value, acid value, and total polyphenols in peanuts. Zhang et al. [72] reported significant increases in the activity of the enzymes superoxide dismutase, peroxidase and catalase in soybean sprouts. Stolarik et al. [52] suggested the induction of faster germination and hormonal activities was related to plant signaling and development during the early growth phase of pea seedlings.

The creation of the glycosylation conjugates of high-temperature peanut protein isolate (HPPI) and lactose, improving the solubility of HPPI in peanuts, was reported by Yu et al. [79]. The increase in the degree of glycosylation and a decrease in the degree of browning by DBD accelerated the glycosylation of HPPI and lactose, increasing the solubility and changing the structure of the L-HPPI conjugates. Also, the analysis of protein surface hydrophobicity indicated that the L-HPPI conjugates had a more hydrophilic, stable, and ordered structure (the consequences are mentioned in the original work).

Moreover, Li et al. [80] investigated the impacts of DBD on soybean trypsin inhibitor, which is considered as one of the most important anti-nutritional factors in soybeans. They found that the soybean trypsin inhibitor activities of soymilk were reduced by 86%, probably due to plasma-induced conformational changes and oxidative modification which might contribute to the inactivation of soybean trypsin inhibitor.

Mehr and Koocheki [81] investigated the structure and emulsifying properties of grass pea (*Lathyrus sativus*) protein isolate after DBD treatment. The content of carbonyl groups, dityrosine cross-linking and free sulfhydryl, secondary and tertiary structures, sodium dodecyl sulphate–polyacrylamide gel electrophoresis, surface charge, surface hydrophobicity and solubility of grass pea protein isolate were followed. Overall, the results indicated that cold plasma treatment had positive effects on the interfacial and emulsifying properties of grass pea protein isolate in terms of thermodynamic stability of protein on interface, globulin dissociation, and increase in oil-droplet surface electrical charge.

The goal of the study [82] was to verify the impact of plasma treatment on DNA damage and the induction of positive adaptive responses in pea seedlings. The positive effect of DCSBD (see above) pre-treatment and the reduction of DNA damage of pea were observed at all exposure times used. The strongest repairing effect was observed at exposure times of 120–240 s.

Finally, Lee et al. [44] studied the influence of PAW on soybean cultivation. Its application increases the amount of ascorbate, asparagine, and γ-aminobutyric acid significantly, in the part of cotyledon and hypocotyl of soybean sprouts during 1 to 4 days of farming.

The response of the seeds to NTP together with the use of nanoparticles was tested in the following experiment by Moghanloo et al. [83]. Seeds of *Astragalus fridae* were treated with DBD cold plasma and grown in hormone-free culture medium manipulated with different concentrations of SiO_2 nanoparticles (nSi). The total dry mass was influenced by different treatments of plasma and nSi, and significant reductions in total dry mass (by 34% and 56%) were observed. The seed treatment with DBD did not cause significant changes in chlorophyll content. Seed treatment with cold plasma and supplementation of rooting medium with nSi led to a severe augmentation in the leaf peroxidase activity when compared to the control. Individual plasma treatments did not produce a significant change in the expression of universal stress protein gene, but the supplementation of culture medium with the different levels of nSi altered the expression rate of this protein. Tissue differentiation patterns (especially vascular system) were affected by the seed treatment with DBD and/or supplementation of rooting medium with nSi.

7. Field Production and Quality Crop Yield

The following three works enhance the studies up to the level of field experiments. In Tarrad et al. [84], seeds of Egyptian clover (*T. alexandrium* cv. Gemmiza 1 and cv. Fahl) were treated by pulsed atmospheric-pressure plasma jet, that increased the final yield. The total dry matter yield increased by about 15% and 9% over the non-treated control for Gemmiza 1 and Fahl cultivars, respectively.

In [85], the authors reported for blue lupine NTP treated seed that due to the decrease of seed infection and stimulation of field germination, to early seedling growth and to plant resistance to pathogens, the yield increased by 27%.

In [64], the authors exposed soybean seeds to NTP in various atmospheres of air, O_2 or N_2. Under greenhouse conditions, dry weight of roots, plant height, stem diameter and yield of plants grown from either healthy or infected seeds were improved. The plant height, stem diameter and root dry weight of plants from plasma-treated seeds showed increases of 3%, 8% and 12%, respectively; the NTP treatment had positive effects on all the monitored parameters, as compared with either infected plants or fungicide control.

8. Miscellaneous Applications

Finally several unclassified but interesting curious works can be mentioned. The first group of works deals with the dry bulk material modulation. In [86], the authors attempted to modify the protein and techno-functional properties of different flour fractions obtained from pea (*P. sativum* cv. Salamanca) seeds. Experiments using a pea protein isolate indicated that the reason for the increase in water and fat binding capacities in protein rich pea flour is based on plasma-induced modifications of the proteins.

Li et al. [80] showed that DBD significantly induced the inactivation of soybean trypsin inhibitor, one of the most important anti-nutritional factors in soybeans, in soymilk and Kunitz-type trypsin inhibitor from soybean in a model system. The soybean trypsin inhibitor activities of soymilk were reduced by 86.1%; the intrinsic fluorescence and surface hydrophobicity of soybean trypsin inhibitor were significantly decreased, while the sulfhydryl contents were increased, so NTP induced conformational changes and oxidative modifications that might contribute to the inactivation of soybean trypsin inhibitor.

Gnapowski et al. [87] also attempted to improve the properties of soybean powder (SBP) suitable as a good food for animals. However, there are two problems with this brew. One is that SBP sinks too fast as parts of SBP are too big and too heavy. Another negative point is a rapid growth of moulds. Their results showed that the NTP is useful to decrease the sinking speed of SBP and no mould growth was observed after the exposure.

The following studies have been devoted to peanuts and their products. Venkataratnam et al. [88] reported the reduction in antigenicity of defatted peanut flour by up to 43% and in whole peanut by up to 9% by the modifications in protein secondary structure caused by NTP. Ji et al. [89,90] exposed peanut protein isolate (PPI) solutions to a DBD plasma. They found a significant improvement in the solubility, emulsion stability, and water holding capacity of PPI, caused by the unfolding of PPI structure, increasing the β-sheet and random coil content and decreasing the α-helix and β-turn content. Moreover, the PPI surface was rougher and more loosely bound, indicating an increase in the PPI specific surface area and exposed protein–water binding sites as well as a marked increase in its oxygen content, suggesting an increase in the hydrophilic groups on the PPI surface. The following study by Ji et al. [91] showed a rapid conjugation between PPI and dextran, caused by changes in the structure of PPI from compact and hydrophobic to loose and hydrophilic.

The last three works present unique and curious topics. In [92], the authors observed that treatment of plant *Mimosa pudica* by NTP induces movements of the pinnules (part of the leaf) and petioles similar to the effects of mechanical stimulation. The gas flow and UV radiation associated with plasma are not the primary reasons for the observed effects, but rather reactive oxygen and nitrogen species (RONS) appeared to be the primary reason for this plasma-induced activation of phytoactuators in plants. Some of these RONS are known to be signaling molecules, which control plants' developmental processes.

Yepez et al. [93] showed that DBD plasma treatment can transform the liquid soybean oil into a solid product and can produce plasma species that may polymerize polyunsaturated triacylglycerols. Finally, the study of [94] evaluated the influence of PAW on the microbial load and food quality of thin sheets of bean curd (tofu, soybean product). Treatment for 30 min with PAW activated for 90 s reduced the microbial count of total aerobic bacteria and total yeasts and moulds on thin sheets of bean curd.

Table 1 provides an overview of the all monitored plant species and the issues that were studied on them.

Table 1. Summary of studies performed on species of the *Fabaceae* family or on products thereof after treatment with non-thermal plasma (NTP). The list is sorted alphabetically by plant species, 19 scientific names are mentioned. Abbreviations: syn. synonymous; sn scientific name; DBD dielectric barrier discharge; PAW plasma activated water; AC alternating current; RF radio frequency; USP universal stress protein; SEM scanning electron microscope.

Plant Species	Plasma Source/Device	Object of Study	References
Alfalfa, sn: *Medicago sativa* L.	AC glow discharge (20–200 W)	germination	[59]
	Rf glow discharge (air + He mixture, 0–280 W, 15 s)	adaptability of alfalfa seeds in different drought environments	[74]
Astragalus fridae, sn: *Astragalus fridae* Rech. F.	DBD (0.84 W/cm², 0–90 s); applications of SiO₂ nanoparticle (0–80 mg/l)	expression of USP gene	[83]
Black gram, sn: *Vigna mungo* (L.) Hepper	PAW generated by air AC discharge (3–6 kV, 3–10 kHz)	SEM analysis + growth parameter + total soluble protein and sugar concentrations + physiological characteristics + enzyme concentration	[57]
	DBD (0.5 atm, 5 kV, 4.5 kHz, 60 mm, 310 K, 45 W, 20–180 s)	seed cultivar germination + early growth + physiological characteristics	[73]
Blue lupine (syn. narrow-leaved lupine), sn: *Lupinus angustifolius* L.	capacitively coupled rf discharge (5.28 MHz, 0.6 W/cm3, 5–20 min)	germination + early growth + surface decontamination	[41,42]
	capacitively coupled rf discharge (5.28 MHz, 0.025 W/cm3, 2–7 min)	microbial reduction + crop yield	[85]
Chickpea, sn: *Cicer arietinum* L.	plasma not specified (20 kV, 1 kHz, 0.5–30 min)	fungal decontamination	[36]
	surface micro-discharge (10 mW/cm²)	microbial reduction + surface of cotyledon + germination	[37]
	plasma not specified (10–50 s)	germination + fungal inactivation	[38]
Common bean, sn: *Phaseolus vulgaris* L.	DBD (8 kV, 510 W, 5–30 min)	fungi decontamination + germination + seed structure	[35]
	plasma not specified (20 kV, 1 kHz, 0.5–30 min)	decontamination of *Aspergillus* sp. and *Penicillium* sp. + seed germination	[36]
	micro-wave discharge (20 W, 15 s–20 min)	wetting properties + imbibition + germination	[47,48]
	capacitevly coupled plasma (13.56-MHz, 0–20 min)	coating of seeds + germination	[70]
Egyptian clover, sn: *Trifolium alexandrinum* L.	pulsed atmospheric-pressure plasma jet (10–20 kV)	morphological characters of two cultivars + fresh and dry yield	[84]
Fenugreek, sn: *Trigonella foenum-graecum* L.	plasma not specified (10–50 s)	germination + fungal reduction	[38]
	plasma jet (30 kV, 30 kHz, 10 s–15 min)	germination + early seedling growth	[60]
	DBD plasma jet (3.5–4 kV, 0–5 min)	elevated expression of diosgenin-related genes and stimulation of the defense system	[77]
Grass pea, sn: *Lathyrus sativus* L.	DBD (9.4 and 18.6 kV, 30 and 60 s)	food hydrocoloid aspects	[81]
Honey clover, sn: *Melilotus albus* Medik.	capacitively coupled rf discharge (5.28 MHz, 0.6 W/cm3, 5–20 min)	laboratory and field germination + decontamination	[41]
Lentil, sn: *Lens culinaris* Medik.	plasma not specified (20 kV, 1 kHz, 0.5–30 min)	decontamination of *Aspergillus* sp. and *Penicillium* sp. + seed germination	[36]
	micro-wave discharge (20 W, 15 s–20 min)	wetting properties + germination	[47]
	inductive RF plasma discharge (13.56 MHz, air, 18 W, 60 s)	seed surface + hydrophilization	[49]
	atmospheric pressure plasma jet (22.1 kV, 12 s)	PAW, germination + early growth + liquid fertilizer	[69]
	DBD (12 kV, 500 Hz, atmospheric ambient air)	PAW irrigation, early growth	[76]
Mimosa, 1st sn: *Mimosa caesalpiniafolia* Benth. 2nd sn: *Mimosa pudica* L.	DBD (17.5 kV, 990 Hz, 3–15 min)	wettability + imbibition + germination	[50]
	Plasma jet (Ar, 10 kV)	movements of pinnules and petioles	[92]

Table 1. *Cont.*

Plant Species	Plasma Source/Device	Object of Study	References
Mung bean (syn. green gram), sn: *Vigna radiata* (L.) R. Wilczek	microwave discharge (2.45 GHz, 1.1 kW) produced PAW (5, 15 or 50 s), PAW sprouts treatment 0–5 min	bacteria inactivation on fresh sprouts by PAW	[45]
	gliding arc (air, 5 kV, 40 kHz)	PAW treatment, sprouts decontamination + antioxidant potential + total phenolic and flavonoid contents + sensory characteristics	[46]
	micro plasma jet (N_2, He, air, and O_2; 20 kV, 9 kHz, 25 W)	germination + early growth + catalase activity	[51]
	atmospheric pressure plasma jet (5 kV, 40 kHz, 750 W), 200 mL of PAW exposed 15–90 s	PAW or plasma treatments of one cultivar, germination + growth characteristics + total phenolic and flavonoid contents	[56]
	PAW atmospheric pressure plasma jet (air, O_2, He, N_2; 30 mA, 30 min)	various PAW (O_2, He, N_2), germination + seedling growth + sterilization + physiological parameters + morphology SEM	[58]
	capacitively coupled RF discharge (13.56 MHz; 40 and 60 W)	germination + early growth + surface change + enzymatic activity	[61]
	DBD (O_2, N_2, air; 18 kV, 500 Hz),	treatment in PAW (direct, indirect), seed germination	[67]
Pea, sn: *Pisum sativum* L.	gliding arc (air, 15 kV)	drought resistance + seed germination + surface decontamination	[39]
	capacitively coupled rf discharge (5.28 MHz, 0.6 W/cm3, 5–20 min)	germination + early growth	[42]
	DBD (20 kV, 14 kHz, 400 W, 60–600 s)	surface + germination + early growth + physiology	[52]
	DBD (20 kV, 14 kHz, 400 W, 60–300 s)	early growth stages	[53]
	DBD (20 kV, 14 kHz, 400 W, 60–300 s)	DNA damage	[66]
	DBD (6 and 12 kV, 3 kHz, 1–10 min)	cultivar, physiology + germination	[62,86]
	capacitevly coupled plasma (13.56-MHz, 0–20 min)	germination	[70]
	DBD (20 kV, 14 kHz, 400 W, 60–300 s)	reduction of DNA damage	[82]
Peanut, sn: *Arachis hypogaea* L.	plasma not specified (air or SF6, 20 kV, 1 kHz, 0–20 min)	antifungal activity of seed surfaces + total aflatoxins	[40]
	plasma jet (4.4 kV, 70–90 kHz, 650 W, 3–5 min)	aflatoxin reduction	[43]
	rf plasma (13.56 MHz, 60–140 W, 15 s)	surface + germination + early growth + yield	[63]
	DBD (10–40 W, 1–15 min)	characteristics of antioxidant properties	[78]
	DBD (90 W, 0–5 min)	glycosylation conjugates of high-temperature peanut protein	[79]
	DBD (80 kV, 0–60 min)	antigenicity for defatted peanut flour and whole peanut	[88]
	DBD (90 W, 1–4 min)	peanut protein characteristics	[89]
	DBD (90 W, 1–10 min)	solubilization of peanut protein isolate	[90]
	DBD (not specified, 0.5–3 min)	glycation of peanut protein isolate and dextran	[91]
Red clover, sn: *Trifolium pratense* L.	capacitively coupled rf discharge (5.28 MHz, 0.6 W/cm3, 5–7 min)	germination + early growth + hormonal and flavonoid contents	[65,75]
Soybean, sn: *Glycine max* (L.) Merr.	plasma not specified (20 kV, 1 kHz, 0.5–30 min)	irrigation water on soybean sprout production.	[36]
	capacitively coupled rf discharge (5.28 MHz, 0.6 W/cm3, 5–20 min)	germination + early growth + surface decontamination	[42]
	PAW from DBD (not specified, air, 0–5 min)	PAW, sprout growth + aerobic microbe decontamination + phytohormone amount	[44]
	inductive RF plasma discharge (13.56 MHz, He, 60–120 W,15 s)	surface changing + seed germination + early growth + seedling physiology	[54]
	DBD (O2, N2, 25 kV, 50 Hz)	root growth + nodule formation + plant growth enhancement	[55]
	DBD (N2 or O2, 25 kV, 50 Hz)	physiological characteristics + vegetative growth + agronomic traits	[64]
	DBD (air, 80 kV, 50 Hz)	PAW, seed germination + plant growth + water uptake	[68]
	capacitevly coupled plasma (13.56-MHz, 0–20 min)	coating of seeds + seed germination	[70]
	DBD (20 kV, 14 kHz, 400 W, 60–300 s)	early growth stages + seedling physiology	[71]
	DBD (Ar, 10.8–22.1 kV, 3.4–15.6 W, 60 Hz, 12 s)	germination + early growth + physiology	[72]
	DBD (90 W, 0–27 min)	inactivation of soybean trypsin inhibitor in soymilk and Kunitz-type trypsin inhibitor in soybean	[80]
	pulsed power plasma discharge (50 Hz, 5 J/pulse, 27 kV, 10 A in pulse maximum, 20 s)	food quality (soybeans powder mixed with water)	[87]
	DBD (80 kV, 0–6 h)	treatment of soybean oil in a hydrogen gas environment at atmospheric pressure	[93]
	PAW gliding arc (air, 5 kV, 40 kHz, 750 W, 30–90 s)	PAW, microbial decontamination + food quality of bean curd (tofu)	[94]

9. Conclusions

Non-thermal plasma (NTP) has become a widely used technique in various fields of biology, medicine, food processing and others. Different methods are used for its preparation, which, however, makes the comparison of different results somewhat difficult. This review provides a current overview of the effects of NTP on species of the *Fabaceae* family (legumes). The text is divided into logical units, which range from influencing the surface of seeds to affecting whole plants or their products: surface seed and sprout decontamination, seed surface properties affection, seed germination, seedling initial growth, seedling metabolite activity affection, field production and quality crop yield, and miscellaneous applications. This overview shows that NTP may also be well established in various agricultural sectors; here special applications for legumes and their products are covered.

From the works included in this review, the following general conclusions can be drawn: the exposure to NTP or to plasma-activated water (PAW) can significantly affect the different properties of legume seeds. Namely, germination starts from water uptake, and the capability of water absorption could be significantly influenced by the action of plasma. Important surface properties and some physiological parameters of seeds could be also modified. Oxidation processes of plasmatic reactive species may increase water adsorption capability by increasing wettability of seed coats and could also be associated to gas exchanges and to electrolyte leakage by the seed. It is likely that NTP can effectively change dormancy of hard seeds by affecting seed permeability and triggering subsequent processes. NTP can positively influence the germination and growth of the seed, and subsequently also the properties of the seedlings. NTP treatment could reduce the hardness associated with mechanical dormancy of many *Fabaceae* species (alfalfa, blue lupine, grass pea, honey clover, *Mimosa* sp., *Trifolium* sp., etc.). NTP can be advantageously used in decontamination of plant seed surfaces or legume products. Legumes tolerate this physico-chemical treatment well, and the mild stress it causes appears to have a positive effect on them. Changes in physiological factors can then have a positive effect on the number of crops in the field and their yield.

The results discussed in the text are summarized in the Table 1, which provides an overview of previous studies performed with NTP on plant seeds of the *Fabaceae* family.

Author Contributions: Conceptualization, B.Š., V.S.; Writing-Original Draft Preparation, B.Š.; Writing-Review & Editing, B.Š., V.S., J.J. (Jana Jirešová), J.K., J.J. (Jaroslav Julák), M.Š. All authors have read and agreed to the published version of the manuscript.

Funding: This study was partially supported by Charles University research program Progress Q25.

Institutional Review Board Statement: Not applicable.

Informed Consent Statement: Not applicable.

Conflicts of Interest: The authors declare no conflict of interest.

References

1. Tendero, C.; Tixier, C.; Tristant, P.; Desmaison, J.; Leprince, P. Atmospheric Pressure Plasmas: A Review. *Spectrochim. Acta B.* **2006**, *61*, 2–30. [CrossRef]
2. Bourke, P.; Ziuzina, D.; Boehm, D.; Cullen, P.J.; Keener, K. The Potential of Cold Plasma for Safe and Sustainable Food Production. *Trends Biotechnol.* **2018**, *36*, 615–626. [CrossRef]
3. Julák, J.; Scholtz, V. The Potential for use of Non-thermal Plasma in Microbiology and Medicine. *Epidemiol. Microbiol. Imunol.* **2020**, *69*, 28–36.
4. Scholtz, V.; Pazlarova, J.; Souskova, H.; Khun, J.; Julák, J. Non-thermal Plasma—A Tool for Decontamination and Disinfection. *Biotechnol. Adv.* **2015**, *33*, 1108–1119. [CrossRef] [PubMed]
5. Zhu, Y.L.; Li, C.Z.; Cui, H.Y.; Lin, L. Feasibility of Cold Plasma for the Control of Biofilms in Food Industry. *Trends Food Sci. Technol.* **2020**, *99*, 142–151. [CrossRef]
6. von Woedtke, T.; Schmidt, A.; Bekeschus, S.; Wende, K.; Weltmann, K.D. Plasma Medicine: A Field of Applied Redox Biology. *In Vivo* **2019**, *33*, 1011–1026. [CrossRef]

7.	Shintani, H.; Sakudo, A. *Gas Plasma Sterilization in Microbiology: Theory, Applications, Pitfalls and New Perspectives*; Caister Academic Press: Poole, UK, 2016. [CrossRef]

8.	Metelmann, H.R.; von Woedtke, T.; Weltmann, K.D. *Comprehensive Clinical Plasma Medicine: Cold Physical Plasma for Medical Application*; Springer: Cham, Switzerland, 2018; ISBN 331967627X.

9.	Holubová, L.; Kyzek, S.; Ďurovcová, I.; Fabová, J.; Horváthová, E.; Ševčovičová, A.; Gálová, E. Non-Thermal Plasma—A New Green Priming Agent for Plants? *Int. J. Mol. Sci.* **2020**, *21*, 9466. [CrossRef]

10.	Scholtz, V.; Šerá, B.; Khun, J.; Šerý, M.; Julák, J. Effects of Nonthermal Plasma on Wheat Grains and Products. *J. Food Qual* **2019**, 1–10. [CrossRef]

11.	Magallanes Lopez, A.M.; Simsek, S. Pathogens Control on Wheat and Wheat Flour: A Review. *Cereal Chem.* **2021**, *98*, 17–30 [CrossRef]

12.	Siddique, S.S.; Hardy, G.S.J.; Bayliss, K.L. Cold Plasma: A Potential New Method to Manage Postharvest Diseases Caused by Fungal Plant Pathogens. *Plant Pathol.* **2018**, *67*, 1011–1021. [CrossRef]

13.	Han, Y.; Cheng, J.H.; Sun, D.W. Activities and Conformation Changes of Food Enzymes Induced by Cold Plasma: A Review. *Crit. Rev. Food Sci. Nutr.* **2019**, *59*, 794–811. [CrossRef] [PubMed]

14.	Ekezie, E.F.G.; Chizoba, F.G.; Sun, D.W.; Cheng, J.H. A Review on Recent Advances in Cold Plasma Technology for the Food Industry: Current Applications and Future Trends. *Trends Food Sci. Technol.* **2017**, *69*, 46–58. [CrossRef]

15.	Pankaj, S.K.; Wan, Z.; Keener, K.M. Effects of Cold Plasma on Food Quality: A Review. *Foods* **2018**, *7*, 4. [CrossRef] [PubMed]

16.	Attri, P.; Ishikawa, K.; Okumura, T.; Koga, K.; Masaharu, S.M. Plasma Agriculture from Laboratory to Farm: A Review. *Processes* **2020**, *8*, 1002. [CrossRef]

17.	Adhikari, B.; Pangomm, K.; Veerana, M.; Mitra, S.; Park, G. Plant Disease Control by Non-Thermal Atmospheric-Pressure Plasma. *Front. Plant Sci.* **2020**, *11*, 77. [CrossRef] [PubMed]

18.	Ten Bosch, L.; Pfohl, K.; Avramidis, G.; Wieneke, S.; Viöl, W.; Karlovsky, P. Plasma-Based Degradation of Mycotoxins Produced by *Fusarium*, *Aspergillus* and *Alternaria* Species. *Toxins* **2017**, *9*, 97. [CrossRef] [PubMed]

19.	Čolović, R.; Puvača, N.; Cheli, F.; Avantaggiato, G.; Greco, D.; Duragič, O.; Kos, J.; Pinotti, L. Decontamination of Mycotoxin-Contaminated Feedstuffs and Compound Feed. *Toxins* **2019**, *11*, 617. [CrossRef] [PubMed]

20.	Graves, D.B. The Emerging Role of Reactive Oxygen and Nitrogen Species in Redox Biology and Some Implications for Plasma Applications to Medicine and Biology. *J. Phys. D Appl. Phys.* **2012**, *45*, 263001. [CrossRef]

21.	Kelly, S.; Turner, M. Atomic oxygen patterning from a biomedical needle-plasma source. *J. Appl. Phys.* **2013**, *114*, 123301. [CrossRef]

22.	Sysolyatina, E.; Mukhachev, A.; Yurova, M.; Grushin, M.; Karalnik, V.; Petryakov, A.; Trushkin, N.; Ermolaeva, S.; Akishev, Y. Role of the Charged Particles in Bacteria Inactivation by Plasma of a Positive and Negative Corona in Ambient Air. *Plasma Process. Polym.* **2014**, *11*, 315–334. [CrossRef]

23.	Liao, X.Y.; Cullen, P.J.; Muhammad, A.I.; Jiang, Z.M.; Ye, X.Q.; Liu, D.H.; Ding, T. Cold Plasma-Based Hurdle Interventions: New Strategies for Improving Food Safety. *Food Eng. Rev.* **2020**, *12*, 321. [CrossRef]

24.	Yousfi, M.; Merbahi, N.; Sarrette, J.P.; Eichwald, O.; Ricard, A.; Gardou, J.P.; Ducasse, O.; Benhenni, M. Non Thermal Plasma Sources of Production of Active Species for Biomedical Uses: Analyses, Optimization and Prospect. In *Biomedical Engineering Frontiers and Challenges*; Reza, F.R., Ed.; Intech Europe: Rijeka, Croatia, 2011; pp. 99–124. [CrossRef]

25.	Ehlbeck, J.; Schnabel, U.; Polak, M.; Winter, J.; von Woedtke, T.; Brandenburg, R.; von dem Hagen, T.; Weltmann, K.D. Low Temperature Atmospheric Pressure Plasma Sources for Microbial Decontamination. *J. Phys. D Appl. Phys.* **2011**, *44*, 013002. [CrossRef]

26.	Simoncicova, J.; Krystofova, S.; Medvecka, V.; Durisova, K.; Kalinakova, B. Technical Applications of Plasma Treatments: Current State and Perspectives. *Appl. Microbiol. Biotechnol.* **2019**, *103*, 5117–5129. [CrossRef]

27.	Khun, J.; Scholtz, V.; Hozák, P.; Fitl, P.; Julák, J. Various DC-driven Point-to-plain Discharges as Non-Thermal Plasma Sources and their Bactericidal Effects. *Plasma Sources Sci. Technol.* **2018**, *27*, 065002. [CrossRef]

28.	Julák, J.; Hujacova, A.; Scholtz, V.; Khun, J.; Holada, K. Contribution to the Chemistry of Plasma-Activated Water. *Plasma Phys. Rep.* **2018**, *44*, 125–136. [CrossRef]

29.	Al-Sharify, Z.T.; Al-Sharify, T.A.; Al-Azawi, A.M. Investigative Study on the Interaction and Applications of Plasma Activated Water (PAW). *IOP Conf. Ser. Mater. Sci. Eng.* **2020**, *870*, 012042. [CrossRef]

30.	Zhou, R.; Zhou, R.; Wang, P.; Xian, Y.; Prochnow, A.M.; Lu, X.; Cullen, J.P.; Ostrikov, K.; Bazaka, K. Plasma-Activated Water: Generation, Origin of Reactive Species and Biological Applications. *J. Phys. D Appl. Phys.* **2020**, *53*, 303001. [CrossRef]

31.	Judd, W.S.; Campbell, C.S.; Kellogg, E.A. *Plant Systematics: A Phylogenetic Approach*; Sinauer Associates: Sunderland, MA, USA, 2008.

32.	Simpson, M.G. *Plant Systematics*; Elsevier-Academic Press: San Diego, CA, USA, 2010; ISBN 0128126280.

33.	Gepts, P.; Beavis, W.D.; Brummer, E.C.; Shoemaker, R.C.; Stalker, H.T.; Weeden, N.F.; Young, N.D. Legumes as a Model Plant Family. Genomics for Food and Feed Report of the Cross-Legume Advances through Genomics Conference. *Plant Physiol.* **2005**, *137*, 1228–1235. [CrossRef]

34.	Sera, B.; Sery, M. Non-Thermal Plasma Treatment as a New Biotechnology in Relation to Seeds, Dryfruits, and Grains. *Plasma Sci. Technol.* **2018**, *20*, 044012. [CrossRef]

35. Runtzel, C.L.; da Silva, J.R.; da Silva, B.A.; Moecke, E.S.; Scussel, V.M. Effect of Cold Plasma on Black Beans (*Phaseolus vulgaris*, L.), Fungi Inactivation and Micro-Structures stability. *Emir. J. Food Agric.* **2019**, *31*, 864–873. [CrossRef]

36. Selcuk, M.; Oksuz, L.; Basaran, P. Decontamination of Grains and Legumes Infected with *Aspergillus* spp. and *Penicillium* spp. by Cold Plasma Treatment. *Bioresour. Technol.* **2008**, *99*, 5104–5109. [CrossRef]

37. Mitra, A.; Li, Y.F.; Klämpfl, T.G.; Shimizu, T.; Jeon, J.; Morfill, G.E.; Zimmermann, J.L. Inactivation of Surface-Borne Microorganisms and Increased Germination of Seed Specimen by Cold Atmospheric Plasma. *Food Bioprocess Technol.* **2014**, *7*, 645–653. [CrossRef]

38. Vejrazka, K.; Hofbauer, J.; Vichova, J. The Influence of Physical Seed Treatment on Minor Crops Germinations and Fungal Pathogen Occurence. In *Seed and Seedlings XI*; CZU: Prague, Czech Republic, 2013; ISBN 978-80-213-2358-2.

39. Khatami, S.; Ahmadinia, A. Increased Germination and Growth Rates of Pea and Zucchini Seed by FSG Plasma. *J. Theor. Appl. Phys.* **2018**, *12*, 33–38. [CrossRef]

40. Basaran, P.; Basaran-Akgul, N.; Oksuz, L. Elimination of Aspergillus Parasiticus from nut Surface with Low Pressure Cold Plasma (LPCP) Treatment. *Food Microbiol.* **2008**, *25*, 626–632. [CrossRef] [PubMed]

41. Filatova, I.; Azharonok, V.; Kadyrov, M.; Beljavsky, V.; Gvozdov, A.; Shik, A.; Antonuk, A. The Effect of Plasma Treatment of Seeds of Some Grain and Legumes on their Sowing Quality and Productivity. *Rom. J. Phys.* **2011**, *56*, 139–143.

42. Filatova, I.; Azharonok, V.; Shik, A.; Antoniuk, A.; Terletskaya, N. Fungicidal Effects of Plasma and Radio-Wave Pre-treatments on Seeds of Grain Crops and Legumes. In *Nanomaterials for Security*; Springer Science and Business Media LLC: Cham, Switzerland, 2011; pp. 469–479.

43. Iqdiam, B.M.; Abuagela, M.O.; Boz, Z.; Marshall, S.M.; Goodrich-Schneider, R.; Sims, C.A.; Marshall, M.R.; MacIntosh, A.J.; Welt, B.A. Effects of Atmospheric Pressure Plasma Jet Treatment on Aflatoxin Level, Physiochemical Quality, and Sensory Attributes of Peanuts. *J. Food Process. Preserv.* **2020**, *44*, e14305. [CrossRef]

44. Lee, E.J.; Khan, M.S.I.; Shim, J.; Kim, Y.J. Roles of Oxides of Nitrogen on Quality Enhancement of Soybean Sprout During Hydroponic Production Using Plasma Discharged Water Recycling Technology. *Sci. Rep.* **2018**, *8*, 16872. [CrossRef] [PubMed]

45. Schnabel, U.; Sydow, D.; Schlüter, O.; Andrasch, M.; Ehlbeck, J. Decontamination of Fresh-cut Iceberg Lettuce and Fresh Mung Bean Sprouts by Non-Thermal Atmospheric Pressure Plasma Processed Water (PPW). *Mod. Agric. Sci. Technol.* **2015**, *1*, 23–39. [CrossRef]

46. Xiang, Q.; Liu, X.; Liu, S.; Ma, Y.; Xu, C.; Bai, Y. Effect of Plasma-Activated Water on Microbial Quality and Physicochemical Characteristics of Mung Bean Sprouts. *Innov. Food Sci. Emerg. Technol.* **2019**, *52*, 49–56. [CrossRef]

47. Bormashenko, E.; Grynyov, R.; Bormashenko, Y.; Drori, E. Cold Radiofrequency Plasma Treatment Modifies Wettability and Germination Speed of Plant Seeds. *Sci. Rep.* **2012**, *2*, 741. [CrossRef]

48. Bormashenko, E.; Shapira, Y.; Grynyov, R.; Whyman, G.; Bormashenko, Y.; Drori, E. Interaction of Cold Radiofrequency Plasma with Seeds of Beans (*Phaseolus vulgaris*). *J. Exp. Bot.* **2015**, *66*, 4013–4021. [CrossRef] [PubMed]

49. Shapira, Y.; Chaniel, G.; Bormashenko, E. Surface Charging by the Cold Plasma Discharge of Lentil and Pepper Seeds in Comparison with Polymers. *Colloids Surf. B Biointerfaces* **2018**, *172*, 541–544. [CrossRef] [PubMed]

50. da Silva, A.R.M.; Farias, M.L.; da Silva, D.L.S.; Vitoriano, J.O.; de Sousa, R.C.; Alves-Junior, C. Using Atmospheric Plasma to Increase Wettability, Imbibition and Germination of Physically Dormant Seeds of *Mimosa caesalpiniafolia*. *Colloids Surf. B Biointerfaces* **2017**, *157*, 280–285. [CrossRef] [PubMed]

51. Zhou, R.; Zhou, R.; Zhang, X.; Zhuang, J.; Yang, S.; Bazaka, K.; Ostrikov, K.K. Effects of Atmospheric-Pressure N 2, He, air, and O 2 Microplasmas on Mung Bean Seed Germination and Seedling Growth. *Sci. Rep.* **2016**, *6*, 32603. [CrossRef] [PubMed]

52. Stolárik, T.; Henselová, M.; Martinka, M.; Novák, O.; Zahoranová, A.; Černák, M. Effect of Low-Temperature Plasma on the Structure of Seeds, Growth and Metabolism of Endogenous Phytohormones in Pea (*Pisum sativum* L.). *Plasma Chem. Plasma Process.* **2015**, *35*, 659–676. [CrossRef]

53. Svubova, R.; Kyzek, S.; Medvecka, V.; Slovakova, L.; Galova, E.; Zahoranova, A. Novel Insight at the Effect of Cold Atmospheric Pressure Plasma on the Activity of Enzymes Essential for the Germination of Pea (*Pisum sativum* L. cv. Prophet) Seeds. *Plasma Chem. Plasma Process.* **2020**, *40*, 1221–1240. [CrossRef]

54. Li, L.; Jiang, J.; Li, J.; Shen, M.; He, X.; Shao, H.; Dong, Y. Effects of Cold Plasma Treatment on Seed Germination and Seedling Growth of Soybean. *Sci. Rep.* **2014**, *4*, 5859. [CrossRef]

55. Perez-Piza, M.C.; Cejas, E.; Zilli, C.; Prevosto, L.; Mancinelli, B.; Santa-Cruz, D.; Yannarelli, G.; Balestrasse, K. Enhancement of Soybean Nodulation by Seed Treatment with Non-Thermal Plasmas. *Sci. Rep.* **2020**, *10*, 4917. [CrossRef]

56. Fan, L.M.; Liu, X.F.; Ma, Y.F.; Xiang, Q.S. Effects of Plasma-Activated Water Treatment on Seed Germination and Growth of Mung Bean Sprouts. *J. Taibah Univ. Sci.* **2020**, *14*, 823–830. [CrossRef]

57. Sajib, S.A.; Billah, M.; Mahmud, S.; Miah, M.; Hossain, F.; Omar, F.B.; Roy, N.C.; Hoque, K.M.F.; Talukder, M.R.; Kabir, A.H.; et al. Plasma Activated Water: The Next Generation Eco-Friendly Stimulant for Enhancing Plant Seed Germination, Vigor and Increased Enzyme Activity, a Study on Black Gram (*Vigna mungo* L.). *Plasma Chem. Plasma Process.* **2020**, *40*, 119–143. [CrossRef]

58. Zhou, R.; Li, J.; Zhou, R.; Zhang, X.; Yang, S. Atmospheric-Pressure Plasma Treated Water for Seed Germination and Seedling Growth of Mung Bean and its Sterilization Effect on Mung Bean Sprouts. *Innov. Food Sci. Emerg. Technol.* **2019**, *53*, 36–44. [CrossRef]

59. Tang, X.; Liang, F.; Zhao, L.; Zhang, L.; Shu, J.; Zheng, H.; Qin, X.; Shao, C.; Feng, J.; Du, K. Stimulating Effect of Low-Temperature Plasma (LTP) on the Germination Rate and Vigor of Alfalfa Seed (*Medicago sativa* L.). In Proceedings of the International Conference on Computer and Computing Technologies in Agriculture IX, Jilin, China, 12–15 August 2016; pp. 522–529. [CrossRef]

60. Fadhlalmawla, S.A.; Mohamed, A.A.H.; Almarashi, J.Q.M.; Boutraa, T. The Impact of Cold Atmospheric Pressure Plasma Jet on Seed Germination and Seedlings Growth of Fenugreek (*Trigonella Foenum-Graecum*). *Plasma Sci. Technol.* **2019**, *21*, 105503. [CrossRef]

61. Sadhu, S.; Thirumdas, R.; Deshmukh, R.R.; Annapure, U.S. Influence of Cold Plasma on the Enzymatic Activity in Germinating Mung Beans (*Vigna radiate*). *LWT Food Sci. Technol.* **2017**, *78*, 97–104. [CrossRef]

62. Bußler, S.; Herppich, W.B.; Neugart, S.; Schreiner, M.; Ehlbeck, J.; Rohn, S.; Schlüter, O. Impact of Cold Atmospheric Pressure Plasma on Physiology and Flavonol Glycoside Profile of Peas (*Pisum sativum* 'Salamanca'). *Food Res. Int.* **2015**, *76*, 132–141. [CrossRef]

63. Ling, L.I.; Jiangang, L.I.; Minchong, S.; Jinfeng, H.; Hanliang, S.; Yuanhua, D.; Jiafeng, J. Improving Seed Germination and Peanut Yields by Cold Plasma Treatment. *Plasma Sci. Technol.* **2016**, *18*, 1027–1033. [CrossRef]

64. Perez-Piza, M.C.; Prevosto, L.; Grijalba, P.E.; Zilli, C.G.; Cejas, E.; Mancinelli, B.; Balestrasse, K.B. Improvement of Growth and Yield of Soybean Plants Through the Application of Non-Thermal Plasmas to Seeds with Different Health Status. *Heliyon* **2019**, *5*, e01495. [CrossRef]

65. Mildaziene, V.; Pauzaite, G.; Nauciene, Z.; Zukiene, R.; Malakauskiene, A.; Norkeviciene, E.; Slepetiene, A.; Stukonis, V.; Olsauskaite, V.; Padarauskas, A.; et al. Effect of Seed Treatment with Cold Plasma and Electromagnetic Field on Red Clover Germination, Growth and Content of Major Isoflavones. *J. Phys. D Appl. Phys.* **2020**, *53*, 264001. [CrossRef]

66. Tomekova, J.; Kyzek, S.; Medvecka, V.; Galova, E.; Zahoranova, A. Influence of Cold Atmospheric Pressure Plasma on Pea Seeds: DNA Damage of Seedlings and Optical Diagnostics of Plasma. *Plasma Chem. Plasma Process.* **2020**, *40*, 1571–1584. [CrossRef]

67. Liu, B.; Honnorat, B.; Yang, H.; Arancibia, J.; Rajjou, L.; Rousseau, A. Non-Thermal DBD Plasma Array on Seed Germination of Different Plant Species. *J. Phys. D Appl. Phys.* **2019**, *52*, 025401. [CrossRef]

68. Lo Porto, C.; Ziuzina, D.; Los, A.; Boehm, D.; Palumbo, F.; Favia, P.; Tiwari, B.; Bourke, P.; Cullen, P.J. Plasma Activated Water and Airborne Ultrasound Treatments for Enhanced Germination and Growth of Soybean. *Innov. Food Sci. Emerg. Technol.* **2018**, *49*, 13–19. [CrossRef]

69. Zhang, S.; Rousseau, A.; Dufour, T. Promoting Lentil Germination and Stem Growth by Plasma Activated Tap Water, Demineralized Water and Liquid Fertilizer. *RSC Adv.* **2017**, *7*, 31244–31251. [CrossRef]

70. Volin, J.C.; Denes, F.S.; Young, R.A.; Park, S.M.T. Modification of Seed Germination Performance through Cold Plasma Chemistry Technology. *Crop Sci.* **2000**, *40*, 1706–1718. [CrossRef]

71. Švubová, R.; Slováková, Ľ.; Holubová, Ľ.; Rovňanová, D.; Gálová, E.; Tomeková, J. Evaluation of the Impact of Cold Atmospheric Pressure Plasma on Soybean Seed Germination. *Plants* **2021**, *10*, 177. [CrossRef]

72. Zhang, J.J.; Jo, J.O.; Mongre, R.K.; Ghosh, M.; Singh, A.K.; Lee, S.B.; Mok, Y.S.; Hyuk, P.; Jeong, D.K. Growth-Inducing Effects of Argon Plasma on Soybean Sprouts Via the Regulation of Demethylation Levels of Energy Metabolism-Related Genes. *Sci. Rep.* **2017**, *7*, 1–12. [CrossRef]

73. Billah, M.; Sajib, S.A.; Roy, N.C.; Rashid, M.M.; Reza, M.A.; Hasan, M.M.; Talukder, M.R. Effects of DBD Air Plasma Treatment on the Enhancement of Black Gram (*Vigna mungo* L.) Seed Germination and Growth. *Arch. Biochem. Biophys.* **2020**, *681*, 108253. [CrossRef]

74. Feng, J.; Wang, D.; Shao, C.; Lili, Z.; Tang, X. Effects of Cold Plasma Treatment on Alfalfa Seed Growth under Simulated Drought Stress. *Plasma Sci. Technol.* **2018**, *20*, 035505. [CrossRef]

75. Mildaziene, V.; Ivanov, A.; Pauzaite, G.; Nauciene, Z.; Zukiene, R.; Degutyte-Fomins, L.; Pukalska, A.; Venskutonis, P.R.; Filatova, I.; Lyushkevich, V. Seed Treatment with Cold Plasma and Electromagnetic Field Induces Changes in Red Clover Root Growth Dynamics, Flavonoid Exudation, and Activates Nodulation. *Plasma Process. Polym.* **2020**, e2000160. [CrossRef]

76. Judée, F.; Simon, S.; Bailly, C.; Dufour, T. Plasma-Activation of Tap Water using DBD for Agronomy Applications: Identification and Quantification of Long Lifetime Chemical Species and Production/Consumption Mechanisms. *Water Res.* **2018**, *133*, 47–59. [CrossRef]

77. Ebrahimibasabi, E.; Ebrahimi, A.; Momeni, M.; Amerian, M.R. Elevated Expression of Diosgenin-Related Genes and Stimulation of the Defense System in *Trigonella Foenum-Graecum* (Fenugreek) by Cold Plasma Treatment. *Sci. Hortic.* **2020**, *271*, 109494. [CrossRef]

78. Gebremical, G.G.; Emire, S.A.; Berhanu, T. Effects of Multihollow Surface Dielectric Barrier Discharge Plasma on Chemical and Antioxidant Properties of Peanut. *J. Food Qual.* **2019**, *2019*, 1–10. [CrossRef]

79. Yu, J.J.; Ji, H.; Chen, Y.; Zhang, Y.F.; Zheng, X.C.; Li, S.H.; Chen, Y. Analysis of the Glycosylation Products of Peanut Protein and Lactose by Cold Plasma Treatment: Solubility and Structural Characteristics. *Int. J. Biol. Macromol.* **2020**, *158*, 1194–1203. [CrossRef]

80. Li, J.; Xiang, Q.; Liu, X.; Ding, T.; Zhang, X.; Zhai, Y.; Bai, Y. Inactivation of Soybean Trypsin Inhibitor by Dielectric-Barrier Discharge (DBD) Plasma. *Food Chem.* **2017**, *232*, 515–522. [CrossRef]

81. Mehr, H.M.; Koocheki, A. Effect of Atmospheric Cold Plasma on Structure, Interfacial and Emulsifying Properties of Grass Pea (*Lathyrus Sativus* L.) Protein Isolate. *Food Hydrocoll.* **2020**, *106*, 105899. [CrossRef]

32. Kyzek, S.; Holubová, L.; Medvecká, V.; Tomeková, J.; Gálová, E.; Zahoranová, A. Cold Atmospheric Pressure Plasma Can Induce Adaptive Response in Pea Seeds. *Plasma Chem. Plasma Process.* **2019**, *39*, 475–486. [CrossRef]

33. Moghanloo, M.; Iranbakhsh, A.; Ebadi, M.; Ardebili, Z.O. Differential Physiology and Expression of Phenylalanine Ammonia Lyase (PAL) and Universal Stress Protein (USP) in the Endangered Species *Astragalus fridae* following Seed Priming with Cold Plasma and Manipulation of Culture Medium with Silica Nanoparticles. *3 Biotech* **2019**, *9*, 1–13. [CrossRef] [PubMed]

34. Tarrad, M.M.; Ahmed, G.; Zayed, E.M. Response of Egyptian Clover Ecotypes to the Non-Thermal Plasma Radiation. *Range Manag. Agrofor.* **2011**, *32*, 9–14.

35. Filatova, I.; Lyushkevich, V.; Goncharik, S.; Zhukovsky, A.; Krupenko, N.; Kalatskaja, J. The Effect of Low-Pressure Plasma Treatment of Seeds on the Plant Resistance to Pathogens and Crop Yields. *J. Phys. D Appl. Phys.* **2020**, *53*, 244001. [CrossRef]

36. Bußler, S.; Steins, V.; Ehlbeck, J.; Schlüter, O. Impact of Thermal Treatment Versus Cold Atmospheric Plasma Processing on the Techno-Functional Protein Properties from Pisum Sativum 'Salamanca'. *J. Food Eng.* **2015**, *167*, 166–174. [CrossRef]

37. Gnapowski, S.; Gnapowski, E.; Duda, A. Inproving of the Quality FOOD for Animals by Pulsed Power Plasma Discharge. *Adv. Sci. Technol. Res. J.* **2015**, *9*, 58–65. [CrossRef]

38. Venkataratnam, H.; Sarangapani, C.; Cahill, O.; Ryan, C.B. Effect of Cold Plasma Treatment on the Antigenicity of Peanut Allergen Ara H 1. *Innov. Food Sci. Emerg. Technol.* **2019**, *52*, 368–375. [CrossRef]

39. Ji, H.; Dong, S.; Han, F.; Li, Y.; Chen, G.; Li, L.; Chen, Y. Effects of Dielectric Barrier Discharge (DBD) Cold Plasma Treatment on Physicochemical and Functional Properties of Peanut Protein. *Food Bioprocess Technol.* **2018**, *11*, 344–354. [CrossRef]

40. Ji, H.; Han, F.; Peng, S.L.; Yu, J.J.; Li, L.; Liu, Y.G.; Chen, Y.; Li, S.H.; Chen, Y. Behavioral Solubilization of Peanut Protein Isolate by Atmospheric Pressure Cold Plasma (ACP) Treatment. *Food Bioprocess Technol.* **2019**, *12*, 2018–2027. [CrossRef]

41. Ji, H.; Tang, X.J.; Li, L.; Peng, S.L.; Gao, C.J.; Chen, Y. Improved Physicochemical Properties of Peanut Protein Isolate Glycated by Atmospheric Pressure Cold Plasma (ACP) Treatment. *Food Hydrocoll.* **2020**, *109*, 106124. [CrossRef]

42. Volkov, A.G.; Xu, K.G.; Kolobov, V.I. Cold Plasma Interactions with Plants: Morphing and Movements of Venus Flytrap and *Mimosa Pudica* Induced by Argon Plasma jet. *Bioelectrochemistry* **2017**, *18*, 100–105. [CrossRef] [PubMed]

43. Yepez, X.V.; Baykara, H.; Xu, L.; Keener, K.M. Cold Plasma Treatment of Soybean Oil with Hydrogen Gas. *J. Am. Oil Chem. Soc.* **2020**, *98*, 103–113. [CrossRef]

44. Zhai, Y.F.; Liu, S.N.; Xiang, Q.S.; Lyu, Y.; Shen, R.L. Effect of Plasma-Activated Water on the Microbial Decontamination and Food Quality of Thin Sheets of Bean Curd. *Appl. Sci.* **2019**, *9*, 4223. [CrossRef]

Review

Effects of Nonthermal Plasma on Morphology, Genetics and Physiology of Seeds: A Review

Pia Starič [1,*], Katarina Vogel-Mikuš [1,2], Miran Mozetič [1] and Ita Junkar [1]

[1] Jožef Stefan Institute, Jamova cesta 39, 1000 Ljubljana, Slovenia; katarina.vogelmikus@bf.uni-lj.si (K.V.-M.); miran.mozetic@ijs.si (M.M.); ita.junkar@ijs.si (I.J.)

[2] Biotechnical Faculty, University of Ljubljana, Jamnikarjeva ulica 101, 1000 Ljubljana, Slovenia

* Correspondence: pia.staric@ijs.si

Received: 17 November 2020; Accepted: 7 December 2020; Published: 9 December 2020

Abstract: Nonthermal plasma (NTP), or cold plasma, has shown many advantages in the agriculture sector as it enables removal of pesticides and contaminants from the seed surface, increases shelf life of crops, improves germination and resistance to abiotic stress. Recent studies show that plasma treatment indeed offers unique and environmentally friendly processing of different seeds, such as wheat, beans, corn, soybeans, barley, peanuts, rice and *Arabidopsis thaliana*, which could reduce the use of agricultural chemicals and has a high potential in ecological farming. This review covers the main concepts and underlying principles of plasma treatment techniques and their interaction with seeds. Different plasma generation methods and setups are presented and the influence of plasma treatment on DNA damage, gene expression, enzymatic activity, morphological and chemical changes, germination and resistance to stress, is explained. Important plasma treatment parameters and interactions of plasma species with the seed surface are presented and critically discussed in correlation with recent advances in this field. Although plasma agriculture is a relatively new field of research, and the complex mechanisms of interactions are not fully understood, it holds great promise for the future. This overview aims to present the advantages and limitations of different nonthermal plasma setups and discuss their possible future applications.

Keywords: cold plasma; seeds; seed priming; plants; germination; yield; gene expression; (de)methylation; growth; ROS; phytohormones; MDA; antioxidant enzymes

---

1. Introduction

Plasma is the fourth fundamental state of matter [1]. It is a mixture of electrons, positively charged ions, radicals, gas atoms, molecules (in excited or basic state) and photons from a range of energies including ultraviolet (UV) and vacuum ultraviolet (VUV) radiation [2,3]. This state is typical of a number of natural phenomena on Earth, such as cloud-to-ground and cloud-to-cloud lightning, the aurora borealis and fire. In space, plasma is the main component of matter, including the solar corona and solar winds. It is found in the tails of comets, in the interstellar and intergalactic media and in the accretion disks around black holes. Under laboratory conditions, it is obtained by supplying energy to a neutral gas, causing the excitation of gaseous molecules and atoms and at least partial dissociation and ionization. The energy can be supplied by heating or exposure to an electromagnetic field [4,5].

Plasma can be categorized according to its thermal equilibrium or nonequilibrium. When generated by heating the gas to a sufficiently high temperature to produce ionized gas it is termed thermal plasma. The constituents of thermal plasma are in thermodynamic equilibrium, possibly at temperatures well above 1000 °C. Thermal plasmas are generated primarily by different gaseous discharges, including direct current (DC), alternate current (AC), radio frequency (RF) and microwave (MW) discharge. A wide range of DC and RF plasma sources with power levels ranging from a few W to several MW levels [6] are available.

In a nonequilibrium plasma or nonthermal plasma (NTP) the excitation, dissociation or ionization of molecules is more effective than increasing the average kinetic energy by heating, resulting in lower temperatures of the heavy particles in comparison to the electron temperature. While the electron temperature is usually of the order of 10,000 K, the temperature of heavy particles is often close to the room temperature. NTP plasma or cold plasma is more suitable for the treatment of thermally unstable biological samples than thermal plasma because the heating of the samples is minimal. The heating can be further suppressed when it comes to short-term NTP exposures, for example, in pulsed modes [2,3,7].

NTP has a wide range of biological applications, from decontamination and sterilization of surfaces [8], food [9], treatment of medical implants for biocompatibility [10], improved wound healing [11], seed germination [12–14] and resistance to certain abiotic stresses [15]. NTP technology is becoming increasingly popular in agriculture, in particular for seed treatments. Studies have shown a positive effect on seed sterilization, presenting an elegant solution for the reduction of the amount of chemical pesticides used, to diminish the high burden on the environment [16] and high risk for human health [17]. NTP has also been shown to improve seed germination and viability, which in turn could help to avoid the use of chemical substances for seed priming and increase the yield of various important crops [18–21].

Interaction of gaseous plasma with the seed surface is schematically presented in Figure 1. Reactive chemical species, the electromagnetic field and the temperature of plasma exert some influences on the seed material. In most cases the seed surface becomes more hydrophilic as the competition between the functionalization of the surface molecules and etching of the surface takes place. Hydrophilicity is mainly related to the removal of waxy structures from the surface of seeds by plasma etching. This can affect the seed morphology, as changes of nanostructures on the seed surface may be observed. The interaction of reactive chemical species in plasma also changes surface chemistry (functionalization of the surface), which may improve surface wettability, water uptake and, consequentially, initiate complex signaling pathways in seeds. Changes in DNA, enzymatic activity and hormone balance have also been detected, depending on the plasma treatment conditions. Summarily plasma-induced changes can affect germination, later growth and development of plants, their resistance to abiotic stress and yield. However, plasma and discharge parameters should be optimized to obtain desired seed responses regarding water uptake, sprouting, growth, etc.

In this review, we focus on physiological and morphological aspects of the effects of NTP on seeds and plants. The present knowledge enables the prediction of possible influences of NTP on seeds in terms of germination and growth, stress resistance, DNA change, transfer of traits to the next generation, enzyme activity, phytohormones and, to some extent, optimization by the choice of the plasma treatment conditions [21–24]. There is already some evidence of the positive effects of NTP on seeds, but it is still unclear what mechanisms are behind the improved seed germination and viability. It is important to emphasize that different research groups use different experimental setups and parameter settings, as well as different types of seeds, which all affect the results of plasma-seed interactions. All results presented in this review and summarized in Table 1 are not necessarily applicable to all types of seeds. However, a basic overview is given.

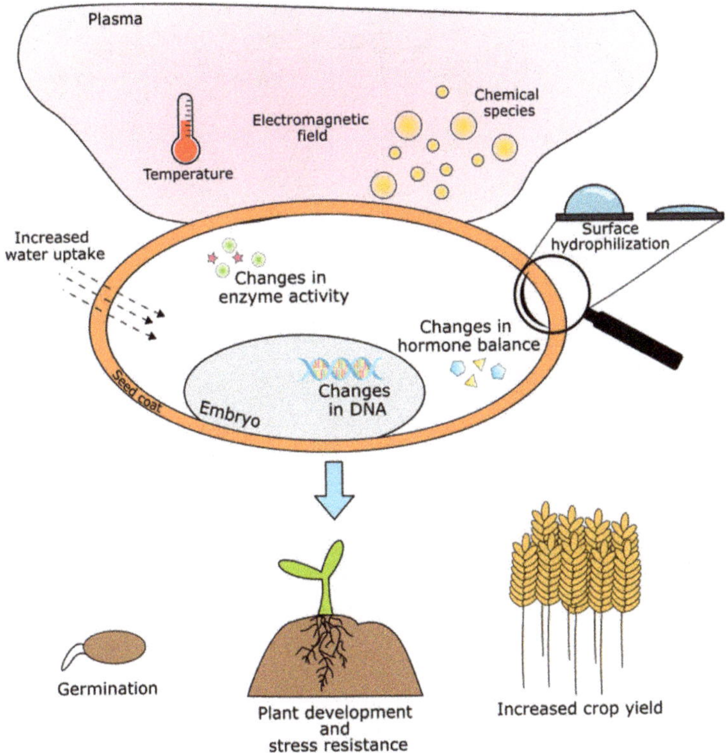

Figure 1. A schematic representation of nonthermal plasma (NTP) effects on seeds.

Table 1. An overview of experimental cold plasma parameters and their effects on different seeds. DBD: dielectric. DCSBD: diffuse coplanar surface barrier discharge. RF-CCP: radio frequency capacitively coupled plasma. RF-EMF: radio frequency electro-magnetic field. LPDBD: low pressure dielectric barrier discharge.

Seed Type	Plasma Parameters	Exposure Time	Result Summary	Ref.
Arabidopsis thaliana	DBD, 10 kHz, 10 kV, atmospheric pressure, air	15 min	Plasma pretreated seeds germinated faster, but the final germination rate was not significantly increased; germination was improved under salinity conditions as germination decrease caused by salinity stress was partially restored.	[25]
Arabidopsis thaliana, *Raphanus sativus*	RF, 13,56 MHz, 60 W, 20–80 Pa, O_2		No significant impact on plant growth, but the gene expression patterns were changed, also when comparing the first and the second generation of seeds.	[26]
Arabidopsis thaliana, *Raphanus sativus*	RF, 13.56 MHz, 60 W, 20–80 Pa, Ar, O_2	5, 15, 30 and 60 min	An enhanced seedling growth (at 80 Pa for 10 or 20 min) and changed gene expression pattern; growth enhancement was not inherited by the second-generation plants.	[27]
Barley (*Hordeum vulgare*)	SDBD (AC), 30 kHz, 400 W, atmospheric pressure, N_2 and air	10, 20, 40 and 80 s	Denser and longer roots and higher shoots in plasma-exposed seedlings; an increase in GABA levels.	[28]
Beans (*Phaseolus vulgaris*)	RF, 10 MHz, 20 W, 6.7×10^{-2} Pa, air	2 min	Faster initial germination, but the final germination rate was the same; hydrophilization of the seed coat surface and accelerated water uptake.	[29]

Table 1. *Cont.*

Seed Type	Plasma Parameters	Exposure Time	Result Summary	Ref.
Bell pepper (*Capsicum annuum*)	DBD (AC), 23 kHz, 11 kV, 80 W, atmospheric pressure, Ar	60, 120 s	Growth-promoting and protecting effects on seedlings; inhibition and delay of nano ZnO toxicity and its negative effects on plant growth and differentiation.	[30]
Brown rice	DC, 1–3 kV, 800 Pa, air	10 min	Increased germination rate, seedling length, water uptake, and GABA levels with optimum treatment at 3kV for 10 min.	[12]
Chickpea (*Cicer arietinum*)	surface microdischarge (SMD), complex 20-ms cycle, atmospheric pressure, air	0.5–5 min	A noticeable increase in germination rate and root and shoot length.	[31]
Maize (*Zea mays*)	DCSBD, 14 kHz, 370 W, atmospheric pressure, air	60, 120 s	No change in the germination rate, noticeably longer roots and bigger wet and dry biomass of plants after plasma treatment.	[32]
Maize (*Zea mays*)	DCSBD, 14 kHz, 20 kV, 400 W, atmospheric pressure, air	30–300 s	No statistically significant changes in the germination rate; at longer exposure times, a decrease in germination rate; an increase in root length and shoot height.	[33]
Mung beans (*Vigna radiata*)	RF-CCP, 13.56 MHz, 40 and 60 W, 20 Pa, air	10, 15 and 20 min	Faster germination, an increased germination rate, higher shoots, an increase in water-soluble sugars and higher amylase and phytase activity.	[22]
Mung beans (*Vigna radiata*)	microplasma, 9 kHz, 0–20 kV, 25 W, atmospheric pressure, N_2, He, air and O_2	10 min	Different effects depending on feed gas, higher germination rate in air and helium plasma, an increase in plant height for air plasma, no significant effects on seeds in O_2 plasma.	[34]
Pea (*Pisum sativum*)	DBD, 14 kHz, 20 kV, 400 W, atmospheric pressure, air/N_2/O_2/N_2 and O_2	60, 180 and 300 s	More DNA damage than in nontreated samples; an ambient air plasma had the least damaging effects on seed DNA, compared to plasma treatment with different mixtures of O_2 and N_2.	[35]
Pea (*Pisum sativum*)	DCSBD, 14 kHz, 20 kV, atmospheric pressure, air	60–300 s	No increase in DNA damage. After the application of DNA-damage agent Zeocin, an increase in DNA damage. The plasma-treated seeds had a lower level of DNA damage than untreated seeds.	[36]
Pea (*Pisum sativum*)	DCSBD, 14 kHz, 10 kV, 370 W, atmospheric pressure, air	60–600 s	Increased biosynthesis of auxin and cytokinins as well as their catabolites and conjugates; a noticeable increase in germination rate and root and shoot length.	[21]
Peanut (*Arachis hypogaea*)	RF-CCP, 13.56 MHz, 60–140 W, 150 Pa, He	15 s	Improved germination rate and peanut yield after plasma treatment of seeds at 120 W.	[37]
Quinoa (*Chenpodium quinoa*)	RF and DBD, 1 kHz, 8.2 kV, 6.4 W, 500 and 0.1 mbar, air	10, 30, 60, 180 and 900 s	A higher germination rate in seeds treated for 10 s with RF plasma or 180 or 900 s by DBD plasma. A drastically changed chemistry of the seed coat outer layer	[13]
Rapeseed (*Brassica napus*)	RF-CCP, 13.56 MHz, 100 W, 150 Pa, He	15 s	Increased germination rate and shoot and root growth. Germination rate, germination index and vigor of seeds exposed to drought stress and treated with plasma was increased, along with plant growth and a decrease in MDA content compared to untreated seeds exposed to drought	[15]
Sunflower (*Helianthus annuus*)	RF-EMF, 5.28 MHz, atmospheric pressure, air	5, 10 and 15 min	No significant changes in the germination rate of plasma and EMF treated seeds; noticeable changes in phytohormone balance; after short exposure of seeds to RF-CP or RF-EMF a long-term effect on gene expression in leaves, mostly stimulating expression of proteins involved in photosynthesis and its regulation.	[38]
	RF-CP, 200 Pa, air	2, 5 or 7 min		
Soybean (*Glycine max*)	60 kHz, 10.8–22.1 kV, 3.4–15.6 W, atmospheric pressure, Ar	12, 24, 48, 60, 120 and 180 s	A slightly higher germination rate and enhanced root and shoot growth; changes in DNA methylation level, an increased SOD, POD and CAT enzyme activity.	[24]
Wheat (*Triticum aestivum*)	SDBD (AC), 50 Hz, atmospheric pressure, air	5, 15 and 20 min	The germination rate unchanged; an enhanced root growth (especially in seedlings treated with plasma for 15 min).	[39]

Table 1. *Cont.*

Seed Type	Plasma Parameters	Exposure Time	Result Summary	Ref.
Wheat (*Triticum aestivum*)	DBD, 50 Hz, 13 kV, atmospheric pressure, air	4 min	Increased germination rate, root length and shoot height; noticeable changes in the expression of genes *LEA1*, *SnRK2* and *P5CS*; increased proline and soluble sugar levels in normal water conditions and in seedlings exposed to drought conditions. A decrease in MDA content in seeds under drought stress. An increase in SOD, POD and CAT enzyme activity.	[40]
Wheat (*Triticum aestivum*)	RF, 3×10^9 MHz, 60, 80 and 100 W, 150 Pa, He	1–5 shots of high voltage nanosecond	No changes in the germination rate and speed; enhanced root and shoot growth and an increased yield.	[41]
Wheat (*Triticum aestivum*)	DBD (AC), 50 kHz, 13 kV, 1.5 W, atmospheric pressure, air	1, 4, 7, 10 and 13 min	Increased root and shoot growth, an increase in proline and soluble sugar levels. With increased exposure time a noticeable decline of MDA content and an increase in SOD and POD enzyme activity; increased germination rate.	[20]
Wheat (*Triticum aestivum*)	DBD, 50 kHz, 80 kV, atmospheric pressure, air	30, 60 and 80 s	Improved seed germination rate and an increase in root growth. Shoots longer only in seeds treated with plasma for 30 s and no retention time, and seeds treated with plasma for 60 s and retention time of 24 h.	[42]
Wheat (*Triticum aestivum*)	DBD (AC), 50 kHz, 13 kV, atmospheric pressure, air, O_2, N_2, Ar		In seeds, treated with air, N_2 or Ar plasma an increase in root length; no increase in seeds treated with O_2 plasma.	[18]
Wheat (*Triticum aestivum*)	LPDBD, 5 kHz, 4.5 kV, 45 W, 10 torr, Ar/O_2 or Ar/air	90 s	Shorter roots and longer shoots compared to the control seedlings; an elevated SOD enzymatic activity in roots and a higher level of H_2O_2 in roots and shoots in seedlings treated with O_2/Ar plasma; with Ar/air plasma an increase in CAT activity in leaves.	[19]
Wheat (*Triticum aestivum*)	RF, 13.56 MHz, 80 W, 0.1 mbar, air	60, 120, 180 and 240 s	With Plasma treatment for 180 s a higher wheat yield and increased plant photosynthesis.	[43]
Maize, wheat, soybeans, tobacco	RF-CCP, 13.56 MHz, 50–1000 W, 30–200 Pa, air	5–90 s	An increase in the germination rate of maize and seedling growth of wheat; SOD and POD enzymatic activity also increased; an increase in the yield of tobacco (20%) and soybean (4%).	[44]

1.1. Nonthermal Plasma Generation Methods

Although authors used a variety of experimental setups, the most common methods for NTP generation used for seed treatment are presented in Figure 2 and are based on atmospheric pressure plasma jet discharge (APPJ) [34,45,46], dielectric barrier discharge (DBD) [13,18–21,23–25,28,32,33,36,39,40,42,44] and radio-frequency discharge (RF) [13,22,26,27,29,37,38,43,47–50]. The type of discharge depends on the frequency of the power source (AC, DC), the ambient gas pressure (low and atmospheric pressure) and the exact shape of the electrodes [51]. The most commonly used methods for seed treatment are the DBD and RF discharges, which are described below in more detail.

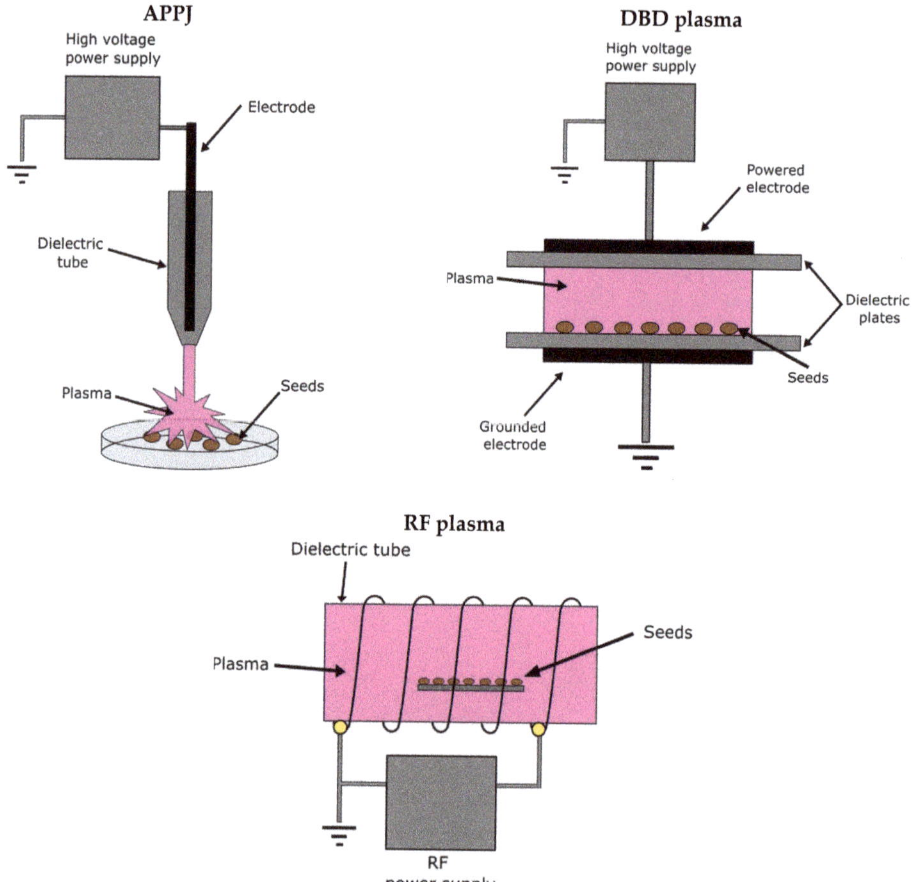

Figure 2. Positioning of seeds in different plasma reactors.

A plasma reactor with DBD consists of two electrodes covered with a solid insulating material (glass, plastic, ceramic, etc.). Such dielectric coating prevents the arc discharge that would cause the generation of thermal plasma. The gas temperature usually does not rise significantly because the DBD plasma runs in streamers of very short time duration, typically in the microsecond range. DBD is generated by applying alternating (AC) high-voltage with a frequency from 50 Hz up to about 100 kHz [52].

RF plasma is normally generated by an RF power supply at 13.56 MHz. The frequency is high enough to provide a sustaining continuous plasma, so no streamers are observed. We can further classify the RF plasma by the shape of the electrodes or the antenna. In capacitively-coupled plasma (CCP), RF power is applied between two electrodes and plasma is generated in the gap between. In inductively-coupled plasma (ICP), an antenna has the shape of a coil to which the RF voltage is applied. The plasma is then generated inside the coil. RF set-up can be used to generate a relatively large volume of rather high-density plasma, with minimal heating under low-pressure conditions [51].

The exact mechanisms that contribute to seed priming by using NTP are not yet well understood. Experiments on seeds to observe the effects of direct (glow mode) and indirect (afterglow mode) plasma treatment could help to reveal the mechanisms by which NTP interactions with seeds affect surface properties and seed germination [51].

Direct and Indirect Effects of Nonthermal Plasma Treatment

The seeds or other biological samples can be exposed to different modes of NTP depending on the plasma system. The treatment can be performed in the so-called direct or glow mode, or indirect (afterglow) mode. Surfaces that are exposed to the area of discharge (direct plasma treatment) are subjected to UV and/or VUV radiation and the plasma particles (ions, electrons and different excited atomic and molecular species). During indirect treatment, however, the samples are exposed to nonequilibrium gas outside of the area of the glowing plasma region but close enough to benefit from interaction with long-lived radicals. Samples, in this case, are not exposed to the radiation and are subjected to lower, less aggressive concentrations of reactive chemical species with lower energy, brought from the discharge area by the gas flow. These samples are thus exposed only to longer-lived or recombined species, which are the result of secondary reactions. These species have marginal kinetic energy and lower potential energy compared to plasma species. Thermal heating of samples in the afterglow mode is marginal but can still functionalize the seed surface [53]. Indirect plasma treatment is usually weaker, and longer times of sample exposure are needed to achieve similar results as in direct plasma treatment [54]. Moreover, the etching of surfaces in this region is significantly lower. Interactions of plasma species in the direct and indirect modes are schematically presented in Figure 3.

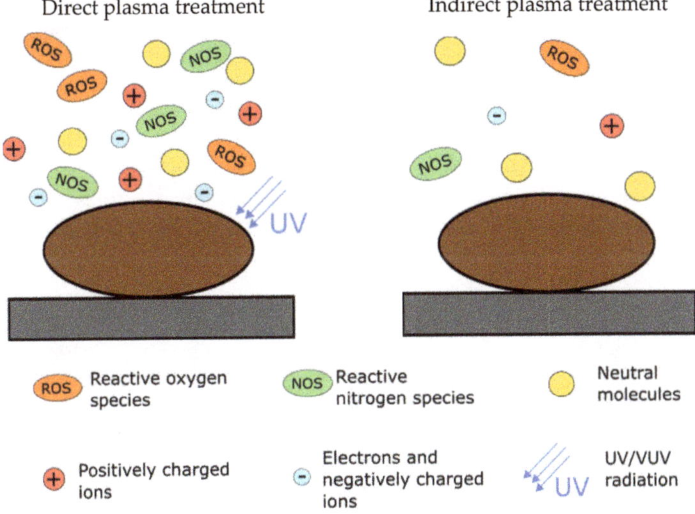

Figure 3. Schematic representation of particles occurring in direct and indirect plasma treatment region. Direct plasma treatment region is a combination of positively and negatively charged particles, neutral chemical species, ROS, NOS and UV radiation. In indirect plasma treatment, however, only lower, less aggressive concentrations of reactive chemical species are present.

2. Nonthermal Plasma Effects on Seeds

2.1. DNA Methylation and Demethylation and DNA Damage

The use of NTP technology on seeds can affect their different physiological and morphological levels. Plasma generation causes formation of different reactive species such as oxygen (ROS) and nitrogen (RNS) reactive species, and UV/VUV radiation that can have some effect on DNA of the embryo. Although there are no reports on DNA damage in seeds, such as double-strand breaks (DSBs) or formations of micronuclei, some experiments showed that NTP treatment caused changes in DNA (de)methylation and gene expression. Experiments conducted by Leduc et al. [55] showed that NTP treatment could potentially cause DSBs and fragmentation of plasmid DNA in case of isolated DNA in phosphate-buffered saline (PBS buffer) or water. It is important to emphasize that the fragmentation of isolated (naked) DNA by NTP does not mean that the treatment of seeds with NTP could result in similar effects, even at the same NTP parameters. Another study conducted by Lazovic et al. [56] showed that atmospheric pressure plasma induced single-strand breaks and base-damage of DNA in the human primary fibroblast cells, depending on the plasma treatment time and power. However, it should be noted that embryonic cells in seeds have cell walls that to a certain degree protect the cells from abiotic stresses and that the seed coat (testa) also acts as a protective barrier. This is the reason that short-term exposure to NTP should not harm the DNA in seeds at least not to the degree of the total DNA fragmentation [55]. Here, it should be stressed that the plasma particles do not penetrate deep into the solid matter (surface treatment technique, where only the top few nanometers of surface are modified), while the radiation can penetrate rather deeply, depending on the photon energy. For example, the UV radiation of the germicidal range of wavelengths (about 230 –280 nm) can penetrate the bacterial cell wall, causing DNA damage [57].

Kyzek et al. [36] showed that NTP treatment of pea seeds did not cause significant DNA damage. When pea seeds were treated with zeocin, that has toxic effects and causes severe DNA damage, seeds pretreated with NTP showed a significantly lower level of DNA damage. Although a low level of DNA damage induced by NTP cannot be excluded, it is possible that NTP triggers DNA repair mechanisms that repair damaged segments of DNA and protect against additional DNA harming agents. Further studies are, however, needed to prove the upregulation of DNA repair mechanisms in seeds treated with NTP.

In recent studies, Tomeková et al. [35], examined the effects of longer exposure times of pea seeds to cold plasma. They used air, nitrogen, oxygen and combinations of O_2 and N_2 in different ratios, as a feed gas for plasma treatment. The study analyzed the composition of the gaseous product of the plasma with FTIR spectroscopy and correlated the spectroscopy results with analysis of DNA damage in pea seeds treated with cold plasma at different conditions. The results indicated that ambient air had the least damaging effects on pea seed DNA. The DNA was more damaged with increasing concentration of nitrogen and exposure time. It was estimated that the optimal combination of the plasma chemical composition and water vapor in ambient air had a positive effect on pea seed germination. However, further research is necessary to confirm this. At the same time, it is necessary to separately investigate the influence of chemically reactive gaseous species and plasma radiation. As mentioned above, the reactive species are unlikely to reach the DNA, whereas the radiation rather easily penetrates into the organic material. The effects may be separated by placing a VUV-transparent window between the glowing plasma and organic matter, which is a practice when studying the effects on polymers [58].

In another study Nakano et al. [27] conducted a microarray analysis on *Arabidopsis thaliana* seeds. The seeds treated with oxygen NTP at a pressure of 20–80 Pa and power of RF generator 60 W displayed an increase in DNA methylation. The NTP treatment also enhanced the growth and germination of seeds. The analysis of microarray results showed a decrease in expression of *AT2G30620*, a gene encoding histone H1.2, which is involved in repression of DNA transcription. Its downregulation implies changes in the chromatin structure. On the other hand, the studies have shown an increase in the RNA-directed DNA methylation 4 gene (*RDM4*) encoding proteins connected with DNA methylation. This observation indicates that NTP treatment may influence the epigenetics of seeds to a certain extent. Treatment with argon NTP exhibited different expression patterns for the before-mentioned genes, with the expression levels of *AT2G30620* remaining the same, but the expression of *RDM4* was lower compared to the oxygen NTP treatment at a low pressure of 20 Pa, indicating different effects of different NTP feed gases. One possible explanation is in the difference in VUV radiation. Low-pressure oxygen plasma is not a significant source of radiation; usually, only the O-line at about 130 nm is observed, whereas atmospheric pressure plasma sustained in pure argon is an extensive source of VUV continuum arising from Ar_2^* dimers [59]. Zhang et al. [24] used an atmospheric pressure discharge and found that argon NTP can reduce methylation and increase gene expression of adenosine triphosphate (ATP) synthase, target of rapamycin (TOR) and growth-regulating factor (GRF) genes in soybeans involved in the biosynthesis of energy-rich molecules and biomass production, and regulate the growth and development of seedlings. A higher concentration of ATP promotes germination and growth of sprouts, which has been shown in germination and growth experiments with soybeans. All these results suggest that NTP has an effect on the epigenetics of seed embryos, but the mechanisms behind its activation are still unknown. Unfortunately, many authors do not report on peculiarities of discharges, in particular the gas purity, which may be the decisive parameter governing the radiation and chemical reactivity of NTPs [60].

2.2. Gene Expression and Protein Synthesis

The effects of NTP described above show that NTP could actually affect seed gene expression. This implies possible changes in protein synthesis in the embryo after NTP treatment. Some research on gene expression was done using microarray and gene ontology analyses. After NTP treatment of seeds, the expression of several genes changed. Hayashi et al. [26] found the upregulation of genes involved in increasing energy production by photosynthesis by upregulating enzymes such as RuBisCO, increasing carbon fixation and production of the plant hormone auxin. Nakano et al. [27] also noted changes in the synthesis of several proteins, but it appears that there were no mutagenic effects on seed embryos in genes associated with plant growth, as the growth increase was not inherited from the first plant generation. NTP treatment of seeds also resulted in upregulation of the expression of the *SnRK2* and *P5CS* genes and, at the same time, downregulation of the expression of the *LEA1* gene. These genes are associated with drought resistance in plants and might be an indicator that the NTP treatment exerted a certain level of stress on seeds. In another study [46], researchers observed an increased expression in redox homeostasis-related genes and in genes responsible for epigenetic changes in tomato seeds exposed to cold plasma treatment for 10 min. All mentioned reports indicate an important activation of gene regulation after plasma treatment of seeds. However, the question is what triggers are produced by plasma and what signals induce changes in gene and protein expression.

2.3. The Effects on Enzyme Activity

Differences in enzyme activities may indicate that the plant has been exposed to some type of stress or change in its natural environment. NTP treatment of seeds could potentially act as a stress agent. To assess this possibility, and to explore the mechanisms of seed response to NTP treatment, researchers examined the activity of different enzyme groups.

For example, malondialdehyde (MDA) is measured as an indicator of membrane lipid peroxidation and membrane damage. Many researchers report that seedlings exposed to NTP treatment exhibit lower concentrations of MDA while showing increased germination and growth. This suggests that NTP exposure of seeds (under appropriate NTP parameters) may reduce membrane damage, possibly by accelerating antioxidant machinery [15,20,40,46].

In the course of evolution, plants have developed an antioxidant system that enables them to combat ROS toxicity and to eliminate its negative effects. Antioxidants have various physiological functions, ranging from converting harmful ROS into less reactive species even if present in relatively low concentrations, to acting as a deactivating agent against oxidation. The antioxidant system comprises both enzymatic and nonenzymatic systems. The enzymatic system consists of superoxide dismutase (SOD), catalase (CAT), ascorbate peroxidase (APX), peroxidase (POD), glutathione reductase (GR), polyphenol oxidase (PPO), among others. Upon deleterious or altered conditions of the plant environment, e.g., extreme oxidative stress, antioxidants scavenge the toxic radicals and help the plants to survive. Under such conditions, the level of antioxidant enzyme activity could increase. However, the increase in antioxidant enzyme activity can not only be correlated with a stressful environment, but also with activation of the energy metabolism during seed germination, which triggers many cellular processes where ROS are actively produced [61].

In many experiments conducted by different researchers, NTP-treated seeds showed an increase in SOD, POD and CAT enzymatic activity [19,20,24,44]. The increased antioxidant enzyme activity was found in a few days-old seedlings. Rahman et al. [19] found no changes in SOD and APX activity in seeds treated with NTP. Increased enzyme activity was found only in older seedlings from seeds treated with NTP, with an increase in SOD activity in the roots and APX activity in the shoots, while there were no detectable changes in the seeds right after the NTP treatment. It appears that the increased antioxidant activity is not the first change in the line of seed response to NTP treatment. It may be a secondary response, but it is not clear what the activation mechanism of increased antioxidant enzyme activity is.

The activity of amylase enzyme secreted by the cells of the aleurone layer of the seed is an important factor in seed germination [62]. Amylase splits the starch reserves into simpler molecules, which are then used for early growth of the seedlings. Seeds exposed to NTP showed a higher increase in the activity of the amylase enzyme after 12 and 24 h of imbibition than untreated imbibed seeds. Together with the increased amylase activity, an increase in soluble sugars, which are a product of amylase degradation of starch was also observed [22]. Chen et al. [12] also observed an increased amylase activity in seedlings of NTP-treated seeds which could be correlated with an increase in germination rate. The effect of NTP was also reflected in the amylase activity of young seedlings. Amylase enzymes are normally activated by plant gibberellin (GA) hormones [63], which indicates that hormones respond to NTP treatment earlier than amylase activity. Another possibility for the activation of amylase activity is by nitrogen species and compounds, more precisely nitric oxide (NO), which may also result from NTP treatment of seeds [13,64].

2.4. Morphological and Chemical Changes of the Seed Coat

The effect of NTP on the seeds can be observed directly after the NTP treatment. By using a scanning electron microscope (SEM), many researchers noticed the etching effect on the seed coat. Li et al. [20] investigated the morphology of wheat seed coat. The conspicuous mesh-like structures seen in nontreated seeds were gradually destroyed with prolonged DBD plasma treatment. After 10 min of treatment, the boundary layer of these mesh-like structures on the seed coat was difficult to identify. Besides, noticeable cracks in the seed coat were observed. Similar effects of NTP treatment were observed by many other researchers [18,19,21,25,28,40,46]. Park et al. [28] reported of an uneven etching of barley seeds, indicating an uneven interaction of the NTP with the seed surface. This can be attributed to specific chemical interactions of the NTP with the seed surface and the energy distribution of the NTP. Stolárik et al. [21] also noted the uneven effect of NTP on the surface of pea seeds with abrasion and disruption of the seed surface. During prolonged NTP treatment, cleavages and cracks were observed in the seed surface. This suggests that a prolonged exposure to NTP treatment causes more severe changes in the seed surface morphology. In contrast to these results, Zahoranová et al. [33] did not observe any surface damage to the seed coat and only a slightly softer surface of corn seeds. It is apparent that the altered morphology of the seed coat depends on the initial seed coat morphology and the NTP parameters used.

Numerous experiments showed a decrease in the water contact angle of the seed surface as shown in Figure 4 [22,29,33,49,65,66]. Bormashenko et al. [49] found a decrease in the water contact angle after 15 s treatment with air RF plasma. At the same time, no morphological changes were detected during SEM analysis. This indicates that the change in water contact angle can be attributed to chemical changes that occur during NTP treatment on the seed surface, turning the normally hydrophobic seed coat to a hydrophilic one. Time-of-flight secondary ion mass spectrometry (TOF-SIMS) analysis of seeds treated with NTP demonstrated a 2.5 to 3 times more intense mass peaks corresponding to the presence of oxygen than on untreated seeds. This shows the enrichment of the seed surface with oxygen species by NTP and the introduction of new chemical species on the surface, resulting in a more hydrophilic nature, a reduced water contact angle and an increase in the wettability of the seeds [13,33,34,42,49], which could, to some extent, be attributed to altered surface morphology (nano-structuring of seed coat).

Figure 4. Plasma treatment of seeds decreases water contact angle. Water contact angle of untreated surface on the left, and decreased water contact angle of the surface after plasma pre-treatment on the right.

An increase in the hydrophilic properties of biological samples by NTP is an effect similar to the hydrophilization of synthetic polymers, which has been carefully studied in terms of altered surface morphology and functionalization [67,68] as well as influence on the preferential etching of amorphous parts of the polymer [69]. NTP-treatment of polymers introduces new functional groups that have a major impact on the physical and chemical properties of the surface, including its wettability. A common phenomenon accompanying plasma treatment of polymers is also a process called hydrophobic recovery in which polymer surface characteristics revert to their original condition over time and regain hydrophobic properties [70]. In seeds, hydrophobic recovery has not been observed so far [71].

Gómez-Ramírez et al. [13] examined the surface of quinoa seeds with XPS (X-ray photoelectron spectroscopy) and found that the surface was largely affected by the NTP treatment. The exposure of the seeds to DBD plasma for 30 s, or to RF plasma for 10 s, induced a significant increase in the oxygen and nitrogen content of the seed surface at the expense of carbon. They also noted a slight increase in potassium content. Similar effects were observed when monitoring the seed surface with SIMS (secondary-ion mass spectrometry) as reported by Bormashenko et al. [47], confirming the results of the XPS analysis. Prolonged exposures of the seeds to NTP corresponded to an even larger increase in oxygen, nitrogen and potassium content. This indicates the ability of the NTP to oxidize the outermost layers of the seeds due to the presence of highly reactive excited species in air NTP. Similar effects have been described for different seeds exposed to the RF plasma and for some polymers where the depth of plasma-induced oxidation down to 100 nm was found using similar NTP parameters [13]. After exposure of the seeds to water, there was a decrease in potassium and nitrogen species (NO_3^-, NO_2^-). Gómez-Ramírez et al. [13] suggest that the mechanism behind this is a diffusion of labile potassium and nitrogen species into the interior of the seed after exposure to water. The diffusion of these molecules promotes seed germination in a similar way the exposure to nitrate-rich water influences seed metabolism and water uptake. Zahoranova et al. [33] reported that there were no detectable morphological changes as observed by SEM after NTP treatment of maize seeds. Attenuated total reflection-Fourier-transform infrared spectroscopy (ATR-FTIR) measurements, however, indicated some changes in the chemical groups of plasma-treated seeds, with a decrease in the lipid group and an increase in polar groups containing oxygen and nitrogen. All these results confirm that NTP reduces the hydrophobic surface of the seeds and transforms it into more hydrophilic one.

2.5. Plant Hormone Balance

Plant hormones, such as auxin, cytokinin, ethylene, gibberellins (GA), abscisic acid (ABA) and brassinosteroids, are molecules that control various physiological and biochemical processes in the plant. Plant hormones can be associated with germination, growth and development, flowering, fruiting, fruit ripening, dormancy and resistance to various abiotic and biotic stresses [64]. During germination, numerous processes are activated in the seeds, such as the release of stored reserves, activation of metabolism and growth of meristems. During germination and the transition from seed to seedling, plants are very sensitive to environmental factors such as light, temperature and water availability. The reaction to these abiotic factors is often mediated by hormones [63].

Researchers indicate that the treatment of seeds with NTP influences the presence and synthesis of hormones in seeds and seedlings [20,21,28,38,40,41]. The effects depend not only on the concentration of a specific hormone but also on the relationship of the specific hormone to other hormones and signaling molecules. By generating low-pressure NTP, seeds are placed in vacuum that could potentially affect their properties and seed germination. Although according to Mildažienė et al. [38], there was no difference in germination and seedling growth between seeds exposed to vacuum and the untreated seeds, there was a noticeable change in the auxin/cytokinin ratio. This indicates the response of the plant to an external factor (vacuum) by activation of signaling pathways related to auxin and cytokinin phytohormones. Additional treatment with RF NTP caused an increase in gibberellin GA_3 concentration, while exposure to an RF electromagnetic field reduced the concentration of ABA. Although seeds in the dehydrated state have a high resistance to abiotic stress, they are still able to respond rapidly to short NTP, electromagnetic field or vacuum treatments by producing plant hormones. Ji et al. [14] also noted an increase in GA_3 concentration in wheat seeds and suggested that hormones and increased activity of hydrolytic enzymes activate seed germination after NTP treatment. Stolárik et al. [21] noted that in NTP-treated seeds, the biosynthesis of auxins and cytokinin, as well as their catabolites and conjugates, was upregulated. Guo et al. [40] reported an increase in ABA content in four-day old wheat seedlings exposed to NTP treatment prior to imbibition. ABA is an important signaling factor in response to dehydration, with the ability to regulate the water status of plants through changing stomatal conductivity and inducing genes involved in dehydration resistance.

There is increasing evidence that the pretreatment of seeds with NTP has an important impact on the hormonal balance. Whether hormones play a fundamental role in the response of seedlings to NTP treatment of seeds, is still not clear and requires additional attention and research.

2.6. Germination and Seedling Growth Parameters

Many researchers noted effects of seed exposure to NTP on germination, and later on the growth and development of the seedlings. In many cases, NTP treatment of seeds could increase the germination rate and induce faster germination in different crops [15,19,22,25,29,31,33,40,42,46]. Nevertheless, in numerous other cases, there was no improvement in the germination rate. The differences between the control group and the NTP treated seeds appeared later as improved root and shoot growth, or root branching of the seedlings [32,33,39,41,72,73], indicating various outcomes dependent on the type of plasma, treatment conditions, plant species and even the plant variety. Thus, more systematic studies correlating treatment conditions with seed types and varieties should be performed. In particular, it is recommended to examine details about the discharge parameters, in particular the concentration of gaseous impurities which may significantly influence the plasma parameters [74].

2.7. Resistance to Stress

Seeds treated with NTP exhibit interesting responses when exposed to abiotic stresses such as drought and salinity. Guo et al. [40] simulated polyethylene glycol (PEG)–induced drought stress on wheat seeds. PEG-induced drought stress caused significant oxidative damage to germinating wheat seedlings, which was confirmed by increased H_2O_2 and O_2 levels, higher MDA concentration and increased proline and soluble sugar content. In contrast, nonthermal DBD plasma treatment mitigated the oxidative damage of drought stress in wheat seeds through the induced expression of the functional gene *LEA1* and regulatory genes such as *SnRK2* and *P5CS*, increasing ABA levels and accelerating the enzymatic activity of antioxidants. The MDA content associated with lipid peroxidation and membrane damage decreased in seeds pretreated with NTP and exposed to drought stress compared to drought-exposed seeds without NTP treatment. Ling et al. [15] came to similar results for NTP DBD plasma treatment of oilseed rape seeds, where statistically significant differences were found between two varieties of oilseed rape. Bafoil et al. [25] conducted experiments on how salinity stress and NTP treatment affect germination of *Arabidopsis thaliana* seeds. Although the germination rate of NTP-treated seeds increased, NTP treatment could, to a certain extent, compensate for the negative effect of salinity stress. Based on the above results, NTP treatment of seeds has a high potential in regions where crop plants are frequently exposed to drought. Furthermore, the initial water uptake of NTP-treated seeds is definitely increased as shown by numerous authors [12,33,48,73,75].

Recently researchers have also investigated how NTP could alleviate the stress induced by potentially toxic metals (Zn) and metal oxide (ZnO) nanoparticles during seed germination and plant development. Iranbakhsh et al. [30] reported that the treatment of seeds with NTP had a great growth-promoting and protective effect. ZnO nanoparticles were proposed to have a less toxic effect, as growth and differentiation of NTP-treated seeds were improved. Promotion of phenylalanine ammonia lyase (PAL) and peroxidase activity after cold plasma treatment of seeds could give insight to the plant's resistance to abiotic stress, caused by plasma compounds such as NO, ozone and UV radiation. Increased activity of PAL and peroxidase enzymes by plasma pretreatment of seeds could thus act as a protective mechanism of the seedlings, involved in the alleviation of toxic effects caused by nano-ZnO. This is an important new aspect of NTP influence on seeds and plants with great potential in the field of bioremediation.

2.8. Transfer of the Traits Induced by Nonthermal Plasma to the Next Generation

As has been demonstrated, NTP treatment affects the seeds and their properties. It is important to know whether the effects have an impact on seeds and plants of future generations. Hayashi et al. [26] reported that the growth of second-generation plants was similar to that of seeds not treated with NTP, indicating that no mutation of growth-related genes occurred during the NTP treatment of first-generation seeds. Gene expression analysis of *Arabidopsis thaliana* seeds, however, showed a different gene expression pattern in the second-generation compared to the first-generation seeds. This confirms that no mutation, but an epigenetic effect occurred during the enhanced germination of the first-generation seeds. In view of these results, it is still unclear whether NTP causes changes in seeds that are passed on to future generations. It is necessary to conduct additional experiments on the possibility that NTP has an effect on the inheritance of NTP-induced traits to next-generation seeds. Due to time consuming experiments, not much work has been conducted in this direction.

3. Conclusions

NTP technology has great potential in agriculture as a new technology for seed priming. It is an ecological, environmentally friendly and rather simple technology. Understanding the mechanisms behind its effects on improved seed germination, plant growth, improved yield and stress tolerance is critical for future exploitation of its benefits and applications. Many researchers have focused on different levels of NTP effects on seeds, ranging from chemical and morphological changes on the seed surface to changes in water contact angle, water absorption, DNA (de)methylation, gene expression patterns, potential DNA damage, the speed and rate of germination, the growth of the seedlings, the activity of various enzymes, the presence of hormones, the transfer of improved properties to the next generation, plant resistance to drought stress and even metal toxicity. However, much more work, focused on systematic studies on the influence of various types of NTP and different crops as well as crop variety, should be conducted. As was already shown, the types of NTP (RF, APPJ, MW, etc.) affects seeds and seedlings, but it is also important to emphasize that researchers use different NTP experimental setups, which may also influence seed modification. Thus, in many cases, controversial results are reported in the literature, and it is hard to provide a general rule on which treatment conditions have the most influence on seeds in terms of their surface properties as well as biological response. Here, it is worth mentioning that the discharge itself does not interact with the seeds, but rather the reactive gaseous species and UV/VUV radiation produced in NTP sustained by the discharge. From this point of view, the scientific challenge remains appropriate plasma characterization, which may not be trivial, in particular for atmospheric-pressure plasma jets where huge gradients in the density of reactive species are common.

The main mechanisms of changes occurring during and after NTP treatment of seeds are still unknown. It is also important to assess which NTP properties are responsible for the changes. It is suggested that oxygen and nitrogen reactive species play a major role in NTP priming of seeds. However, the presence of UV and/or VUV radiation, electromagnetic field, possible temperature fluctuations and other products and by-products of NTP generation can also contribute to changes in seeds. The response of seeds to NTP could additionally be species and variety-dependent. Furthermore, the size of seeds of the same species could also play a decisive role, as the distribution of NTP energy on the seed surface could be different.

Author Contributions: Conceptualization, P.S. and K.V.-M.; validation, I.J., M.M. and K.V.-M.; investigation, P.S.; writing—original draft preparation, P.S.; writing—review and editing, I.J., M.M. and K.V.-M.; visualization, P.S.; supervision, M.M., I.J. and K.V.-M.; All authors have read and agreed to the published version of the manuscript.

Funding: This research was funded by Slovenian Research Agency, (Javna Agencija za Raziskovalno Dejavnost RS–ARRS), young researcher grant number PR-09756, core fund P2-0082(Thin film structures and plasma surface engineering), program group P1-02012 (Plant biology). We acknowledge also Slovenian Research Agency and Ministry of Agriculture, Forestry and Food of Slovenia for funding the project under grant number V4-2001 and projects N1-0105, N1-0090, J7-9398 and J7-9418.

Conflicts of Interest: The authors declare no conflict of interest.

References

1. Crookes, W. Experiments on the Dark Space in Vacuum Tubes. *Proc. R. Soc. A Math. Phys. Eng. Sci.* **1907**, *79*, 98–117. [CrossRef]

2. Tendero, C.; Tixier, C.; Tristant, P.; Desmaison, J.; Leprince, P. Atmospheric pressure plasmas: A review. *Spectrochim. Acta Part B At. Spectrosc.* **2006**, *61*, 2–30. [CrossRef]

3. Randeniya, L.K.; de Groot, G.J.J.B. Non-Thermal Plasma Treatment of Agricultural Seeds for Stimulation of Germination, Removal of Surface Contamination and Other Benefits: A Review. *Plasma Process. Polym.* **2015**, *12*, 608–623. [CrossRef]

4. Misra, N.N.; Schluter, O.; Cullen, P.J. *Cold Plasma in Food and Agriculture*, 1st ed.; Academic Press: Cambridge, MA, USA, 2016.

5. Frank-Kamenetskii, D.A.; Frank-Kamenetskii, D.A. Introduction. In *Plasma: The Fourth State of Matter*; Springer: Berlin, Germany, 1972; pp. 1–8.

6. Samal, S. Thermal plasma technology: The prospective future in material processing. *J. Clean. Prod.* **2017**, *142*, 3131–3150. [CrossRef]

7. Moreau, M.; Orange, N.; Feuilloley, M.G.J. Non-thermal plasma technologies: New tools for bio-decontamination. *Biotechnol. Adv.* **2008**, *26*, 610–617. [CrossRef] [PubMed]

8. Niemira, B.A.; Boyd, G.; Sites, J. Cold Plasma Rapid Decontamination of Food Contact Surfaces Contaminated with *Salmonella* Biofilms. *J. Food Sci.* **2014**, *79*, M917–M922. [CrossRef] [PubMed]

9. Niemira, B.A. Cold Plasma Decontamination of Foods. *Annu. Rev. Food Sci. Technol.* **2012**, *3*, 125–142. [CrossRef] [PubMed]

10. Weltmann, K.-D.; Polak, M.; Masur, K.; von Woedtke, T.; Winter, J.; Reuter, S. Plasma Processes and Plasma Sources in Medicine. *Contrib. Plasma Phys.* **2012**, *52*, 644–654. [CrossRef]

11. Haertel, B.; von Woedtke, T.; Weltmann, K.D.; Lindequist, U. Non-thermal atmospheric-pressure plasma possible application in wound healing. *Biomol. Ther.* **2014**, *22*, 477–490. [CrossRef]

12. Chen, H.H.; Chang, H.C.; Chen, Y.K.; Hung, C.L.; Lin, S.Y.; Chen, Y.S. An improved process for high nutrition of germinated brown rice production: Low-pressure plasma. *Food Chem.* **2016**, *191*, 120–127. [CrossRef]

13. Gómez-Ramírez, A.; López-Santos, C.; Cantos, M.; García, J.L.; Molina, R.; Cotrino, J.; Espinós, J.P.; González-Elipe, A.R. Surface chemistry and germination improvement of Quinoa seeds subjected to plasma activation. *Sci. Rep.* **2017**, *7*, 12. [CrossRef] [PubMed]

14. Ji, S.H.; Choi, K.H.; Pengkit, A.; Im, J.S.; Kim, J.S.; Kim, Y.H.; Park, Y.; Hong, E.J.; Jung, S.K.; Choi, E.H.; et al. Effects of high voltage nanosecond pulsed plasma and micro DBD plasma on seed germination, growth development and physiological activities in spinach. *Arch. Biochem. Biophys.* **2016**, *605*, 117–128. [CrossRef] [PubMed]

15. Ling, L.; Jiangang, L.; Minchong, S.; Chunlei, Z.; Yuanhua, D. Cold plasma treatment enhances oilseed rape seed germination under drought stress. *Sci. Rep.* **2015**, *5*, 1–10. [CrossRef] [PubMed]

16. Selcuk, M.; Oksuz, L.; Basaran, P. Decontamination of grains and legumes infected with Aspergillus spp. and Penicillum spp. by cold plasma treatment. *Bioresour. Technol.* **2008**, *99*, 5104–5109. [CrossRef]

17. Kim, K.H.; Kabir, E.; Jahan, S.A. Exposure to pesticides and the associated human health effects. *Sci. Total Environ.* **2017**, *575*, 525–535. [CrossRef]

18. Meng, Y.; Qu, G.; Wang, T.; Sun, Q.; Liang, D.; Hu, S. Enhancement of Germination and Seedling Growth of Wheat Seed Using Dielectric Barrier Discharge Plasma with Various Gas Sources. *Plasma Chem. Plasma Process.* **2017**, *37*, 1105–1119. [CrossRef]

19. Rahman, M.M.; Sajib, S.A.; Rahi, M.S.; Tahura, S.; Roy, N.C.; Parvez, S.; Reza, M.A.; Talukder, M.R.; Kabir, A.H. Mechanisms and Signaling Associated with LPDBD Plasma Mediated Growth Improvement in Wheat. *Sci. Rep.* **2018**, *8*, 1–11. [CrossRef]

20. Li, Y.; Wang, T.; Meng, Y.; Qu, G.; Sun, Q.; Liang, D.; Hu, S. Air Atmospheric Dielectric Barrier Discharge Plasma Induced Germination and Growth Enhancement of Wheat Seed. *Plasma Chem. Plasma Process.* **2017**, *37*, 1621–1634. [CrossRef]

21. Stolárik, T.; Henselová, M.; Martinka, M.; Novák, O.; Zahoranová, A.; Černák, M. Effect of Low-Temperature Plasma on the Structure of Seeds, Growth and Metabolism of Endogenous Phytohormones in Pea (*Pisum sativum* L.). *Plasma Chem. Plasma Process.* **2015**, *35*, 659–676. [CrossRef]

22. Sadhu, S.; Thirumdas, R.; Deshmukh, R.R.; Annapure, U.S. Influence of cold plasma on the enzymatic activity in germinating mung beans (Vigna radiate). *LWT Food Sci. Technol.* **2017**, *78*, 97–104. [CrossRef]

23. Iranbakhsh, A.; Ardebili, N.O.; Ardebili, Z.O.; Shafaati, M.; Ghoranneviss, M. Non-thermal Plasma Induced Expression of Heat Shock Factor A4A and Improved Wheat (*Triticum aestivum* L.) Growth and Resistance Against Salt Stress. *Plasma Chem. Plasma Process.* **2018**, *38*, 29–44. [CrossRef]

24. Zhang, J.J.; Jo, J.O.; Huynh, D.L.; Mongre, R.K.; Ghosh, M.; Singh, A.K.; Lee, S.B.; Mok, Y.S.; Hyuk, P.; Jeong, D.K. Growth-inducing effects of argon plasma on soybean sprouts via the regulation of demethylation levels of energy metabolism-related genes. *Sci. Rep.* **2017**, *7*, 41917. [CrossRef] [PubMed]

25. Bafoil, M.; Le Ru, A.; Merbahi, N.; Eichwald, O.; Dunand, C.; Yousfi, M. New insights of low-temperature plasma effects on germination of three genotypes of Arabidopsis thaliana seeds under osmotic and saline stresses. *Sci. Rep.* **2019**, *9*, 1–10. [CrossRef] [PubMed]

26. Hayashi, N.; Ono, R.; Nakano, R.; Shiratani, M.; Tashiro, K.; Kuhara, S.; Yasuda, K.; Hagiwara, H. DNA Microarray Analysis of Plant Seeds Irradiated by Active Oxygen Species in Oxygen Plasma. *Plasma Med.* **2016**, *6*, 459–471. [CrossRef]

27. Nakano, R.; Tashiro, K.; Aijima, R.; Hayashi, N. Effect of Oxygen Plasma Irradiation on Gene Expression in Plant Seeds Induced by Active Oxygen Species. *Plasma Med.* **2016**, *6*, 303–313. [CrossRef]

28. Park, Y.; Oh, K.S.; Oh, J.; Seok, D.C.; Kim, S.B.; Yoo, S.J.; Lee, M.J. The biological effects of surface dielectric barrier discharge on seed germination and plant growth with barley. *Plasma Process. Polym.* **2018**, *15*, 1600056. [CrossRef]

29. Bormashenko, E.; Shapira, Y.; Grynyov, R.; Whyman, G.; Bormashenko, Y.; Drori, E. Interaction of cold radiofrequency plasma with seeds of beans (*Phaseolus vulgaris*). *J. Exp. Bot.* **2015**, *66*, 4013–4021. [CrossRef]

30. Iranbakhsh, A.; Ardebili, Z.O.; Ardebili, N.O.; Ghoranneviss, M.; Safari, N. Cold plasma relieved toxicity signs of nano zinc oxide in Capsicum annuum cayenne via modifying growth, differentiation, and physiology. *Acta Physiol. Plant.* **2018**, *40*, 154. [CrossRef]

31. Mitra, A.; Li, Y.-F.; Klämpfl, T.G.; Shimizu, T.; Jeon, J.; Morfill, G.E.; Zimmermann, J.L. Inactivation of Surface-Borne Microorganisms and Increased Germination of Seed Specimen by Cold Atmospheric Plasma. *Food Bioprocess Technol.* **2014**, *7*, 645–653. [CrossRef]

32. Henselová, M.; Slováková, Ľ.; Martinka, M.; Zahoranová, A. Growth, anatomy and enzyme activity changes in maize roots induced by treatment of seeds with low-temperature plasma. *Biologia* **2012**, *67*, 490–497. [CrossRef]

33. Zahoranová, A.; Hoppanová, L.; Šimončicová, J.; Tučeková, Z.; Medvecká, V.; Hudecová, D.; Kaliňáková, B.; Kováčik, D.; Černák, M. Effect of Cold Atmospheric Pressure Plasma on Maize Seeds: Enhancement of Seedlings Growth and Surface Microorganisms Inactivation. *Plasma Chem. Plasma Process.* **2018**, *38*, 969–988. [CrossRef]

34. Zhou, R.; Zhou, R.; Zhang, X.; Zhuang, J.; Yang, S.; Bazaka, K.; Ostrikov, K. Effects of Atmospheric-Pressure N_2, He, Air, and O_2 Microplasmas on Mung Bean Seed Germination and Seedling Growth. *Nat. Publ. Gr.* **2016**. [CrossRef] [PubMed]

35. Tomeková, J.; Kyzek, S.; Medvecká, V.; Gálová, E.; Zahoranová, A. Influence of Cold Atmospheric Pressure Plasma on Pea Seeds: DNA Damage of Seedlings and Optical Diagnostics of Plasma. *Plasma Chem. Plasma Process.* **2020**, 1–14. [CrossRef]

36. Kyzek, S.; Holubová, Ľ.; Medvecká, V.; Tomeková, J.; Gálová, E.; Zahoranová, A. Cold Atmospheric Pressure Plasma Can Induce Adaptive Response in Pea Seeds. *Plasma Chem. Plasma Process.* **2019**, *39*, 475–486. [CrossRef]

37. Li, L.; Li, J.; Shen, M.; Hou, J.; Shao, H.; Dong, Y.; Jiang, J. Improving Seed Germination and Peanut Yields by Cold Plasma Treatment. *Plasma Sci. Technol.* **2016**, *18*, 1027–1033. [CrossRef]

38. Mildažienė, V.; Aleknavičiūtė, V.; Žūkienė, R.; Paužaitė, G.; Naučienė, Z.; Filatova, I.; Lyushkevich, V.; Haimi, P.; Tamošiūnė, I.; Baniulis, D. Treatment of Common Sunflower (*Helianthus annus* L.) Seeds with Radio-frequency Electromagnetic Field and Cold Plasma Induces Changes in Seed Phytohormone Balance, Seedling Development and Leaf Protein Expression. *Sci. Rep.* **2019**, *9*, 6437. [CrossRef]

39. Dobrin, D.; Magureanu, M.; Mandache, N.B.; Ionita, M.D. The effect of non-thermal plasma treatment on wheat germination and early growth. *Innov. Food Sci. Emerg. Technol.* **2015**, *29*, 255–260. [CrossRef]

40. Guo, Q.; Wang, Y.; Zhang, H.; Qu, G.; Wang, T.; Sun, Q.; Liang, D. Alleviation of adverse effects of drought stress on wheat seed germination using atmospheric dielectric barrier discharge plasma treatment. *Sci. Rep.* **2017**, *7*, 1–14. [CrossRef]

41. Jiang, J.; He, X.; Li, L.; Li, J.; Shao, H.; Xu, Q.; Ye, R.; Dong, Y. Effect of cold plasma treatment on seed germination and growth of wheat. *Plasma Sci. Technol.* **2014**, *16*, 54–58. [CrossRef]

42. Los, A.; Ziuzina, D.; Boehm, D.; Cullen, P.J.; Bourke, P. Investigation of mechanisms involved in germination enhancement of wheat (*Triticum aestivum*) by cold plasma: Effects on seed surface chemistry and characteristics. *Plasma Process. Polym.* **2019**, *16*, 1800148. [CrossRef]

43. Saberi, M.; Modarres-Sanavy, S.A.M.; Zare, R.; Ghomi, H. Amelioration of Photosynthesis and Quality of Wheat under Non-thermal Radio Frequency Plasma Treatment. *Sci. Rep.* **2018**, *8*, 1–8. [CrossRef] [PubMed]

44. Zhang, B.; Li, R.; Yan, J. Study on activation and improvement of crop seeds by the application of plasma treating seeds equipment. *Arch. Biochem. Biophys.* **2018**, *655*, 37–42. [CrossRef] [PubMed]

45. De Groot, G.J.J.B.; Hundt, A.; Murphy, A.B.; Bange, M.P.; Mai-Prochnow, A. Cold plasma treatment for cotton seed germination improvement. *Sci. Rep.* **2018**, *8*, 1–10. [CrossRef] [PubMed]

46. Adhikari, B.; Adhikari, M.; Ghimire, B.; Adhikari, B.C.; Park, G.; Choi, E.H. Cold plasma seed priming modulates growth, redox homeostasis and stress response by inducing reactive species in tomato (*Solanum lycopersicum*). *Free Radic. Biol. Med.* **2020**, *156*, 57–69. [CrossRef] [PubMed]

47. Pauzaite, G.; Malakauskiene, A.; Nauciene, Z.; Zukiene, R.; Filatova, I.; Lyushkevich, V.; Azarko, I.; Mildaziene, V. Changes in Norway spruce germination and growth induced by pre-sowing seed treatment with cold plasma and electromagnetic field: Short-term versus long-term effects. *Plasma Process. Polym.* **2018**, *15*, 1700068. [CrossRef]

48. Shapira, Y.; Bormashenko, E.; Drori, E. Pre-germination plasma treatment of seeds does not alter cotyledon DNA structure, nor phenotype and phenology of tomato and pepper plants. *Biochem. Biophys. Res. Commun.* **2019**, *519*, 512–517. [CrossRef]

49. Bormashenko, E.; Grynyov, R.; Bormashenko, Y.; Drori, E. Cold radiofrequency plasma treatment modifies wettability and germination speed of plant seeds. *Sci. Rep.* **2012**, *2*, 3–10. [CrossRef]

50. Ling, L.; Jiafeng, J.; Li, J.; Minchong, S.; Xin, H.; Hanliang, S.; Yuanhua, D. Effects of cold plasma treatment on seed germination and seedling growth of soybean. *Sci. Rep.* **2014**, *4*, 5859. [CrossRef]

51. Bermúdez-Aguirre, D. *Advances in Cold Plasma Applications for Food Safety and Preservation*, 1st ed.; Bermudez-Aguirre, D., Ed.; Academic Press: Cambridge, MA, USA, 2019; ISBN 9780128149218.

52. Conrads, H.; Schmidt, M. Plasma generation and plasma sources. *Plasma Sources Sci. Technol.* **2000**, *9*, 441–454. [CrossRef]

53. Han, L.; Patil, S.; Boehm, D.; Milosavljević, V.; Cullen, P.J.; Bourke, P. Mechanisms of inactivation by high-voltage atmospheric cold plasma differ for Escherichia coli and Staphylococcus aureus. *Appl. Environ. Microbiol.* **2016**, *82*, 450–458. [CrossRef]

54. Fridman, G.; Brooks, A.D.; Balasubramanian, M.; Fridman, A.; Gutsol, A.; Vasilets, V.N.; Ayan, H.; Friedman, G. Comparison of Direct and Indirect Effects of Non-Thermal Atmospheric-Pressure Plasma on Bacteria. *Plasma Process. Polym.* **2007**, *4*, 370–375. [CrossRef]

55. Leduc, M.; Guay, D.; Coulombe, S.; Leask, R.L. Effects of non-thermal plasmas on DNA and mammalian cells. *Plasma Process. Polym.* **2010**, *7*, 899–909. [CrossRef]

56. Lazović, S.; Maletić, D.; Leskovac, A.; Filipović, J.; Puač, N.; Malović, G.; Joksić, G.; Petrović, Z.L. Plasma induced DNA damage: Comparison with the effects of ionizing radiation. *Appl. Phys. Lett.* **2014**, *105*, 124101. [CrossRef]

57. Rastogi, R.P.; Richa; Kumar, A.; Tyagi, M.B.; Sinha, R.P. Molecular mechanisms of ultraviolet radiation-induced DNA damage and repair. *J. Nucleic Acids* **2010**, *2010*, 32. [CrossRef]

58. Zaplotnik, R.; Vesel, A. Effect of VUV radiation on surface modification of polystyrene exposed to atmospheric pressure plasma jet. *Polymers* **2020**, *12*, 1136. [CrossRef]

59. Golda, J.; Biskup, B.; Layes, V.; Winzer, T.; Benedikt, J. Vacuum ultraviolet spectroscopy of cold atmospheric pressure plasma jets. *Plasma Process. Polym.* **2020**, *17*, 1900216. [CrossRef]

60. Sarani, A.; Nikiforov, A.Y.; Leys, C. Atmospheric pressure plasma jet in Ar and Ar/H_2O mixtures: Optical emission spectroscopy and temperature measurements. *Phys. Plasmas* **2010**, *17*. [CrossRef]

61. Pehlivan, F.E. Free Radicals and Antioxidant System in Seed Biology. In *Advances in Seed Biology*; InTech: London, UK, 2017; p. 167.

62. Ranki, H.; Sopanen, T. Secretion of α-Amylase by the Aleurone Layer and the Scutellum of Germinating Barley Grain. *Plant Physiol.* **1984**, *75*, 710–715. [CrossRef]

63. Finkelstein, R.R. The role of hormones during seed development and germination. *Plant Horm. Biosynth. Signal Transduct. Action* **2010**, 549–573. [CrossRef]

64. Miransari, M.; Smith, D.L. Plant hormones and seed germination. *Environ. Exp. Bot.* **2014**, *99*, 110–121. [CrossRef]

65. Kurek, K.; Plitta-Michalak, B.; Ratajczak, E. Reactive oxygen species as potential drivers of the seed aging process. *Plants* **2019**, *8*, 174. [CrossRef] [PubMed]

66. Los, A.; Ziuzina, D.; Akkermans, S.; Boehm, D.; Cullen, P.J.; Van Impe, J.; Bourke, P. Improving microbiological safety and quality characteristics of wheat and barley by high voltage atmospheric cold plasma closed processing. *Food Res. Int.* **2018**, *106*, 509–521. [CrossRef] [PubMed]

67. Fritz, J.L.; Owen, M.J. Hydrophobic recovery of plasma-treated polydimethylsiloxane. *J. Adhes.* **1995**, *54*, 33–45. [CrossRef]

68. Junkar, I.; Modic, M.; Mozetič, M.; Vesel, A.; Hauptman, N.; Cvelbar, U. Modification of PET surface properties using extremely non-equilibrium oxygen plasma. *Plasma Process. Polym.* **2009**, *6*, 667–675. [CrossRef]

69. Junkar, I. Plasma treatment of amorphous and semicrystalline polymers for improved biocompatibility. *Vacuum* **2013**, *98*, 111–115. [CrossRef]

70. Modic, M.; Junkar, I.; Vesel, A.; Mozetic, M. Aging of plasma treated surfaces and their effects on platelet adhesion and activation. *Surf. Coatings Technol.* **2012**, *213*, 98–104. [CrossRef]

71. Bormashenko, E. Progress in understanding wetting transitions on rough surfaces. *Adv. Colloid Interface Sci.* **2015**, *222*, 92–103. [CrossRef]

72. Šerá, B.; Špatenka, P.; Šerý, M.; Vrchotová, N.; Hrušková, I. Influence of plasma treatment on wheat and oat germination and early growth. *IEEE Trans. Plasma Sci.* **2010**, *38*, 2963–2968. [CrossRef]

73. Holc, M.; Primc, G.; Iskra, J.; Titan, P.; Kovač, J.; Mozetič, M.; Junkar, I. Effect of oxygen plasma on sprout and root growth, surface morphology and yield of garlic. *Plants* **2019**, *8*, 462. [CrossRef]

74. Nikiforov, A.Y.; Sarani, A.; Leys, C. The influence of water vapor content on electrical and spectral properties of an atmospheric pressure plasma jet. *Plasma Sources Sci. Technol.* **2011**, *20*. [CrossRef]

75. Alves Junior, C.; de Oliveira Vitoriano, J.; Layza Souza da Silva, D.; de Lima Farias, M.; Bandeira de Lima Dantas, N. Water uptake mechanism and germination of Erythrina velutina seeds treated with atmospheric plasma. *Sci. Rep.* **2016**, 33722. [CrossRef]

Publisher's Note: MDPI stays neutral with regard to jurisdictional claims in published maps and institutional affiliations.

MDPI
St. Alban-Anlage 66
4052 Basel
Switzerland
Tel. +41 61 683 77 34
Fax +41 61 302 89 18
www.mdpi.com

Plants Editorial Office
E-mail: plants@mdpi.com
www.mdpi.com/journal/plants